Design and Operation for Air Pollution Control

Design and Operation for Air Pollution Control

Contributors

Litao Wang, Jing Yang et al.

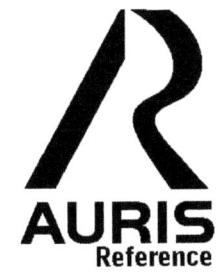

AURIS
Reference

www.aurisreference.com

Design and Operation for Air Pollution Control

Contributors: Litao Wang, Jing Yang et al.

Published by Auris Reference Limited

www.aurisreference.com

United Kingdom

Design and Operation for Air Pollution Control

ISBN: 978-1-78154-834-9

British Library Cataloguing in Publication Data
A CIP record for this book is available from the British Library

Printed in the United Kingdom

Exclusively distributed by CBS Publishers & Distributors Pvt. Ltd.

Sales & Distribution Rights only for India, Pakistan, Bangladesh, Sri Lanka, Nepal and Bhutan. This book is not to be sold outside these territories.

Contents

List of Abbreviations

ACGIH	American Conference of Governmental Industrial Hygienists
ACR	Air Change Rate
AHU	Air Handling Unit
AIHA	American Industrial Hygiene Association
BAT	Best Available Technology
BF	Bio-Filters
BRC	Building Related Health Complaints
BREC	Building-Related Environmental Complaints
BRI	Building-Related Illness
BRS	Building-Related Symptoms
BS	Bio-Scrubbers
BTF	Bio-Trickling Filters
CDC	Centers for Disease Control
DBAC	Desiccant Based Air Conditioning
EBRT	Empty Bed Residence Time
EC	Elimination Capacity
ECA	European Collaborative Action
ETS	Environmental Tobacco Smoke
GC	Gas Chromatography
GDP	Gross Domestic Product
HVAC	Heating, Ventilating, and Air Conditioning
IAQ	Indoor Air Quality
IL	Inlet Load
LOD	Limits of Detection
LOQ	Limits of Quantification
MCS	Multiple Chemical Sensitivity
MEK	Methyl Ethyl Ketone
MMEF	Maximum Mid Expiratory Flow
MMMF	Man-Made Mineral Fibres
MS	Mass Spectrometer
NI	Nosocomial Infection
OSHA	Occupational Safety and Health Administration
PAH	Polycyclic Aromatic Hydrocarbons
PEF	Peak Expiratory Flow
POP	Persistent Organic Pollutant
PVF	Poly-Vinyl Formal
RE	Removal Efficiency
SBS	Sick Building Syndrome
TSP	Total Suspended Particle

| VOC | Volatile Organic Compound |
| WHO | World Health Organization |

List of Contributors

Litao Wang
Department of Environmental Engineering, Hebei University of Engineering, Handan, China

Jing Yang
Department of Environmental Engineering, Hebei University of Engineering, Handan, China

Pu Zhang
Department of Environmental Engineering, Hebei University of Engineering, Handan, China

Xiujuan Zhao
Department of Environmental Engineering, Hebei University of Engineering, Handan, China

Zhe Wei
Department of Environmental Engineering, Hebei University of Engineering, Handan, China

Fenfen Zhang
Department of Environmental Engineering, Hebei University of Engineering, Handan, China

Jie Su
Department of Environmental Engineering, Hebei University of Engineering, Handan, China

Chenchen Meng
Department of Environmental Engineering, Hebei University of Engineering, Handan, China

Marios. P. Tsakas
Laboratory of Environmental Chemistry, National and Kapodestrian University of Athens, Greece

Apostolos P. Siskos
Laboratory of Environmental Chemistry, National and Kapodestrian Univer-

sity of Athens, Greece

Panayotis A. Siskos
Laboratory of Environmental Chemistry, National and Kapodestrian University of Athens, Greece

Detlef Laussmann
Robert Koch Institute, Germany

Dieter Helm
Robert Koch Institute, Germany

M. D. Larrañaga
Oklahoma State University, USA

E. Karunasena
Texas Tech University, USA

H.W. Holder
SWK, LLC, USA

E. D. Althouse
Air Intellect, LLC, USA

D.C. Straus
Texas Tech University Health Sciences Center, USA

M.G. Beruvides
Texas Tech University, USA

Moinuddin Sarker
Natural States Research, Inc., USA

F. Javier Álvarez-Hornos
GI2AM Research Group, Department of Chemical Engineering, University of Valencia, Burjassot, Spain

Feliu Sempere
GI2AM Research Group, Department of Chemical Engineering, University of Valencia, Burjassot, Spain

Marta Izquierdo
GI2AM Research Group, Department of Chemical Engineering, University of Valencia, Burjassot, Spain

Carmen Gabaldón
GI2AM Research Group, Department of Chemical Engineering, University of Valencia, Burjassot, Spain

José A. Orosa
University of A Coruña, Spain

Alessandro Bacaloni
Department of Chemistry, University of Rome "La Sapienza", Italy

Susanna Insogna
Department of Chemistry, University of Rome "La Sapienza", Italy

Lelio Zoccolillo
Department of Chemistry, University of Rome "La Sapienza", Italy

Preface

The control of air pollution is one of the principal areas of pollution control, along with wastewater treatment, solid-waste management, and hazardous-waste management. The text *Design and Operation for Air Pollution Control* discusses the operation and maintenance of air pollution control equipment and explores the problems connected with their use. First chapter focuses on the air quality history and status according to both the governmental reports and the relative academic studies, and also review the air pollution control measurements pursued within Hebei area and give suggestions on future pollution control. Second chapter describes the method which can be used to determine the air change rate with tracer gas measurements. The objectives of third chapter are to highlight the necessity for multiple (replicate) air samples per sample location to conduct valid assessments of the airborne concentrations of bioaerosols. The aim of fourth chapter is to quantify the removal capabilities of a rotary wheel (honeycomb) solid-desiccant dehumidifier at removing selected indoor air quality (IAQ)-related fungal organisms from the airstream. Fifth chapter focuses on the conversion of municipal waste plastics to liquid hydrocarbon fuel which has been carried out in thermal degradation process with/without catalyst. Sixth chapter presents studies conducted to assess environmentally friendly biotechnologies, such as biofilters and biotrickling filters, for volatile organic compounds (VOC) abatement in air at two scales. In seventh chapter, a new methodology based on the statistical study of One-Way ANOVA has been developed to test bioindicators, such as fungi, in real case studies. Eighth chapter presents the way to handle the problem in indoor air quality (IAQ) assessment, and some practical applications, in order to provide the logical pathway to face the majority of actual cases. Last chapter focuses on those species common to indoor and outdoor air pollution and their impact on human health.

Chapter 1

A REVIEW OF AIR POLLUTION AND CONTROL IN HEBEI PROVINCE, CHINA

Litao Wang, Jing Yang, Pu Zhang, Xiujuan Zhao, Zhe Wei, Fenfen Zhang, Jie Su, and Chenchen Meng

Department of Environmental Engineering, Hebei University of Engineering, Handan, China

ABSTRACT

Hebei is one of the most air polluted provinces in China. According to the Ministry of Environmental Protection (MEP) for the severe fog-haze month of Jan. 2013, seven of the top ten most polluted cities in China are located in Hebei Province. In this study, the air pollution history and status of the Hebei Province are reviewed and discussed, using the governmental published Air Pollution Index (API), the academic observations by various scientific research groups and the long-term statistics of visibility and haze frequencies. It is found that within the Hebei Province, the air pollution in the southern cities is much more severe than the northern cities. Particulate matter (PM) is undoubtedly the major air pollutant, sulfur dioxide (SO_2) and nitrogen oxides (NO_x) pollutions are also unnegligible. Ozone (O_3) pollution in larger cities, such as Shijiazhuang, is significant. Air pollution control history from 1998 is discussed as well. Although Hebei Province has made a great effort on air quality, the pollutant emissions, such as SO_2 and fly ash, showed a notable increase in 2001 to 2006. However, after 2006 the emissions started to decrease due to the strict implementation of the national 11th Five Year Plan (FYP). In addition, regional jointly air pollution control and prevention strategies are expected in the future to substantially change the severe air pollution status in Hebei Province.

INTRODUCTION

In Jan. 2013, continuous, severe haze pollution happened in east and central China, attracting the most public attention. In Beijing, only five days were not fog and haze days during Jan. 2013. It is reported that the daily fine particulate matter ($PM_{2.5}$) concentrations in Beijing and Shijiazhuang has been over 500

$\mu g \cdot m^{-3}$, which is 6.7 times of the new China National Ambient Air Quality Standard (CNAAQS) [1]. In the statistics of the Ministry of Environmental Protection of China (MEP), during this month, the ten most polluted cities are Xingtai, Shijiazhuang, Baoding, Handan, Langfang, Hengshui, Jinan, Tangshan, Beijing and Zhengzhou city, out of the reported 74 key cities all over China (http://hebei.sina. com.cn/news/yz/2013-02-06/075733562.html). It should be noted that in these ten top polluted cities, seven cities are within Hebei Province and five of them are located in the southern area of Hebei (see Figure 1). The air pollution in Hebei Province has aroused wide public concern. Hebei Province, located in the north-east of China, is east to the Taihang Mountains and north to the Yellow River (see Figure 1). It encloses two municipal cities, Beijing and Tianjin. The neighboring provinces (in clockwise direction) are Shandong Province, Henan Province, Shanxi Province, Inner Mongolia Autonomous Region and Liaoning Province. It has the area of 187,700 sq·km and population of 71.85 million (http://www.stats.gov.cn/ tjgb/ rkpcgb/dfrkpcgb/t20120228_402804324.htm). Most of the northwest area of Hebei is mountainous or hilly, while the central and south areas belong to North China Plain. Hebei has a monsoon climate of medium latitudes, which has dry and windy springs, hot and rainy summers and dry-cold winters.

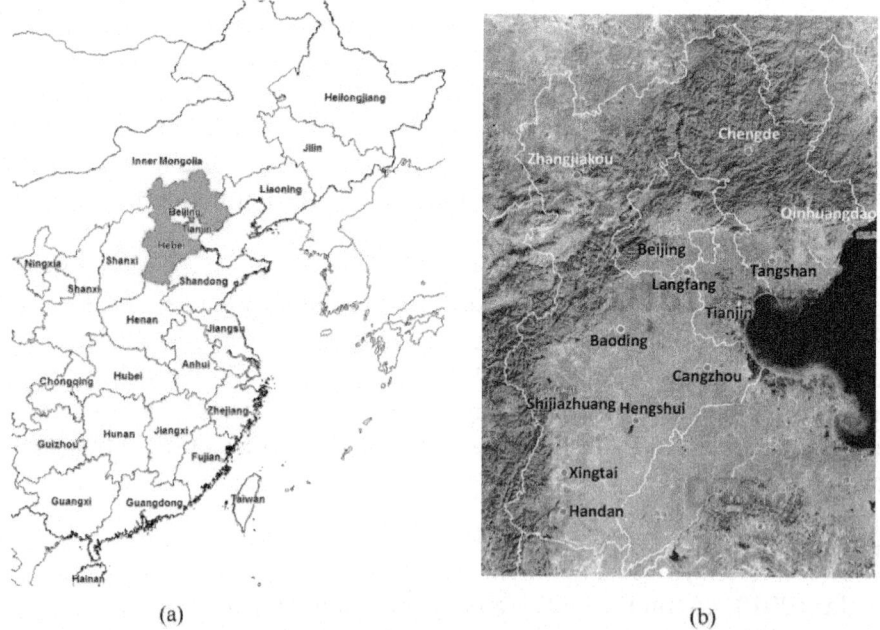

(a)   (b)

Figure 1. Location of Hebei Province and the cities: (a) locations of Hebei Province in China; (b) eleven cities in Hebei Province.

In 2011, Hebei's Gross Domestic Product (GDP) is 2.45 trillion RMB, accounting for 5.18% of national GDP and ranking 6th in China [2]. The major industries in Hebei are iron, steel, coke and cement. In 2011, 45.5% of the steel in the world was produced in China, out of which 24.0% was produced in Hebei [2]. China's coke production accounted for more than 60% of the world, of which 14.5% was produced in Hebei [3]. Hebei's cement production was 6.9% of the national total amount [2]. The air pollution burden of the southern Hebei area is particularly heavy because of its special location. It is surrounded by the other three populated and industrialized provinces, Shandong, Henan and Shanxi.

The steel, coke and cement productions of the four neighbored provinces are as large as 40.8%, 50.1% and 22.6% of the national total amount. That is to say, 18.6% of steel production in the world, 30.0% of the coke and 13.6% of the cement were yielded in this area.

The large industrial productions induce huge quantities of pollutants emission. In the widely-used Asian INTEXB emission inventory [4], the $PM_{2.5}$ emissions from the four provinces accounted for 28% of the national total emission in 2006. The percentages for SO_2, NO_X, CO, VOC, BC and OC were 28%, 25%, 28%, 24%, 30% and 24%. It gives us a clue why Hebei Province, especially the southern area, has the most severe air pollution all over China.

Hebei is somewhat overshadowed by its two neighbors, Beijing and Tianjin. Like other aspects, its air pollution problems haven't been paid enough attention for a long time. Very few studies have focused on the air pollution status in Hebei area, analyzed the present control strategies and given relative suggestions. In the 12th FYP of Air Pollution Control in the Key Regions by MEP at the end of 2012 (http://www.mep.gov.cn/gkml/hbb/bwj/2012 12/ t20121205_243271.htm), Beijing, Tianjin and Hebei area is one the three key air pollution control regions and will be pursued regional jointly air pollution prevention and control. Better strategies and more effective actions should be expected in Hebei area. In this paper, we summarize the air quality history and status according to both the governmental reports and the relative academic studies, and also review the air pollution control measurements pursued within Hebei area and give suggestions on future pollution control.

AIR POLLUTION HISTORY AND STATUS

API and AQI

The API is a non-dimensional number calculated according to the urban daily average concentrations of three pollutants: SO_2, nitrogen dioxide (NO_2) and coarse particulate matter (PM$_{10}$). Besides of the simplicity, it provided the only publicly accessible urban air quality data before the real time concentrations of the three pollutants of the national sites were started to be published online in 2011. The API record of the key environmental protecttion cities from 2000 is on the website of the Ministry of Environmental Protection (http://datacenter. mep.gov.cn/). The detailed introduction of API system can be found on in [5].

Before 2003, only Shijiazhuang, the capital city of Hebei was listed in the China's key environmental protection cities and had the API record on the website of MEP from 2000. In Feb. 2011, three other Hebei cities, Tangshan, Baoding and Handan started to publish their APIs. Figure 2 provides the distribution of the APIs in Shijiazhuang from 2001 to 2012. It can be seen that in general, air quality in Shijiazhuang city was visibly improved in 2001 to 2009. The number of the days with the APIs less than or equal to 50 (which is also "no key pollutant" day, see [5]) and between 50 - 100 was notably increased from 0 and 93 in 2001, to 43 and 274 in 2012, respectively. And the days having the APIs in 100 - 150 and 150 - 200 decreased, respectively, from 168 and 79 in 2001, to 44 and 2 in 2009. The severely polluted days, in which the API was higher than 300, decline from 12 to 0 in 2001 to 2009.

After 2009, the pollution level kept relatively stable. The frequencies of the APIs within 150 - 200 even increased from 2 days in 2009 to 10 days in 2012 (note that the API less than 100 means the city's air quality reach the CNAAQS [6]. The annual average APIs were 77.5, 74.6, 75.5, and 74.1 in 2009, 2010, 2011 and 2012, respectively. This number is 138.3 in 2001, 91.7 in 2005, and 83.5 in 2008.

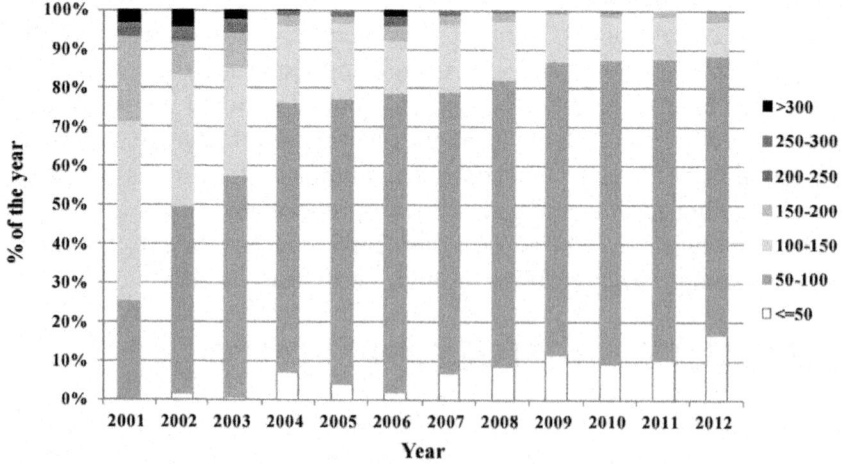

Figure 2. API distribution in Shijiazhuang city in 2001-2012.

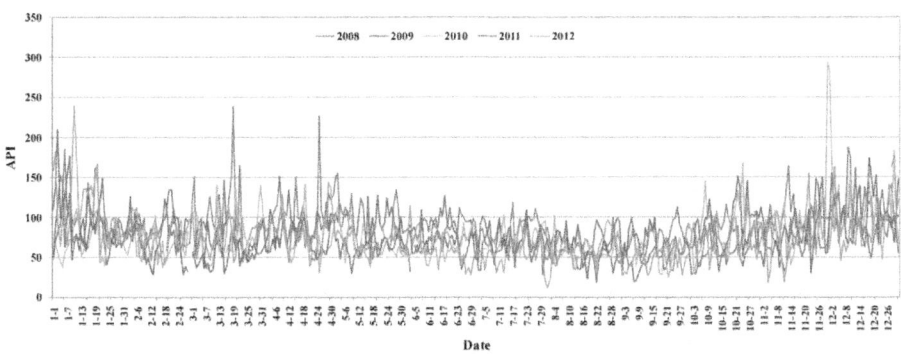

Figure 3. Time series of API in Shijiazhuang in 2008-2012.

Figure 3 shows the time series of API in Shijiazhuang city in the past five years. It is shown that the daily variations of APIs were consistent for the five years, that the best air quality appeared in Sep., Aug. and Jul., with the three-month-average APIs of 55.5 (2012) to 66.9 (2008). Then followed Jun., May and Feb. Winter was the worst season that the monthly average APIs were within 69.6 (2010) to 114.0 (2012) for Jan., 78.8 (2011) to 87.8 (2008) for Nov., and 89.1 (2009) to 111.1 (2010) for Dec. In Mar. and Apr. 2008, the APIs indicated two pollution episodes induced by sand storms. Despite the two episodes, the average APIs were 55.9 (2011) to 76.0 (2010) for Mar., and 72.2 (2011) to 89.3 (2009) for Apr. Other information indicated by Figure 3 is that the air quality in Shijiazhuang didn't show noticeable improvement or deterioration during the recent five years, which is consistent with the above discussions.

Since February 2011, three other cities, Tangshan, Baoding and Handan started to release their APIs on the website of MEP. To compare with Shijiazhuang city, Figure 4 presents the time series of APIs for the four cities from 2010 to Jan. 14, 2013. In general, Shijiazhuang city had the worst air quality that the average API in Feb. 2012 to Jan. 2013 was 83.3. Then was Handan city, with the average API of 73.6 during the same period. Baoding and Tangshan city had the average numbers of 70.8 and 69.4, respectively.

At the end of 2010, there was a severe pollution episode in Shijiazhuang city, with the API of near 300 (as 420 $\mu g \cdot m^{-3}$ of daily average PM_{10}). At the end of 2011, Shijiazhuang city didn't show big difference in pollutant character from 2010, but a highly polluted episode happened in Handan city that the largest API was as high as 348 (as 458 $\mu g \cdot m^{-3}$ of daily average PM_{10}). In this month (Dec. 2011), the average APIs were 118.1, 102.9, 100.1 and 92.1 for Handan, Baoding, Tangshan and Shijiazhuang, respectively.

As discussed above, very severe haze pollution happened over east China in the winter of 2012, in which Beijing-Tianjin-Hebei area was one of the most polluted regions. It is clearly indicated in Figure 4. The APIs reached the top limit, 500 (representing 600 $\mu g \cdot m^{-3}$ of daily average PM_{10}). It is four times of the CNAAQS. One of the most possible reasons might be the special meteorological conditions, comparing with the former two years. Lots of investigations are needed before drawing a convictive conclusion of this episode.

In Feb. 2012, the MEP released the new CNAAQS [7], which will be implemented in 2016, and the new technical regulations on air quality daily reports [8]. In the new system, more pollutants, such as CO, O_3 and $PM_{2.5}$, are involved and API is replaced by Air Quality Index (AQI).

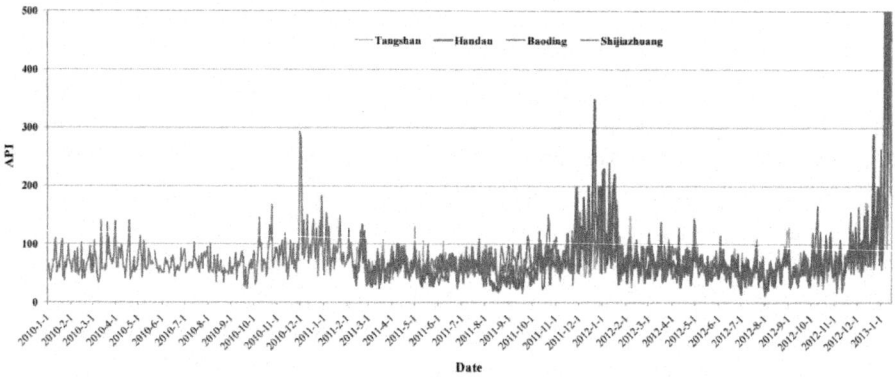

Figure 4. Time series of API in Tangshan, Handan, Baoding and Shijiazhuang in 2010-2012.

From Jan. 2013, the MEP started to publish the realtime concentrations of SO_2, NO_2, CO, 1-hr and 8-hr O_3, PM_{10} and $PM_{2.5}$, as well as their AQIs, of all the national monitoring sites in the 74 cities including the capitals, major cities in Beijing-Tianjin-Hebei area, Yangtze River Delta, and Pearl River Delta (http://113.108.142.147: 20035/emcpublish/). At the same time the old daily APIs for those cities were stopped to update after Jan. 14, 2013. It is a progress that more detailed information on air quality can be accessed on the real time system, but it is a pity that up to now, the users could not access any longer history data except for the data for the past 24 hours. More understanding on the past special winter might be obtained when the historical real time data are accessible from the system.

Academic Observations

The largest scale of air quality observations over the Beijing-Tianjin-Hebei area were pursued by the Institute of Atmospheric Physics, Chinese Academy of Science (IAP, CAS) [9]. 25 monitoring stations were established in northern China by IAP since 2009, out of which 16 were located in Hebei area [9], involving nine cities: Shijiazhuang, Baoding, Tangshan, Hengshui, Cangzhou, Langfang, Chengde, Zhangjiakou, and Qinhuangdao. PM_{10}, SO_2 and NO_X were observed in these stations. Table 1 summarizes the results of these studies and the observations carried out by the other researchers. It is listed according to the locations of the observation sites, from the north to the south of Hebei.

Table 1. Air quality observations in Hebei Province (unit: $\mu g \cdot m^{-3}$).

Site	Period	Environment	City size	Region	Daily PM₁₀	Daily PM₂.₅	Daily SO₂	Daily NO₂	Daily NOₓ	Daily O₃	Hourly O₃	Ref.
Xinglong	Aug. 15-Sep. 15 2006	Rural and background	County-level city	Chengde	69.4 (7.0 - 141.2)[a]	-	4.4 (0.2 - 17.2)[a]	-	15.1 (5.9 - 22.1)[a]	-	137 (21 - 291)[a]	[10]
Xianghe	Jun.-Sep. 2008	Rural	County-level city	Langfang	113 ± 52 (248)[b]	76 ± 42 (184)[b]	13.4 ± 15.2 (84.4)[b]	14.5 ± 8.6 (41.0)[b]	15.9 ± 9.1 (43.3)[b]	82 ± 38 (230)[b]	-	[11]
Tangshan	Jun.-Aug. 2007	Cultural and educational	City	Tangshan	-	109.8 ± 26.4	-	41.7 ± 7.6	-	66.7 ± 16.8	157.8 ± 42.5	[12]
Tangshan	Sep.-Oct, 2007	Cultural and educational	City	Tangshan	-	150.9 ± 58.1	-	46.0 ± 8.9	-	46.4 ± 25.8	123.5 ± 61.4	[12]
Tangshan	Jun.-Aug. 2008	Cultural and educational	City	Tangshan	-	103.9 ± 50.3	44.8 ± 31.1	38.6 ± 10.5	-	74.6 ± 25.8	153.0 ± 52.8	[12]
Tangshan	Sep.-Oct. 2008	Cultural and educational	City	Tangshan	-	91.9 ± 62.4	52.2 ± 25.2	40.4 ± 12.6	-	42.4 ± 21.4	108.3 ± 52.5	[12]
Zhuozhou	Jul. 2009-Feb. 2011	Cultural and educational	County-level city	Baoding	153	-	27	-	68	44	-	[13]
Baoding	Jul. 2009-Feb. 2011	Cultural and educational	City	Baoding	189	-	72	-	83	48	-	[13]
Shijiazhuang	Aug. 12-Sep. 25 2007	Cultural and educational	Capital city	Shijiazhuang	-	114.5 ± 45.3	-	51.9 ± 11.0	-	81.3 ± 35.6	166.7 ± 67.1	[14]
Shijiazhuang	Jun.-Sep. 2008	Cultural and educational	Capital city	Shijiazhuang	150.7 ± 62.1	99.4 ± 48.6	80.2 ± 36.6	43.9 ± 19.8	-	64.0 ± 33.9	148.6 ± 71.5	[14]
Shijiazhuang	Jul. 2009-Feb. 2011	Cultural and educational	City	Shijiazhuang	204	-	87	-	74	34	-	[13]
Handan	Aug.-Oct. 2012	Cultural and educational	City	Handan	196.5 ± 107.1	96.7 ± 55.2	69.2 ± 31.4	37.0 ± 26.3	66.3 ± 31.4	67.5 ± 22.9	135.8 ± 53.4	[15]
Handan	Nov.-Dec. 2012	Cultural and educational	City	Handan	248.6 ± 102.4	121.5 ± 60.4	142.4± 56.4	51.9 ± 12.5	130.0 ± 53.9	14. ± 8.6	40.5 ± 16.7	[15]
Handan	Jan. 2013	Cultural and educational	City	Handan	347.2 ± 174.6	233.4 ± 144.4	222.6± 86.5	93.3 ± 25.2	215.6 ± 72.7	8.8 ± 5.2	22.0 ± 14.8	[15]

[a]The data is "average(minimum-maximum)", [b]The data is "average ± standard deviation (maximum)". Other data is "average" or "average ± standard deviation".

The northest site is Xinglong, located in the Chengde city. It is only the rural and background site in the table. The daily average concentrations of PM_{10}, SO_2, NO_X and O_3 didn't exceed the present CNAAQS [6], neither the maximum daily concentrations. Only the maximum hourly O_3 concentrations exceeded the standard (291 $\mu g \cdot m^{-3}$, 46% higher than the CNAAQS). O_3 pollution at this background site is unnegligible according to the observations [10], which may be induced by the large quantities of pollutant emissions for the near huge cities.

Xianghe site is in Langfang city, between Beijing and Tianjin city. The average PM_{10}, SO_2 and NO_X didn't exceed the present daily standard during the summer of 2008, but the $PM_{2.5}$ concentration were slightly higher than the new CNAAQS for daily $PM_{2.5}$ concentrations (75 $\mu g \cdot m^{-3}$). Note that this was the Beijing Olympic period, during which air quality were much better than the normal situation.

Tangshan is one of the heavy industry bases in China. From the Table 1 we can found that PM was the key pollutant in Tangshan. Even in the Olympic period (Jun.- Aug. 2008), the average $PM_{2.5}$ concentration were 39% higher than the new standard for daily $PM_{2.5}$. Zhuozhou and Baoding site showed the same characteristics that PM was the most important pollutant. Their average PM_{10} concentrations during Jul. 2009 to Feb. 2011 were 153 and 189 $\mu g \cdot m^{-3}$, respectively, much higher than the annual limit of PM_{10} (100 $\mu g \cdot m^{-3}$). SO_2 in Baoding site exceeded the national annual standard (60 $\mu g \cdot m^{-3}$), and NO_X in both Zhuozhou and Baoding site exceeded the new national annual standard (50 $\mu g \cdot m^{-3}$).

According to Table 1, Shijiazhuang, the capital city of Hebei, and Handan, the southest city in Hebei, were the two most polluted cities. The PM_{10} and $PM_{2.5}$ concentrations in all observed periods were exceeding the CNAAQS. In Shijiazhuang, the average PM_{10} concentrations were 204 $\mu g \cdot m^{-3}$ for Jul. 2009 to Feb. 2011, which was about two times of the present annual limit. $PM_{2.5}$ concentration, even in Olympic period, was 1.3 times of new CNAAQS. Observation studies in Handan were pursued since Aug., 2012. The average PM_{10} and $PM_{2.5}$ concentrations could reach 248.6 and 121.5 $\mu g \cdot m^{-3}$ in the winter of 2012, which is about 1.6 times of the present PM_{10} limit and the new standard for $PM_{2.5}$. As discussed above, Jan. 2013 was quite special and highly polluted. During this month, the average PM_{10} and $PM_{2.5}$ concentrations were 347.2 and 233.4 $\mu g \cdot m^{-3}$, respectively, which were 2.3 and 3.1 times of the daily limit. SO_2 and NO_X in wintertime were also unnegligible in Shijiazhuang and Handan city. SO_2 pollution in Shijiazhuang was more severe than in Baoding, and NO_X was on the contrary. Comparing with Handan, O_3 pollution in Shijiazhuang city was more serious according to the limited data. Long-term observations are needed to understand the air pollution characteristics in the southern Hebei cities.

In intercomparison with other cities outside Hebei, Wu's study [13] compared the PM_{10}, SO_2, NO_X and O_3 concentrations in Beijing, Zhuozhou (a county-level city within Baoding), Baoding and Shijiazhuang and concluded that during the monitoring period (Jul. 2009-Feb. 2011), Shijiazhuang city had highest concentrations of PM_{10} and SO_2. The highest NO_X concentrations appeared in Beijing and the most severe O_3 pollution happened in Baoding. And during the three years, NO_X concentrations were increasing and SO_2's were decreasing. O_3 and PM_{10} concentrations kept stable.

In [9], two severe pollution episodes over North China were reported in the period of Oct. 27 to Nov. 10 in 2009. In the comparison of Beijing, Tianjin and nine cities in Hebei (except Xingtai and Handan city), the highest daily PM_{10} concentration appeared in Shijiazhuang of 600 $\mu g \cdot m^{-3}$, four times of the national limit of 150 $\mu g \cdot m^{-3}$. Another important conclusion is that the heavy pollution episodes were characterized by nearly uniform concentrations over northern China and directly related to the strength and duration of the southern flows. And the meteorological conditions of light wind, temperature reversion and low mixed layer were important contributors to the increase of PM.

In Liu et al.'s study [16], PM_{10} were sampled during Sep. to Oct. 2005 all over Hebei Province, Beijing and Tianjin City, to analysis the PAH pollution. They found that the highest PM_{10} concentration appeared in Handan and Shijiazhuang and the PAH pollution were most serious in Handan, Shijiazhuang and Tangshan.

In general, air pollution in the southern cities is more severe than in the northern cities in Hebei. PM is undoubtedly the key pollutant. SO_2 and NO_X are unnegligible and O_3 pollution in larger cities is also significant, indicating that the both the coal-burning emissions and the mobile sources should be considered in the air pollution controls.

Visibility and Haze Frequencies

Visibility might be seen, to some extent, as an indicator of air quality. It has a longer history data for analysis. Che et al. gathered the visibility data from 1981 to 2005 of 615 meteorological stations in mainland China and found that 71% of these stations observed a visible deterioration and this trend became more clear after 1990 [17]. The highest haze frequencies happen in three areas: North China, the Yangtze River Delta and the Pearl River Delta. And the rapid increase in haze frequencies occurs in the middle and southern areas of North China Plain, the middle and lower reaches of the Yangtze River, and South China. The North China area has both the highest number of haze days and the most rapid growth in haze frequency [18].

Within North China area, Zhao [19] analyzed the data from 100 stations in Beijing-Tianjin-Hebei area from 1980 to 2008 and found that the southern cities in Hebei, such as Shijiazhuang, Xingtai and Handan had the lowest visibility of 10 - 14 km on annual average (Beijing was 15 - 20 km) since 1990. In comparison with other cities using the data from 743 stations all over China, the haze frequency in Xingtai City ranked second on average from 1951 to 2005, and became the first after the mid-1990s [20].

Wang et al. [18] analyzed the haze frequencies in 2001 to 2010 of the seven typical cities in North China, Beijing, Tianjin, Shijiazhuang, Xingtai, Taiyuan, Zhengzhou and Jinan. It was found that 2007 was the worst year and the haze frequencies from highest to lowest were Taiyuan, Shijiazhuang, Zhengzhou, Xingtai, Jinan, Beijing, and Tianjin.

AIR POLLUTION CONTROL HISTORY AND EMISSIONS

Hebei Province made a great effort on air pollution control since 1998 [21]. In 1998, air pollution control measurements focused on the key corporations, key industries and key regions, such as tourist regions and areas along the high way. In 1999, eleven cities were all required to make the comprehensive air pollution control action plan, and the mobile emission controls were strengthened as well.

In 2002, besides the continuous emission controls in major industries, the energy using in cities was paid more attention to and the central heating was pushed to spread in urban areas. In 2003, the provincial total amount control of SO_2 was started, according to the national control plan. But it was found that the pollution emissions didn't decrease in the following three years, partially because of the unexpected rapid increase in energy consumption [22,23].

In 2007, Hebei government released the Action Plan of the Comprehensive Controls of Flue Gas Emissions in Hebei Province. It required all the emission instruments reached the national emission standards before Jun. 2008. The explosive increase in vehicle population in Hebei Province was noticed and its pollution control was strengthened as well.

In 2008, the objectives of the Action Plan were accomplished and urban air quality was improved due to the flue gas cleaning, fugitive dust control and mobile source control. During the 2008 Beijing Olympics, lots of small industries, high-pollution plants were phase out or shutdown to ensure the good air quality in Beijing. It brought a better air quality in Hebei Province as well in this year. In 2009, the SO_2 total amount control, energy optimizing in cities, moving high-pollution plants from our urban area were continuous pushing forward in Hebei Province.

In 2010, the national SO_2 emission control objective was successfully accomplished [23]. Regional air quality jointly control and prevention were brought forward by MEP. Hebei government published the regional air pollution control guideline to accelerate the regional scale air quality improvement.

In the national 12th FYP of jointly air pollution prevention and control published in 2012, Beijing-TianjinHebei was listed in the three key regions. More effective controls could be expected under this action structure.

Figure 5 gives the provincial emissions of SO_2, fly ash and dust emissions from 1998 to 2010. The data is come from the Report on the Environmental State of Hebei Province published every year by Hebei Environmental Protection Bureau (HBEPB) [21]. It can be seen that the total SO_2 emissions decreased in 1998 to 2002 by 9.0% from the 1405 kt·y^{-1} to 1279 kt·y^{-1}. Then it began to increase to 1545.5 kt·y^{-1} in 2006, which could partially attribute to the rapid increase of energy use during these five years [22,24]. After that it kept decreasing again, due to the effective national-scale SO_2 emission controls during the 11th FYP (2006-2010) [23]. In 2010 the total SO_2 emission was 1233.8 kt·y^{-1}, which is 20.2% lower than that in 2006.

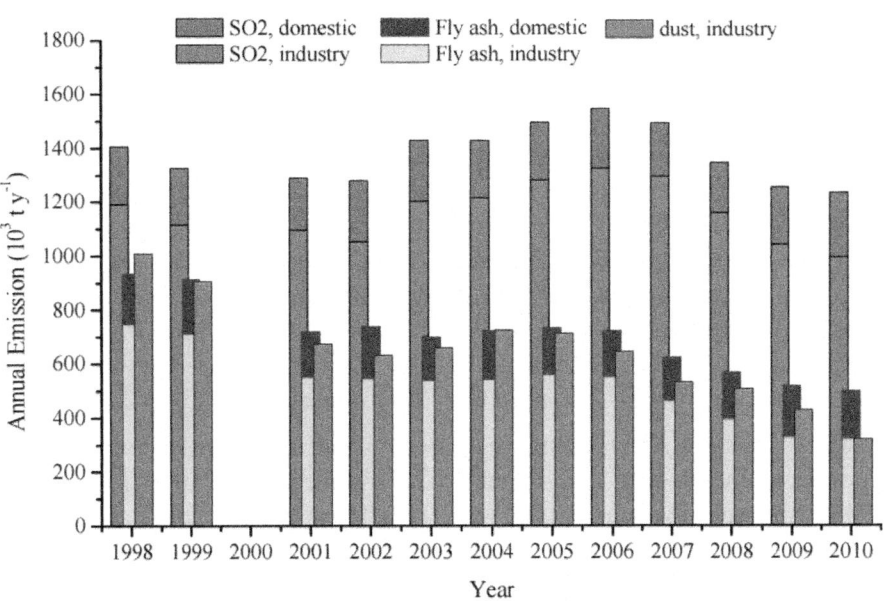

Figure 5. SO_2, fly ash, and dust emissions in Hebei Province in 1998-2010.

Figure 5 also indicated that industry contributed most of the SO_2 emissions in Hebei, which the fractions were between 80.6% (2010) to 86.7% (2007). The domestic emissions were relatively stable that kept between 186.4 kt·y^{-1} (2008) to 239.6 kt·y^{-1} (2010).

As to fly ash emissions, the general trend was decreasing, from 934 kt·y^{-1} in 1998 to 499.7 kt·y^{-1} in 2010. During 2001 to 2006 it kept relatively stable between 699 kt·y^{-1} to 738 kt·y^{-1}. After that the decreasing trend was obvious from 723.2 kt·y^{-1} in 2006 by 30.9% to 499.7 kt·y^{-1} in 2010. The industrial emission also accounted for a big fraction of 63.6% (2009) to 80.0% (1998).

The industrial dust emissions showed the similar variation as the SO$_2$ emissions. The general trend was decreasing, from 1007 kt·y^{-1} in 1998 by 68.1% to 320.9 kt·y^{-1} in 2010. The rapid decrease happened during the 11th FYP (2006-2010) as well.

CONCLUSIONS

Air pollution in Hebei Province has aroused a wide public concern, partially because of the severe fog-haze period happening in the beginning of 2013. It was reported that during this period, seven out of the top ten polluted cities in China were within Hebei Province. But, most of the previous studies involving Hebei focused on Beijing's air quality, the impact of Hebei's emissions to the air quality in Beijing, etc. Very few studies were pursued focusing on the severe air pollution within Hebei.

In this study, we reviewed and analyzed the air pollution history and status in Hebei Province, according to the API data and relative academic observations. It is concluded that air pollution in southern cities are much more severe than in northern cities. PM is the most important pollutant in Hebei cities, and SO$_2$ and NO$_X$ pollution are unnegligible as well. O$_3$ pollution in larger cities is significant, indicating that Hebei's cities are on the way from the coal-burning pollution to the mixed-source pollution. Visibility and haze frequencies in Hebei cities are discussed, that Hebei has both the highest number of haze days and the most rapid growth in haze frequency in recent years.

Hebei made a great effort on the air pollution control since 1998. The major air pollutant, such as SO$_2$ and fly ash, showed a trend of increase in 2001 to 2006 and decrease since 2006. In 2012 MEP published national plan of the regional jointly air pollution prevention and control, more effective control strategies and measurement could be expected to improve the air quality in Hebei Province.

ACKNOWLEDGEMENTS

This study was sponsored by the National Natural Science Foundation of China (No. 41105105) and the Natural Science Foundation of Hebei Province (No. D2011 402019).

REFERENCES

1. MEP, "China National Ambient Air Quality Standards," GB3095-2012, MEP, Beijing, 2012.

2. NBS (Natural Bureau of Statistics), "China Statistical Yearbook 2012," China Statistical Press, Beijing, 2012.

3. NBS (Natural Bureau of Statistics), "China Energy Statistical Yearbook 2012," China Statistical Press, Beijing, 2012.

4. Q. Zhang, D. G. Streets, G. R. Carmichael, K. B. He, H. Huo, A. Kannari, Z. Klimont, I. S. Park, S. Reddy, J. S. Fu, D. Chen, L. Duan, Y. Lei, L. T. Wang and Z. L. Yao. "Asian Emissions in 2006 for the NASA INTEX-B Mission," Atmospheric Chemistry and Physics, Vol. 9, 2009, pp. 5131-5153. doi:10.5194/acp-9-5131-2009

5. L. T. Wang, P. Zhang, S. B. Tan, X. J. Zhao, D. D. Cheng, W. Wei, J. Su and X. M. Pan, "Assessment of Urban Air Quality in China Using Air Pollution Indices (APIs)," Journal of the Air & Waste Management Association, Vol. 63, No. 2, 2013, pp. 170-178. doi:10.1080/10962247.2012.739583

6. State Environmental Protection Agency of China, "China National Ambient Air Quality Standard (in Chinese)," GB3095-1996, State Environmental Protection Agency of China, Beijing, 1996.

7. State Environmental Protection Agency of China, "China National Ambient Air Quality Standard (in Chinese)," GB3095-2012, State Environmental Protection Agency of China, Beijing, 2012.

8. State Environmental Protection Agency of China, "Technical Regulations on Ambient Air Quality Index (in Chinese)," HJ633-2012, State Environmental Protection Agency of China, Beijing, 2012.

9. D. S. Ji, Y. S. Wang, L. L. Wang, L. F. Chen, B. Hu, G. Q. Tang, J. Y. Xin, T. Song , T. X. Wen, Y. Sun, Y. P. Pan and Z. R. Liu, "Analysis of Heavy Pollution Episodes in Selected Cities of Northern China," Atmospheric Environment, Vol. 50, 2012, pp. 338-348. doi:10.1016/j.atmosenv.2011.11.053

10. T. X. Wen, Y. S. Wang, H. H. Xu, Z. Q. Ma and S. D. Ji. "Comparison of the Characteristics of Ambient Pollutant s in Urban and Background Region in Beijing during August and September (in Chinese)," Research of Environmental Sciences, Vol. 20, No. 5, 2007, pp. 7-11.

11. Y. P. Pan, Y. S. Wang, B. Hu, Q. Liu, Y. H. Wang and W. D. Nan, "Observation on Atmospheric Pollution in Xianghe During Beijing 2008

Olympic Games (in Chinese)," Environmental Science, Vol. 31, No. 1, 2010, pp. 1-9.

12.   X. Y. Wang, J. Y. Xin, Y. S. Wang, X. X. Feng and Y. P. Zhang, "Observation and Analysis of Air Pollution in Tangshan During Summer and Autumn Time," Environmental Science, Vol. 31, No. 4, 2010, pp. 877-885.

13.   Y. Wu, "The Network Observation and Research on the Atmospheric Pollutants in Beijing, Zhuozhou, Baoding and Shijiazhuang," Master's Thesis, Nanjing University of Information Science &Technology, 2011.

14.   W. P. Du, Y. S. Wang, T. Song, J. Y. Xin, Y. S. Cheng and D. S. Ji, "Characteristics of Atmospheric Pollutants during the Period of Summer and Autumn in Shijiazhuang," Environmental Science, Vol. 31, No. 7, 2010, pp. 1-8.

15.   P. Zhang, S. B. Tan, L. T. Wang, X. J. Zhao, J. Su, F. F. Zhang, Z. Wei, W. Wei and D. D. Cheng. "Characteristics of Atmospheric Particulate Matter Pollution in Handan City," Acta Scientiae Circumstantiae, (in Press).

16.   S. Z. Liu, X. Q. Li, S. Tao, Z. F. Tian, J. F. Wang and Y. Gao, "PAHs in the Atmosphere from the Area of Hebei, Beijing, and Tianjin during the Non-Heating Season," Acta Scientiae Circumstantiae, Vol. 28, No. 10, 2008, pp. 2105-2110. (in Chinese)

17.   H. Z. Che, X. Y. Zhang, Y. Li, Z. J. Zhou, J. J. Qu and X. J. Hao, "Haze Trends over the Capital Cities of 31 Provinces in China, 1981-2005," Theoretical and Applied Climatology, Vol. 97, No. 3-4, 2009, pp. 235-242. doi:10.1007/s00704-008-0059-8

18.   L. T. Wang, J. Xu, J. Yang, X. J. Zhao, W. Wei, D. D. Cheng, X. M. Pan and J. Su, "Understanding Haze Pollution over the Southern Hebei Area of China Using the CMAQ Model," Atmospheric Environment, Vol. 56, 2012, pp. 69-79. doi:10.1016/j.atmosenv.2012.04.013

19.   P. S. Zhao, X. L Zhang, X. F. Xu and X. J. Zhao, "Long-Term Visibility Trends and Characteristics in the Region of Beijing, Tianjin, and Hebei, China," Atmospheric Research, Vol. 101, No. 3, 2011, pp. 711-718,. doi:10.1016/j.atmosres.2011.04.019

20.   D. Wu, X. J. Wu, F. Li, H. B. Tan, J. Chen, Z. Q. Cao, X. Sun, H. H. Chen, H. Y. Chen, "Temporal and spatial variation of haze during 1951-2005 in Chinese mainland (in Chinese)," Meteorologica Sinica, Vol. 68, 2010, pp. 680-688.

21.   HBEPB, "Report on the Environmental State in Hebei Province, 1998-2010," 2013. http://www.hb12369.net:8000/hjzlzkgb/

22.   L. T. Wang, C. Jang, Y. Zhang, K. Wang, Q. Zhang, D. Streets, J. Fu, Y.

Lei, J. Schreifels, K. B. He, J. M. Hao, Y. F. Lam, J. Lin, N. Meskhidze, S. Voorhees, D. Evarts and S. Phillips, "Assessment of air Quality Benefits from National Air Pollution Control Policies in China. Part I: Background, Emission Scenarios and Evaluation of Meteorological Predictions," Atmospheric Environment, Vol. 44, No. 28, 2010, pp. 3442-3448. doi:10.1016/j.atmosenv.2010.05.051

23.   W. B. Xue, J. N. Wang, H. Niu, J. T. Yang, B. P. Han, Y. Lei, H. L. Chen and C. L. Jiang, "Assessment of Air Quality Improvement Effect under the National Total Emission Control Program during the Twelfth National Five-Year Plan in China," Atmospheric Environment, Vol. 68, No., 2013, pp. 74-81. doi:10.1016/j.atmosenv.2012.11.053

24.   L. T. Wang, C. Jang, Y. Zhang, K. Wang, Q. Zhang, D. Streets, J. Fu, Y. Lei, J. Schreifels, K. B. He, J. M. Hao, Y. F. Lam, J. Lin, N. Meskhidze, S. Voorhees, D. Evarts and S. Phillips, "Assessment of Air Quality Benefits from National Air Pollution Control Policies in China. Part II: Evaluation of Air Quality Predictions and Air Quality Benefits Assessment," Atmospheric Environment, Vol. 44, No. 28, 2010, pp. 3449-3457. doi:10.1016/j.atmosenv.2010.05.058

Chapter 2

AIR CHANGE MEASUREMENTS USING TRACER GASES

Detlef Laussmann and Dieter Helm

[1] Robert Koch Institute, Germany

INTRODUCTION

Both comfortable and healthy indoor climate conditions can only be achieved by constant fresh air supply. However, the minimum of air change required to reach this goal is depending upon different perspectives. Measures for thermal insulation and energy savings are difficult to bring in line with air quality requirements that result from findings of epidemiological studies (Seppänen et al., 1999; Seppänen& Fisk, 2004; Wargocki et al., 2002), and supply air facilities considering the technical construction of buildings and safety aspects (Erhorn & Gertis, 1986). Moreover, construction deficits impairing the integrity of the building envelope, meteorological conditions (thermal and flow induced pressure differences) and, not least, the behavior of the residents, affect the air change in a variety of ways (Heidt, 1987).

For a number of different reasons it is necessary and desirable to examine the real fresh air flow between indoor and outdoor climate under given circumstances. Air change processes are of particular importance in studies focusing on their relationship to indoor air pollutants.

For manifold reasons the air quality inside buildings has been intensively investigated since quite a long time. Of particular relevance are the following aspects (Seifert & Salthammer, 2003):

- In countries with a cold or temperate climate, inhabitants spend more than 50% of the time in their homes. In certain population groups (e.g. infants and the elderly), this proportion is even exceeding 90%.

- Indoor air contains a wide range of different organic and inorganic components. Therefore, the carbon dioxide concentration alone cannot always be regarded as an indicator of air quality (Fanger, 1988; Persily,

1997). Above all, organic compounds which are released from the building, furnishings, household and hobby devices, as well as by daily activities of the inhabitants, such as cooking, baking or frying, and especially smoking, altogether contribute to air pollution caused by volatile organic compounds.

The energy crisis in the 1970s caused an increase in energy costs, evoking an urgent need to reduce the consumption of heat energy. The prevention of heat loss in homes is a very effective way to save energy and related costs. For climate protection and the reduction of global CO_2 emissions the economical use of energy resources is of outstanding significance. Thermal insulation of the building envelope protects efficiently against heat loss, but enhanced tightness includes the disadvantage of reduced air change, thereby increasing the indoor pollutant concentration. To evaluate the indoor air quality under conditions of natural ventilation, the air change rate (ACR) can be determined through standard tracer gas measurement.

For this purpose, a small amount of the tracer gas is released in the room (or building) under study and its concentration is then recorded as a function of time. Subsequently, using appropriate evaluation algorithms the ACR can be calculated from the data obtained.

The first section is introducing basic physical principles of air change processes between indoor and outdoor environment, followed by a description of tracer gases which are frequently used in daily routine, and methods determining the ACR. Moreover, the applicability of carbon dioxide as a tracer gas has been compared with results obtained by hexafluorobenzene or sulphur-hexafluoride. In addition, we are discussing the impact of weather conditions on ACR data obtained under natural ventilation. This section is followed by a review on ACR due to passive ventilation through facades in selected residences in Berlin. Finally, the relationship between air change and the concentration of selected volatile compounds under worst case conditions will be discussed.

MODELLING AIR CHANGE IN INDOOR ROOMS

According to VDI (2001) 4300, part 7, air change is defined as the ratio of air supply Q(t) into a zone (i.e. a room or space) in relation to the volume of this zone V_R (room volume) and is generally expressed as air change per hour [h^{-1}] or [ACH]. The following equation expresses this definition:

$$\lambda(t) = Q(t)/V_R \tag{1}$$

$\lambda(t)$ is the ventilation rate or air change rate [h^{-1}],

Q(t) is the air supply into a room [m^3/h],

V_R is the room's volume [m³], and t = time [h].

Model Assumptions

For the model described here, which is designed to calculate the time course of the tracer gas concentration, the following simplifying assumptions has been made:

- The tracer gas is considered to be chemically stable and inert; i.e. there will be no chemical reactions capable to alter the concentration of the tracer gas in the room.

- There will be no adsorption processes on walls, ceiling or furnishings of the room that may lower the concentration of the tracer gas in the room.

- The air is considered to be completely mixed throughout the measurements. Inside the room there are no concentration gradients, i.e. the concentration of the tracer gas at a given time is the same for the whole room.

- An exchange of tracer gas-containing air with ambient air only occur in those areas that are in contact with the outside, i.e. air change with other interior spaces is considered to be negligible. The room in which the tracer gas was released is considered a single zone system.

- The exchange processes that take place during the measurement period are assumed to be temporally invariant. The air supply rate Q(t) and, thus, the air change rate λ(t) are constant. Q(t) and λ(t) can be replaced by Q and λ.

Model Equations

The basis for the description of the relationship between the mass or concentration of a gaseous substance in a space as a function of time is the mass balance equation. This equation expresses that the mass and – in a fixed volume – the concentration of a tracer gas can only change when either more tracer gas is added to the original amount or tracer gas is removed by elimination processes. Considering the above assumptions, the following supply and removal processes are significant for the tracer gas concentration in the room air (physical dimensions of these variables are given in brackets):

- Transport of tracer gas from the room air to the outside: $Q*C_i$ [mass per time unit]

- Transport of tracer gas from the outside air into the room air $Q*C_a$ [mass per time unit]

- (Constant) emission E of tracer gas into the space by a tracer gas source

[mass per time unit]

Thus, the mass balance equation can be formulated as the following differential equation (Heidt& Werner, 1986):

$V_R*dC_i(t)/dt = -C_i*Q + C_a*Q + E$

or

$$V_R*dC_i(t)/dt = -(C_i - C_a)*Q + E$$

(2a)

After dividing both sides of the equation by the volume V_R, we obtain an ordinary differential equation which describes the concentration change of a tracer gas in the room per time unit:

$$dC_i(t)/dt = -(C_i - C_a)*Q/V_R + E/V_R$$

(2b)

C_a: tracer gas concentration in outside air [mass / volume]

C_i: tracer gas concentration in the indoor air [mass / volume]

Q: exchange air flow between room and outside [volume / time unit]

E: amount of tracer gas emitted per unit time [mass / time unit]

V_R: room volume

t: time

As stated above, Q/V_R is defined as the air change rate. When Q/V_R in Eq. 2b is replaced by λ (Eq. 1) we obtain Eq. 2c:

$$dC_i(t)/dt = -(C_i - C_a)*\lambda + E/V_R$$

(2c)

Equation 2c expresses that under constant homogeneous mixing the concentration change of the tracer gas is proportional to the concentration difference between indoor and outdoor spaces $(C_i - C_a)$ at time t, the air change rate λ, and the amount of tracer gas emitted per time unit (emission rate E).

By integration we obtain the starting conditions at time 0 $(C(t=0) = C_0)$:

$$C_i(t) = C_a + E/(\lambda*V_R) + [C_0 - C_a - E/(\lambda*V_R)] \exp(-\lambda t_i)$$

(3a)

Expanding and transposing yields:

$$C_i(t) = (C_0 - C_a) \exp(-\lambda t_i) + C_a + E/(\lambda*V_R) [1 - \exp(-\lambda t_i)]$$

(3b)

Integration of the differential equation thus leads to an exponential function with the air change rate λ in the exponent. Equation 3b is the basis for the mathematical analysis of tracer gas measurements that are recorded as concentration-time curves. This function is characterized by the following features: If no tracer gas is emitted into the room (i.e. E = 0), and there is already a non-zero tracer gas concentration C_0 present at the time t = 0 which is higher than the outdoor air concentration C_a, then the expression $(C_0 - C_a)$ $\exp(-\lambda t) + C_a$ describes the elimination of the tracer gas out of the room. The curve starts with the initial concentration C_0 and decays exponentially until the ambient tracer gas concentration C_a or any other constant background concentration is reached. If a tracer gas is used, which does not occur in the outside air (i.e. $C_a = 0$) the concentration decreases over time to the value zero. Equation 3b can be simplified to:

$$C_i(t) = C_0 \exp(-\lambda t_i)$$

(3c)

If a source in the room is emitting tracer gas with a constant rate E and if the initial tracer gas concentration in the room at the beginning of the measurement C_0 is equal to the outside concentration C_a ($C_0 = C_a$), then the tracer gas will accumulate within the room. Starting with the concentration C_a, the mathematical function grows with increasing t asymptotically towards C_{eq}, which reaches the final value of $C_a + E/(\lambda*V_R)$ at infinity.

$$C(t) = C_a + [E/(\lambda*V_R)][1-\exp(-\lambda t)]$$

(4a)

When the outdoor air concentration can be neglected, then:

$$C(t) = [E/(\lambda*V_R)][1-\exp(-\lambda t)]$$

(4b)

For $t \gg 1/\lambda$ Equation 4b can be simplified to:

$$C_{eq} = E/(\lambda*V_R)$$

(4c)

C_{eq}: equilibrium concentration

Equation 4c reflects the fact that under equilibrium conditions (i.e. emission equals elimination) the tracer gas concentration C_{eq} is proportionally dependent on the emission rate E but inversely proportionally dependent on both the air change rate λ and the volume of the room V_R. This relationship forms the basis for the determination of the air change rate with tracer gas measurements made under equilibrium conditions.

DETERMINATION OF THE AIR CHANGE RATE (ACR) USING TRACER GASES

In addition to the physical properties discussed in Section 2.1 (model assumptions) tracer gases should fulfil some other requirements in regard to their practical suitability. These include health safety aspects, low environmental burden, high availability and good handling in practical use at the lowest costs possible and, not least, tracer gases should be well recordable with established measurement techniques over a wide concentration range and with high selectivity (Raatschen, 1995).

Requirements for Ideal and Commonly Used Tracer Gases

In the past, a number of gaseous substances such as helium, hydrogen, oxygen, carbon monoxide, methane, acetone, and the radioactive noble gases argon-41 and krypton-85, have been studied and tested for the determination of ACR, mostly in comparison with other tracer gases. Reviews are given by Grimsrud et al. (1980), Shaw (1984), and Sherman (1990). Until the early 1990s, krypton-85 was still used for air change measurement (Schulze &Schuschke, 1990), but later skipped for safety reasons (radiation protection). Nowadays, the following tracer gases are mainly used in practice (Raatschen, 1995):

- Nitrous oxide (N_2O)
- Sulphur hexafluoride (SF_6)
- Halogenated hydrocarbons, such as hexafluorobenzene (C_{6F6} and perfluorocarbons (PFC).

Nitrous Oxide

Formerly, nitrous oxide (N_2O) was widely used for air change measurement in buildings, primarily in Europe (Heidt& Werner, 1986; Keller &Beckert, 1994; Salthammer, 1994; Wegner, 1983, 1984). In the U.S. it was rarely used because of its low TLV (threshold limit value) of 50 ppm (Lagus& Grot, 1997). Germany's equivalent of TLV, the MAK, is however 100 ppm. For precautionary reasons it should not be used in occupied buildings (Raatschen, 1995). Other disadvantages of N_2O are the ease of adsorption on surfaces at concentrations below 1000 ppm, and its high solubility in water, which means that the air change rate can be substantially overestimated in very airtight rooms (Schulze &Schuschke, 1990).

Sulphur Hexafluoride (SF₆)

After Gregory's observation in 1962, that sulphur hexafluoride (SF_6) can be measured reliably on the nanogram scale with electron capture detection – ECD (Gregory, 1962), it has been widely used for air infiltration measurement in buildings since the early 1970s (Drivas et al., 1972; Hunt & Burch, 1975). Of all candidates, the characteristics of SF_6 are nearest to the ideal of a tracer gas. Today, SF_6 is the most frequently used tracer gas worldwide, which is confirmed by the number of relevant publications. In this chapter, only on a very limited selection of publications could be considered (Chuah et al., 1997; Howard-Reed et al., 2002; Kumar et al., 1979; Lagus& Grot, 1997; Raatschen, 1995; Shaw, 1984; Walker & Forest, 1995; Wilson et al., 1996). SF_6 is very stable and only decomposes above 550 C. Background concentration of SF_6

in ambient air is ≈ 1 ppt(6 ng/m³) (Raatschen, 1995). Although its density is about five times higher than that of air, this difference causes no systematically distorting effects on the results of air change measurements with concentrations usually applied in practice (Niemelä et al., 1991; Shaw, 1984). SF_6 can be used in occupied buildings. It can be recorded with high accuracy within a wide concentration range. According to Raatschen (1995) the concentrations commonly used for indoor air change measurement are not exceeding one hundredth of the German MAK value which was defined to be 1000 ppm or 6100 mg/m³ (TRGS 900, 1999). It should be noted that this value is not a toxicity limit. It is simply defined as the still manageable analytical upper limit for gases which are not imminently toxic (BIA-Report, 2001). Due to its high stability SF_6 is only very slowly degraded in the atmosphere and it belongs, like the perfluorinated hydrocarbons, to the climatic relevant greenhouse gases. Therefore, to avoid unnecessary environment hazard it should be used carefully and sparingly in concentrations as low as possible, like other tracer gases.

Hexafluorobenzene (C_6F_6) and Perfluorocarbon-Hydrocarbon Tracers (PFT)

Hexafluorobenzene, and perfluorocarbon-hydrocarbon tracers (PFT) such as perfluorodimethylcyclobutane, perfluoromethylcyclobutane, perfluorodimethylcyclo-hexane, and perfluoromethylcyclohexane are also appropriate as tracer gases and are preferably used in the determination of ACR with the constant injection method applying the passive sampler technique. Enrichment for active sampling is also possible (Cheong &Riffat, 1995; Dietz & Cote, 1982; Dietz et al., 1986; Krooß et al., 1997; Mailahn et al., 1989; Salmon et al., 2000). Since these compounds, unlike SF_6, adsorb well on activated carbon or Tenax they are frequently used in field studies (surveys) to determine indoor air change rates (Andersen et al., 1997; Bornehag et al., 2005; Hirsch et al., 2000; Lembrechts et al., 2001; Øie et al., 1997, 1998; Parker, 1986; Pandian et al., 1993; Ruotsalainen et al., 1992; Sakaguchi & Akabayashi, 2003). The disadvantage of these compounds, however, is that they tend to attach to room surfaces and the emission rates of these gases are strongly temperature dependent (Hill et al., 2000). Thus, a sufficiently long conditioning period is necessary as well as an accurate temperature control of the storage vessels.

Methods for the Determination of ACR

Basically, three appropriate methods exist for the determination of ACR using tracer gases. According to VDI (2001) 4300, part 7, these are the concentration decay method, the constant injection method, and the constant concentration method.

Concentration Decay Method

Tracer gas is injected into the room for a short period of time, either from a gas bottle with pressure reducer or manually from filled gas tanks. After mixing with the room air the tracer gas concentration is measured at regular time intervals. Because the decay curve of the tracer gas concentration C follows an exponential course when completely mixed with the room air (see Eq. 3c) each sampling will not only remove old air but also a certain amount of fresh air supply as well. This essentially means that there will be still $\approx$37% (100/e^1) of the originally added tracer gas (37% old room air), after a complete air change cycle has occurred. The time (1/λ), after which the air change cycle is completed is known as the nominal time constant τ (Maas, 1997; Sherman, 1990). After 3τ (3/λ), 4τ (4/λ) and 4.6τ (4.6/λ), the tracer gas concentration in the room volume under study is 5%, 2% and 1% of its initial value, respectively (Fig. 1a).

By using the concentration decay function

$$C(t) = C_0 * e^{-\lambda * t}$$

(5)

and applying non linear regression analysis we can determine the ACRλ (C_0: tracer gas concentration at time t=0 (Sherman, 1990)). However, in most cases logarithmic concentration values are used to obtain a linear relationship between the logarithm of the tracer gas concentration $C_i(t)$ and the time t (Fig. 1b):

$$\ln C_i(t) = \ln C_0 - \lambda * (t_i)$$

(5a)

The ACR λ is then calculated via linear regression analysis according to Eq. 5a. Both evaluation options – the linear and the non-linear regression analysis – are particularly well suited for the examination of concentration-time curves recorded over a longer period of time. If only a few measurements are available, e.g. when sampling is performed with syringes or other appropriate devices, then the ACR can be determined using the following relationship:

$$[\ln C_i(t=t_i) - \ln C_i(t=t_{i+1})] / (t_{i+1} - t_i) = \lambda.$$

(5b)

$C_i(t_i)$ is the tracer gas concentration at time t_i

$t_{i+1} - t_i$ is the time interval between two measurements

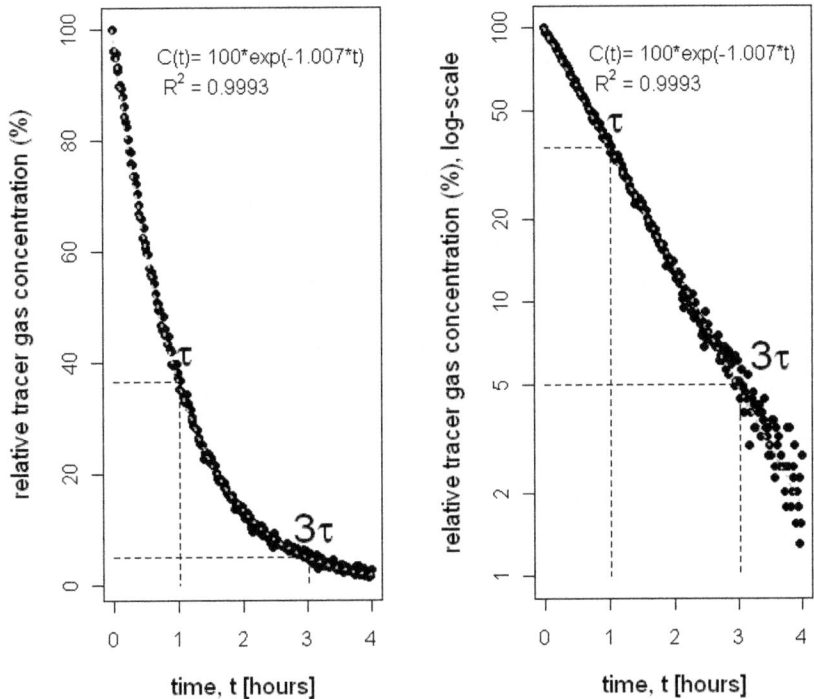

Figure 1. Relationship between tracer gas concentration and time; a, untransformed ordinate; b, after logarithmic transformation.

An illustration of the procedure is given in Fig. 2a. Sulphur hexafluoride (SF_6) was released in an office space (volume 48 m³, equipped with two double glazed box windows). Initial concentration of the tracer gas was adjusted to ≈5 mg/m³. The concentration was measured for 7 hours in 15 minutes intervals using a photo-acoustic infrared detector with selective filter (Bruel&Kjaer Single Gas Monitor). During the measurement the office was unoccupied; door and windows were kept closed. Tracer gas and room air were not continuously mixed with a fan or similar device. For the analysis, SF_6 concentration values were chosen in such a way that each concentration-time pair – $c(t_{i+1})$, $c(t_i)$ – covered exactly a time interval of one hour, meaning that the denominator was equal to 1 when determining the air change rate (ACR). After choosing the six measurement points depicted in Fig. 2awe calculated an average ACR of 0.327

h⁻¹. The standard deviation was 0.038 h⁻¹, which corresponds to a coefficient of variation of ≈12%. When the same measuring points were analysed by linear regression analysis we obtained an ACR of 0.323 h⁻¹ (Fig. 2b). Both values are almost equal. The determination of ACR using concentration-time pairs is also known as the two-point method (Sherman, 1990). For this method, a number of at least five concentration-time pairs are recommended which should be evenly distributed over the entire measurement period (ASTM E 741 – 00; VDI, 2001). In order to reduce the statistical error to ≤10% the time interval between the first and the last measuring point should be in the order of magnitude of the nominal time constant (ASTM E 741 – 00;Heidt& Werner, 1986; Maas, 1997; Sherman, 1990). This prerequisite is, however, hardly fulfilled for ACR between 0.1 and 0.2 h⁻¹ or even lower, since the required minimum decay time would be between 5 and 10 hours or more. Most probably, in the case of very low ACR a statistical error of more than 10% must be accepted. Recommended minimum time intervals between the first and the last measurement and the corresponding measurement intervals are summarised in Table 1, according to ASTM E 741 – 00 and VDI (2001).

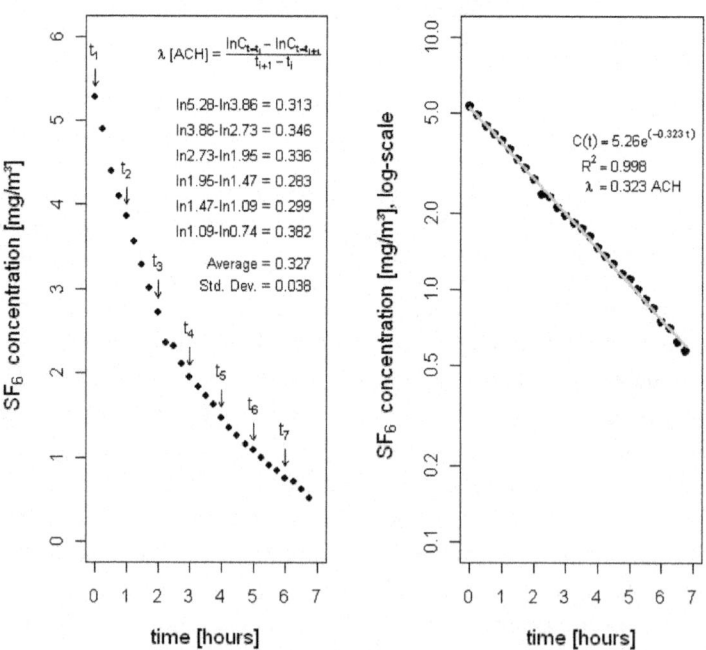

Figure 2. ACR determination from concentration decay with the two-point method,

time interval $t_{i+1} - t_i$ between two measurements: one hour; a, linear scale; b, logarithmic scale, linear regression.

Table 1. Examples for minimum time spans between first and last sampling based on recommendations of ASTM E 471 – 00 and recommended sampling intervals (VDI, 2001) for air change measurement with the decay method.

Air change rate [ACH]	Minimum time span of tracer gas measurement [hours]	Air change rate [ACH]	Sampling interval [min]
0.05	20		
0.125	8		
0.25	4	< 0.5	30 - 40
1	1	0.5 to 1	20 - 30
2	0.5	1 to 2	10
4	0.25	2 to 5	5
10	0.1	> 10	< 2

The concentration decay method is the most commonly used one in practice. It is particularly well suited for the determination of ACR up to 10 h^{-1} (10 ACH) in indoor rooms with a volume below 500 m^3 (VDI, 2001). If data loggers are used to record the concentration decay, the measurement intervals can be shortened down to seconds and in consequence ACR >10 h^{-1} can be determined.

Constant Injection Method

To determine ACR with this method a diffusion tube containing C_6F_6 or PFT is frequently used as tracer gas source. A defined amount of tracer gas is constantly emitted over a certain period of time. Thus, the tracer gas concentration increases with time and reaches a stable value (equilibrium concentration) which depends on the room volume V_R, the air change rate λ, and the emission rate E. At that time point, one or more air samples are taken and the tracer gas concentration is determined for each sample. The ACR can be calculated after solving Eq. 4c for λ. Fig. 3a depicts an example of the constant injection method.

Crucial for this method is, however, that sampling can only be started when the tracer gas concentration is near the equilibrium. If sampling starts too soon, an overestimation of the ACR will be the consequence. The time to reach approximate equilibrium conditions depends on λ (see Fig. 3b).

According to ASTM E 741 – 00, the tracer gas concentration should have reached at least 95% of the equilibrium concentration before measurements can be started. The time until approximate equilibrium is reached can be estimated applying Eqs. 4b, 4c. However, under worst case conditions in rooms with very low air changes, it will be difficult to correctly measure ACRs with this method, because the build-up time to reach nearly equilibrium concentrations is too long (Table 2).

The time intervals for the tracer gas concentration to reach 95%, 98% or 99% of the target equilibrium value (C_{eq}) are $3/\lambda$, $4/\lambda$ and $4.6/\lambda$, respectively (cf. Table 2).

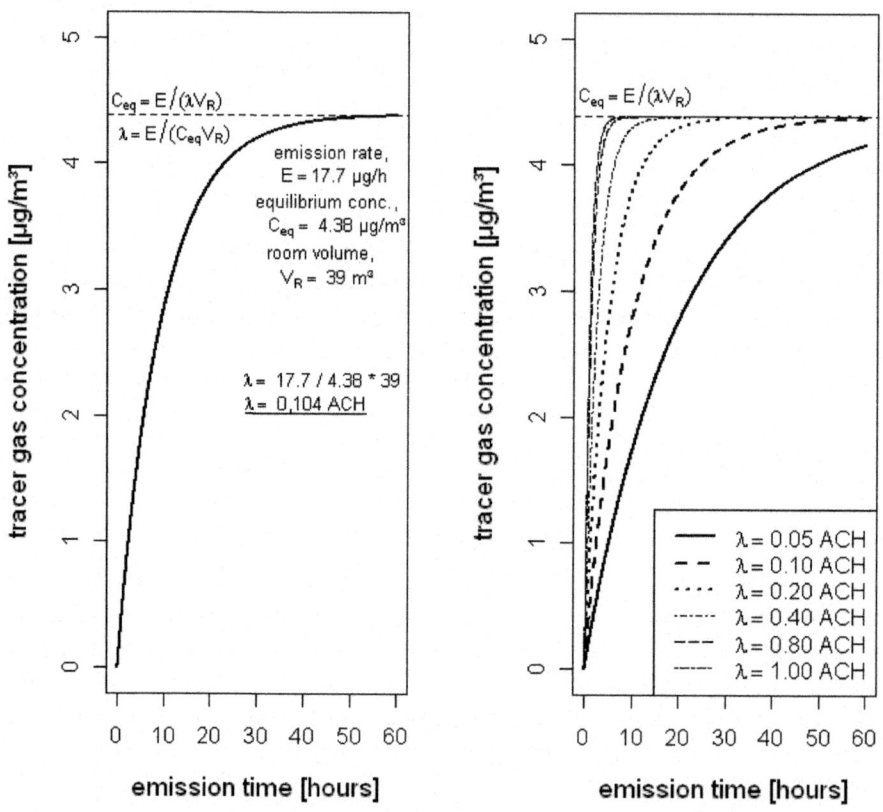

Figure 3. Determination of ACR; a, constant injection method; b, concentrations after constant injection for different ACR.

As Table 2 shows, the time periods needed to reach approximate equilibrium can be several days when the ACR is only 0.1 h^{-1} or even lower. For this reason, the constant injection method is suitable only for short-term measurements of ACR above 0.3 h^{-1} when diffusion tubes are used as tracer gas source. Under usual "worst case" conditions (i.e. last active window ventilation 10-12 hrs before sampling) and ACR <0.3 h^{-1} the tracer gas concentrations are still too far away from the equilibrium. The advantage of the constant injection method is that it can be used for indoor hygiene studies addressing the relationship between pollution and air change in daily used rooms. In this case, tracer gas measurements can be performed with passive samplers and exposure times range from days to weeks.

Table 2. Time spans [hours] needed to reach 95%, 98% or 99% of equilibrium concentration (C_{eq}) for a wide range of air change rates.

	Time span [hours] to reach nearly equilibrium concentration C$_{eq}$		
Air change rate [hours]	95% C$_{eq}$	98% C$_{eq}$	99% C$_{eq}$
0	-	-	-
0.01	300	400	460
0.05	60	80	92
0.1	30	40	46
0.2	15	20	23
0.3	10	13	15
0.4	7.5	10	11.5
0.5	6	8	9.2
1	3	4	4.6
1.5	2	2.7	3
3	1	1.3	1.5

Constant Concentration Method

During constant and thorough mixing with the indoor air, tracer gas is released in the room until a predefined concentration is reached. During the entire measurement the tracer gas concentration is kept constant with an automated dosing and control system. Under the condition of constant tracer gas concentration the air supply is proportional to the tracer gas supply rate. The air supply rate can then be calculated from the ratio of the tracer gas supply to the tracer gas concentration. If the room volume is known, the ACR can be calculated from this ratio (Chao et al., 2004; Kumar et al., 1979; Maas, 1997). An advantage of the constant concentration method is that even short-term changes of air supply can be detected. Compared to the previously described methods the technical equipment required for this method is, however, rather expensive and thus this method is comparatively rarely used for indoor air quality evaluation.

The Use of Carbon Dioxide as a Tracer Gas

Carbon dioxide (CO_2) is one of the gaseous organic compounds always detectable in the indoor air. Since humans exhale metabolic carbon dioxide in considerable quantities, its concentration can increase to several thousand ppm (ml/m^3 room air) within a short time. CO_2 concentration is often used to assess the air quality of occupied rooms. In this context we remind of Pettenkofer's reference concentration. Already in 1858 the German chemist and hygienist pointed out that a CO_2 concentration of 1000 ppm (0.1 vol %) is the upper tolerable limit in indoor environments. Nowadays CO_2measurements are often used for the determination of the indoor ACR, because it can be easily quantified and the required devices are reasonably priced and easy to operate. Moreover, CO_2 fulfils a number of the above mentioned specifications of a good tracer gas.

A huge number of studies are published testing the feasibility of exhaled human carbon dioxide as tracer gas in air change settings. Next to the already mentioned work of Pettenkofer (1858) we would like to allude to the studies done by Penman (1980), Penman & Rashid (1982), and Smith (1988)which are of special importance. Results of Dols&Persily (1992), Nabinger et al. (1994), and Persily (1997) have, however, demonstrated that ACR cannot be reliably determined from spot, peak or average values of the CO_2 concentration inside buldings, because these values are strongly influenced by the number of occupants in the rooms, their times of stay, and, hence, the incessantly changing carbon dioxide supply rates. Depending on the amount of natural ventilation the ACR is sometimes over-estimated up to 2-fold of the real value. The reason is that the air-tightness of modern buildings and the usual sojourn times of the occupants prevent in most cases that the equilibrium concentration can be approached. In practice it is much better to derive the ACR from the decay or build-up curve as was already shown (Barankova, 2005; Bekö et al., 2010; Chao et al., 1997; Chung & Hsu, 2001;Guo& Lewis, 2007; Menzies et al., 1995; Roulet&Foradini, 2002; Schulze &Schuschke, 1990; Sekhar, 2004; Shaw, 1984).

Statistical Evaluation of CO_2 Decay Curves (With Examples)

After CO_2 is released in a room, cither as exhaled breath or via a gas container, its concentration will decay exponentially, if no further CO_2 supply occurs. To exemplify this, the CO_2 decay recorded in a bedroom of an older building (built 1908) is depicted in Fig. 4. In 1990 this bedroom was equipped with a double box window; the room volume is 30 m^3. The room was doped by a person with exhaled carbon dioxide, then the measurement was started and the room was left. During the entire procedure the room was unoccupied – door and window were kept closed. Measurement of CO_2 concentration was done by a CO_2 sensitive probe with infrared absorption. A Testo 400 device (Testo,

Lenzkirch, Germany) was used for data logging and as control unit.

Unlike other tracer gases, CO_2 has the particularity that there is always a certain amount of CO_2 in the outdoor air meaning that the background concentration cannot be neglected. In general the outdoor air concentration of CO_2 is between 350 and 450 ppm or even higher, depending on the season. Thus, the CO_2 decay curve will not decline to zero. Instead, the CO_2 concentration decreases to values which are near to that of the outdoor air. This must be taken into account when analysing CO_2 decay curves. Eq. 5 cannot be applied; the air change rate (ACR) must be determined using the term derived from Eq. 3 b:

$$C(t) = (C_0 - C_a)*e^{-\lambda*t} + C_a \tag{6}$$

C_0: initial concentration, C_a: concentration in the ambient air (i.e. background concentration)

It is possible to examine the linearised curve when the background concentration is subtracted before linearising. Therefore, the background concentration must be determined by an additional measurement.

$$\ln(C(t) - C_a) = \ln(C_0 - C_a) - \lambda*t \tag{6a}$$

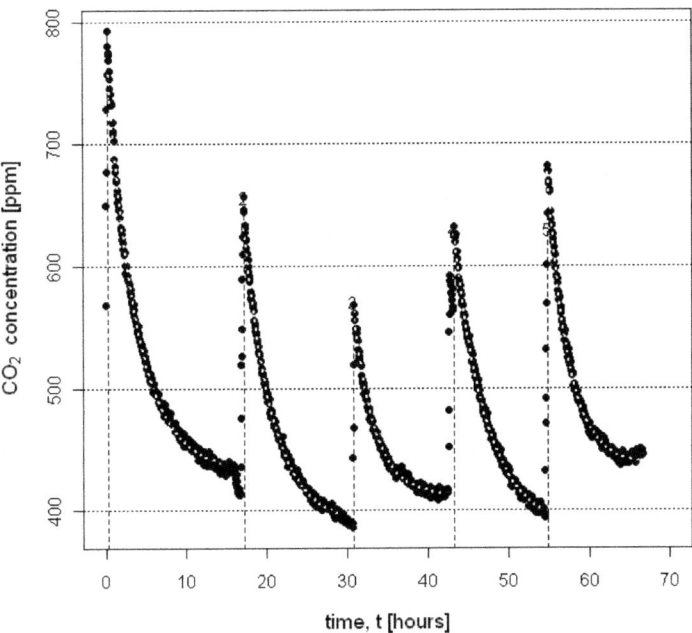

Figure 4. Decay curves of CO_2 (black glyphs) in a bedroom equipped with double box

window and located in a building built in 1908, estimated with nonlinear regression (white curves).

With this relation the ACR can be determined by linear regression analysis. In practice, however, this will be difficult in most cases since it is not feasible to record the CO_2-concentration until the decay curve reaches background level. Modern buildings are in general designed to achieve high airtightness resulting in a very slow decay, and hence the linearisation of the decay curve using the logarithm of the concentration differences is not possible. Furthermore, the background concentrations are often unknown or not accessible to direct measurement. In these cases, the air change rate λ can be determined only by non-linear regression with iterative calculation methods according to the model as given in Eq. 6. Iterative calculation methods start with initial values (raw values) specified by the user for the function parameters, which are then improved iteratively by using the method of least squares until the model function is fitted optimally to the measured curve.

The results obtained from the CO_2 decay curves depicted in Figure 4 were evaluated using the method of nonlinear regression and are shown in Table 3. The calculated ACR ranged from 0.25 to 0.4 h^{-1}(mean 0.32 h^{-1}). The goodness of fit of the regression model is very high, since more than 99% of the variability of the CO_2 concentration can be explained by the respective regression functions. This is also reflected by high accuracy of the estimates for the various ACR.

Table 3. Air change rates in a bedroom located in an older building (built in 1908) equipped in 1990 with wooden framed double box window, decay method and CO_2 as tracer gas. Avg.: Average, CI: confidence interval

curve	time [hours]	air change λ [ACH] (95% CI)	background conc. [ppm] (95% CI)	goodness of curve fit (R^2)
1	14.5	0.250 (0.246 – 0.254)	428 (427 – 429)	0.997
2	12.8	0.301 (0.295 – 0.306)	391 (390 – 392)	0.997
3	11.3	0.394 (0.383 – 0.406)	412 (411 – 413)	0.992
4	11.0	0.249 (0.243 – 0.254)	384 (382 – 385)	0.998
5	10.8	0.400 (0.393 – 0.407)	438 (437 – 439)	0.998
Avg. 95% CI		0.319 (0.226 – 0.411)		

The 95% confidence intervals are very narrow and deviate only about 2 – 3% downward and upward from each individual value. These results were due to both the low variance of the CO_2 values from the fitted curves (white curves in Fig. 4) and the extensive measurement periods (several hours) which yielded high numbers of nodes (concentration-time data pairs) at measuring intervals of 3 minutes. However, the differences between day-to-day measurements are much larger. The 95% confidence interval for the mean ACR ranges from 0.226 to 0.411 h^{-1}. A possible explanation for this relatively high day-to-day variation is given below (see section 3.5). ACR, which were determined under the terms of exclusive joint ventilation (i.e. windows and doors closed), vary considerably and show extremes that span two orders of magnitude. This is shown in Figures 5a and 5b where six examples of CO_2 decay curves are depicted which can typically be recorded indoors. For illustrative reasons decay curves were selected which start at about almost the same initial concentration of 1500 – 1800 ppm and were recorded over a measurement period of more than 10 hours (Fig. 5a). Another selection criterion was the existence of stable weather conditions over the entire measurement period. Both figures demonstrate that CO_2 decay curves can decrease exponentially over a period of 50 hours and longer under stable weather conditions (Fig. 5a).

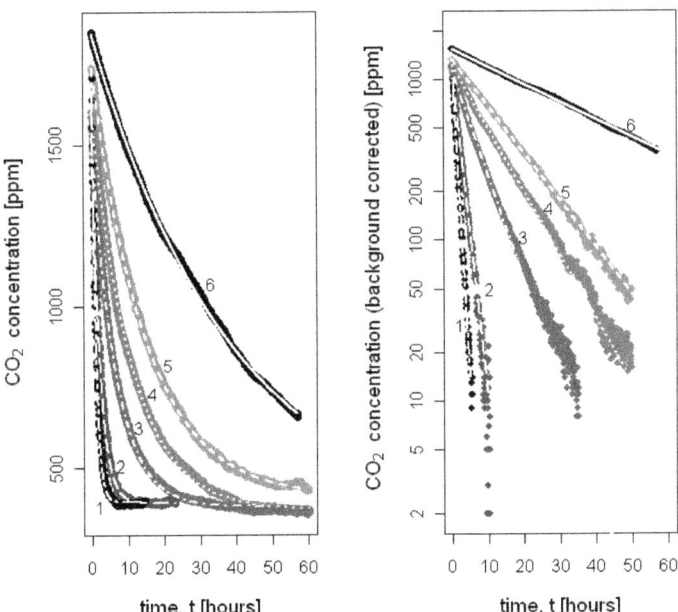

Figure 5. CO_2 decay curves (black) in 6 rooms of different air tightness. White curves show results of regression analyses; a, nonlinear regression; b, linear regression of logarithmic values.

This is an indication that tracer gas and room air can remain homogeneously mixed over such long periods without the need of a fan. The ACR estimated from the decay curves are given in Table 4, and the analysis of the curves was done with non-linear regression analysis (see Eq. 6). In addition to the air change rate λ, the background concentration C_a was also determined with the regression model and is presented in Table 4. After subtracting the respective background concentration from the measured CO_2 concentration values, the logarithmic plot of the concentration differences shows a nearly perfect linear concentration-time relationship (Fig. 5b).

Table 4. ACR and background concentrations with their 95% confidence intervals (95% CI) measured in 6 rooms with differing natural ventilation (windows and doors closed); estimated from CO_2 decay curves with the statistical method of nonlinear regression (time interval between two concentration values: 3 min).

curve	time [hours]	air change, [ACH] (95% CI)	background conc. [ppm]	goodness of fit (R^2)
1	14.5	0.942 (0.931 – 0.953)	393 (391 – 394)	0.998
2	23.0	0.465 (0.463 – 0.468)	392 (391 – 393)	0.999
3	48.0	0.154 0.153 – 0.155)	380 (379 – 381)	0.998
4	60.3	0.091 (0.091 – 0.092)	367 (366 – 368)	0.999
5	60.0	0.067 (0.066 – 0.067)	406 (405 – 407)	0.9996
6	57.0	0.0255 (0.0253 – 0.0257)	357 (350 – 365)	0.999

Determination of ACR from the Concentration Increase of Carbon Dioxide

When CO_2 is supplied at a constant rate, it is possible to determine the ACR from the increasing concentration values (Figure 6). Evaluation of the concentration curve is the same as with the constant injection method. The air change rate λ is calculated with the non-linear regression model approach $y = a_0 + a_1*[1-\exp(-a_2*t)]$ (cf. Eq. 4a), where $a_0 = C_a$, $a_1 = E/(\lambda*V_R)$, and $a_2 = \lambda$. When the ACR is small, the build-up curve will approach the equilibrium only slowly, and this will lead to a quasi-linear concentration curve if the measurement period is too short. Therefore, the ACR calculated from this curve can be afflicted with large uncertainties. For this reason, to obtain a more reliable estimate of the ACR the concentration build-up must be recorded over a period as long as several hours, with intervals of a few minutes. When metabolic CO_2 is used, measurements should be done best during sleep (of the occupants), because this comes nearest to the requirement of a constant CO_2 supply. Figure 6 also shows the ACR determined from the analysis of the CO_2 respective build-up curves depicted here.

These CO_2 build-up curves were recorded in the same room as the decay curves in Figure 4. The measurement of the CO_2 increase occurred, however, four years before, when the room was occupied by 3 persons (two adults, one child). The ACR of both series of measurements, decay and build-up curves, differ only slightly from each other taking daily variation into account.

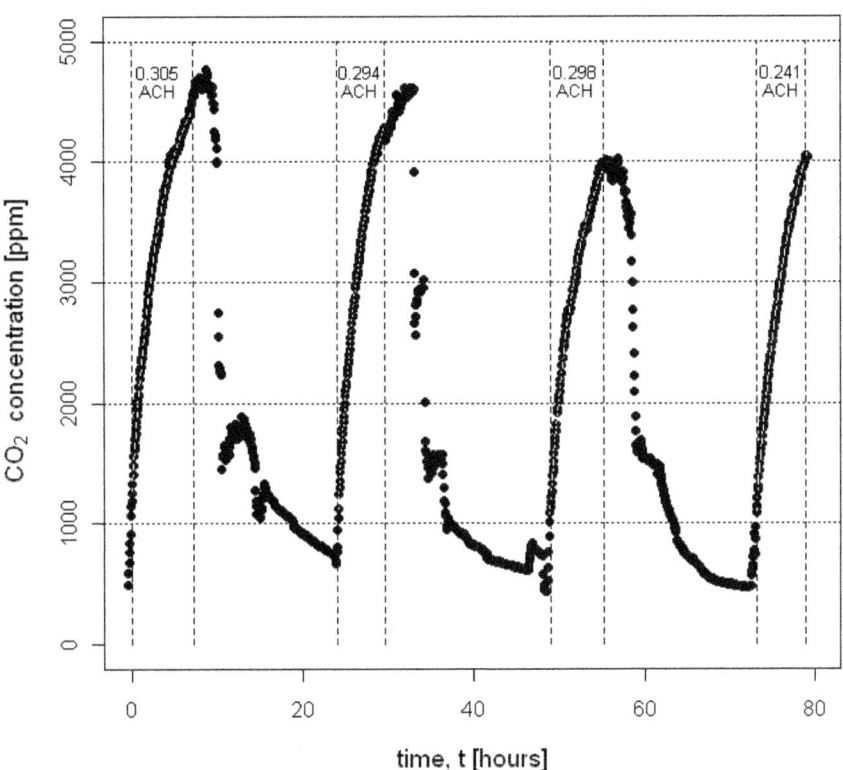

Figure 6. CO_2 concentration build-up curves in a bedroom equipped with a double box window and located in a building built in 1908. Corresponding ACR, estimated with nonlinear regression (fitted values: white curves). Constant-injection method, room volume 30 m³, occupancy: 2 adults, one child.

Comparison of Different Tracer Gases for the Determination Air Change Rates

The above examples have shown the applicability of determining ACR from CO_2 concentration curves, when either the influence of the background concentration can be mathematically eliminated or a constant supply of CO_2 is provided. It can, however, not be concluded that conventional and established tracer gases are dispensable. Compared to CO_2 these have some indisputable advantages, namely a much wider range of possible applications and virtually negligible background concentrations. Both of which greatly simplifies the determination of ACR.

Under certain circumstances the use of established tracer gases is not possible, and in these cases we have to resort to the CO_2 method. This may be necessary in patients with environmentally related health problems, who will not accept the use of tracer gases in their homes because they fear health hazards. In such cases, the CO_2 method is a valuable alternative. It is therefore necessary to examine the extent to which results obtained with CO_2 differ from those obtained with conventional and established tracer gases. To achieve this we performed a number of comparative studies by which an established tracer gas like hexafluorobenzene (C_6F_6) or sulphur hexafluoride (SF_6) was used in parallel to CO_2.

Determination of Air Change Rates with CO_2 and C_6F_6

The measurements described below were done in two different rooms, firstly, in a bedroom (24 m³effective volume) of a terraced house (built December 1997) equipped with double glazed windows and insulating rubber seals, conducting measurements over a period of 5 days, and secondly, in an office space (72 m³ effective volume) of an old brick building (built ≈1900) equipped with two double-box windows. Windows and doors were closed during the experiment. A total of two measurement cycles were performed at intervals of two months.

At the start of the conditioning phase a diffusion tube which contained hexafluorobenzene was placed in both rooms. First sampling on Tenax-tubes occurred about 72 hours after the windows and doors were closed, further samples (two samples each) were taken every 24 hours. Room air was collected using sampling tubes (1 liter per tube) with a bellows pump, type Accuro (Dräger, Lübeck, Germany). ACR were calculated from the known emission rate of the C_6F_6-tubes, the equilibrium concentration and the effective room volume, according to the procedure for the constant injection method as described in VDI (2001), using Eq. 4c. In the bedroom, CO_2 concentration was measured continuously over the entire study period with measuring intervals of 3 minutes using the Test to 400-device equipped with CO_2 probe, and the measured values were stored in the internal data logger. In the office space, the CO_2 concentration was measured with a second device in a 24hour rhythm according to the American standard ASTMD6245-98. Measurements occurred mainly at nighttime in the then unoccupied office building. Prior to the start of the measurement, the CO_2 concentration in the office space was set to ≈600 ppm by the experimenter. The calculation of the ACR was achieved with non-linear regression analysis using the statistical package SPSS, where the measured concentration values were fitted to exponential terms with consideration of the calculated back ground concentrations.

Results of the CO_2 measurements:

The CO_2 time-concentration curve recorded for the bedroom is depicted in Figure 7. The bold black line represents the measured concentration values, whereas the grey line describes the graph which was obtained by means of section-wise non-linear regression analysis. Above the individual curve sections, the calculated ACR are plotted. Given the high air-tightness of the bedroom ACR as low as 0.05 to0.1h^{-1} were determined. It is obvious that the CO_2 concentration increases very quickly during the night (one adult) to levels greater than 4000 ppm and decay exponentially during the day to values of 1500 - 2000 ppm. Due to the low ACR, the equilibriumconcentrationofCO2isnotcompassed during a 7 hours' sleep. The statistical analysis of the CO_2 curves obtained for the office space equipped with double box windows resulted, however, in significantly higher ACR, which ranged between 0.2 and 1.4 h^{-1} because of different window and door features.

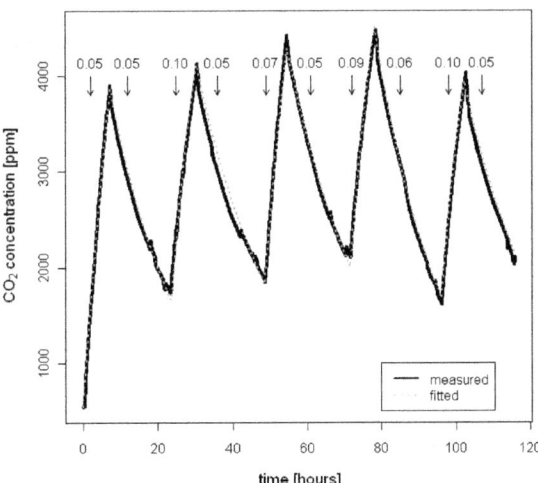

Figure 7. Course of CO_2 concentration in a bedroom equipped with an insulating glass

window with rubber sealing. Window and door closed. Measured values (black) with corresponding air change rates [in ACH], dotted curve: theoretical values. Room volume:24 m³, occupancy: one adult.

Results of the hexafluorobenzene measurements:

In Figure 8, the values of the hexafluorobenzene concentrations from a series of measurements in both rooms are shown together with calculated curves. The solid lines represent the concentration curves which are calculated using Eq. 4b on the basis of the known emission rate E, the room volume V_R and the given air change rate λ, whereas the points represent the concentrations that have been measured (bedroom: triangles; office space: squares). It is evident that the measurements in the bedroom oscillate about values which would be expected for an ACR of about 0.05 h⁻¹. The reason for the large deviation of the two first measurements from the curve is probably due to largely differing flow resistances of the Tenax tubes used. In comparison to the bedroom, the CO_2 values in the office (lower curve) show a much lower level despite the double emission rate. On the one hand, the lower concentration increase is caused by the larger room volume, but another reason is the significantly higher ACR of 0.45 h⁻¹. The calculation of the ACR from the individual concentration values measured led to values ranging from 0.05 to 0.14 h⁻¹ for the bedroom and from 0.2 to 1.7 h⁻¹ for the office space, respectively.

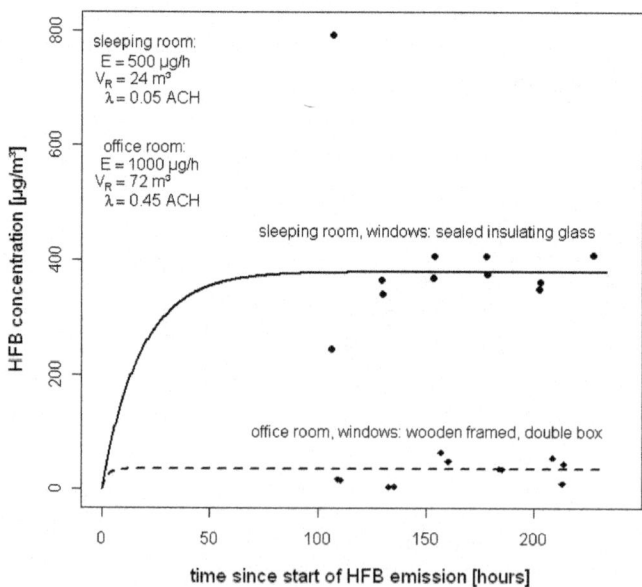

Figure 8. Course of hexafluorobenzene concentration (HFB) in two rooms with dif-

ferent volumes and different window types, theoretical curves calculated from given emission rates E, air change rates λ, known room volumes V_R. Black glyphs: measured values. The results of the tracer gas comparison are shown in Table 5 for both test series. At large, a very good agreement was found for ACR determined with the two different tracer gases.

Table 5. ACR results of comparative measurements with CO_2 and hexafluorobenzene as tracer gases. Windows and doors closed.

building	Window construction	CO_2 (ACH)		Hexafluorobenzene (ACH)	
		serie 1	serie 2	serie 1	serie 2
new	Insulating glass with rubber sealing	0.05 to 0.08	0.05 to 0.1	0.05 to 0.06	0.09 to 0.14
		n = 10	n = 10	n = 5	n = 6
old	wooden framed double box	0.3 to 0.5	0.2 to 1.4	0.2 to 0.9	0.3 to 1.7
		n = 10	n = 10	n = 5	n = 6

Parallel Testing of CO_2 and SF_6

In another series of investigations an office space with a volume of 48 m³ and two double box windows was doped with CO_2 and SF_6 in parallel on 6 consecutive days, and the ACR were determined from the decay curves. Measurement of SF_6 was done in the same way as described in the above section "decay curve". An example of the parallel concentration decay of both tracer gases is given in Figure 9; results of the 6 experiments, which differ only slightly, are compiled in Table 6.

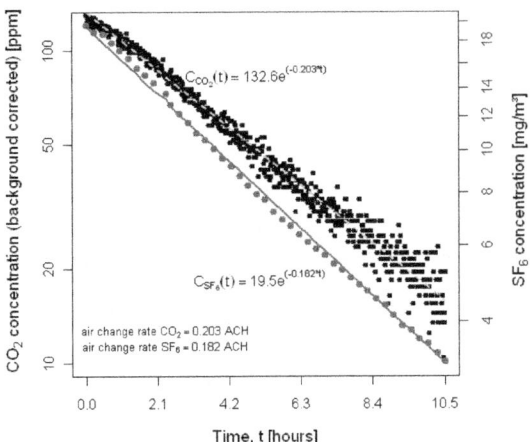

Figure 9. Decay curves of CO_2 and sulphur hexafluoride (SF_6) and curve fits; simultaneous measurements in the same room.

Table 6. Comparative measurements with CO_2 und SF_6; office space (volume 48 m³, two double box windows); COV (%): coefficient of variation = 100*standard deviation (std.dev.) / average.

measurement	air change rate [ACH]		Quotient of air changes
	CO_2	SF_6	ACR_{CO2} / ACR_{SF6}
1	0.211	0.240	0.879
2	0.447	0.438	1.021
3	0.330	0.327	1.009
4	0.203	0.182	1.115
5	0.330	0.320	1.031
6	0.300	0.280	1.071
average	0.304	0.298	1.021
std.dev.	0.090	0.087	0.080
COV (%)	29.7	29.2	8.0

The average ACR was 0.304 h⁻¹ for CO_2 (SD ± 0.09 h⁻¹) and 0.298 h⁻¹ for SF_6 (SD ± 0.09 h⁻¹). To check for systematic differences between the individual determinations we calculated the ratio of the two air change rates and got a mean ratio of 1.02 with a standard deviation of ± 0.08. Thus, no statistically significant systematic differences between the compared tracer gases could be detected. Since our sample size was rather small (only 6 determinations for each tracer gas) we included results from other studies (Guo and Lewis 2007; Roulet and Foradini 2002; Shaw, 1984) for comparison. After inclusion of these data the difference was statistically significant, albeit the systematic deviation between the ACR, determined under the same conditions with either CO_2 or SF_6, was very small (Fig. 10). A similarly good agreement was found by Stavova (Baránková) et al. (2006)comparing CO_2 and Freon 134a.

The mean ratio of the two air change rates (CO_2/SF_6) is about 1.06 (95% CI: 1.01 – 1.12). Standard deviation ranged from 0.93 to 1.21. The ≈6% overestimation of the ACR, as obtainedwithCO_2, is yet much smaller than the 50percentoverestimation obtained by Schulze and Schuschke (1990) who used krypton 85. On the basis of existent data it is so far not possible to conclude with certainty whether or not very small ACR (i.e. 0.1 h⁻¹) are responsible for higher systematic deviations.

Wind and Temperature Effects on the Natural Air Flow in Case of Joint Ventilation

Natural ventilation is driven by the air-pressure difference between the interior and external environment which prevails on the room's opening areas like windows and doors. The air change rate is also influenced by the tightness of windows and doors versus joints and cracks in the building envelope. The air-pressure difference is predominantly caused by temperature differences between indoor and outdoor climate, and by local wind effects. On the windward side of a building, the wind is trapped reaching its highest dynamic power when the flow angle equals 90. On the sides of a building that parallel the wind direction and on the far side, negative air-pressures occur regularly. These air-pressure differences result in raised rim hole rates, particularly in case of leaks or open windows on opposite sides of a building, due to suction and pressure cycles of the wind(i.e. cross ventilation). The entire air-pressure on to the façade is dependent on both the wind speed and squalls, as well as the building's constructional quality, namely the tightness of the façade.

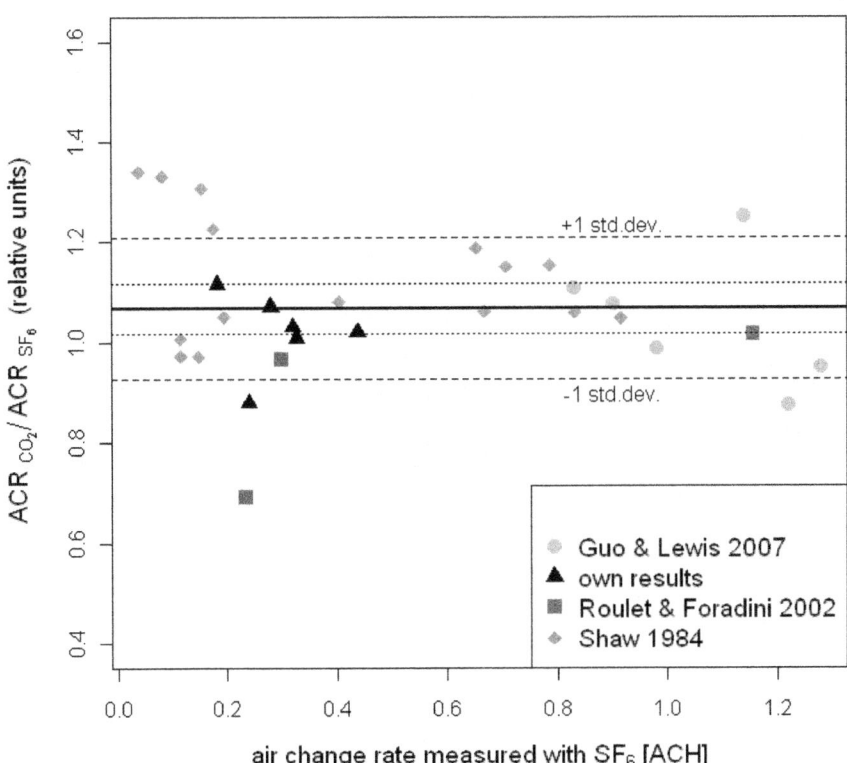

Figure 10. Comparative air change determinations with CO_2 and SF_6 as tracer gases. Results from other authors and own results. Solid line: mean, dotted lines: 95% confidence interval of the mean, dashed lines: interval of $\pm$ 1 standard deviation.

Wind Influences

To study the effects of weather conditions on the ACR, measurements were conducted in two different office buildings of the Robert Koch Institute, Berlin, Germany. Building #1 is an older brick building (built ca. 1900), surrounded by allotment gardens (to the north and south) and large open spaces. On the east side is a three-storey row house, and on the west side a four-storey office building. Due to the location of the building, unhindered wind flow is limited to south-east or south-west directions. All of the five rooms which have been selected in building #1 were located on the south side and equipped with double box windows (rooms A– D and F; room volumes between 30and 90 m³).

The rooms in building #2 constructed in the late 1970s were equipped with tightly closing windows featuring single glazing and insulating rubber seals. The room selected for the measurements had a large window sill with a couple of windows in north-east direction and a room volume of 40 m³ (room E). In front of the windows was an open space (length ≈150 m), limiting wind flow to solely from north-east to south-east directions. On the south side, room E was protected by a cross-building located in a distance of ≈10 m. Likewise, on the north side, the building was protected by a north-east to south-west oriented wing of the same building (in a distance of about 70 m from room E).

Between January 1999 and December 2010 a total of 611 air change measurements were performed in the five rooms of building #1 (A – D and F) using the CO_2 decay method. The decay curves were mainly recorded at night or on weekends according to ASTMD6245-98, when the building was unoccupied. Analogously, 388 air change measurements were carried out in room E of building #2 between July 2002 and September 2008. During the measurements all doors and windows were kept closed. Data about meteorological conditions during the tests, such as outside air temperature, wind speed and wind direction were obtained from the web sites of local Berlin weather stations. In room E of building #2 the indoor temperature was recorded during the measurements. Table7provides an overview of the air change rates determined. For rooms with double box windows individual values of ACR varied from <0.05 h^{-1} to 1.8 h^{-1}(i.e. a factor of ≈40), whereas the medians (50th percentile) obtained for the rooms A – D and F varied only between 0.3 and 0.6h^{-1}. Significant differences between the individual rooms could not be detected. Significantly lower ACR were determined for room E in building #2 equipped with double glazed windows for which the median was 0.08h^{-1}. The individual values varied between 0.016 and 0.40h^{-1}, i.e. maximum and minimum differ more than a magnitude. Wind influences, especially wind speed, were identified as the main reason for the variability of the ACR in the two buildings (Fig. 11).

Table 7. Air change rates [ACH] in office spaces with windows of different construction, rooms A-D, F: wooden double box windows, room E: insulating glass window with plastic frame and rubber sealing

room	sample size	Minimum [ACH]	Median (95% CI)	Maximum [ACH]
A	15	0.209	0.457 (0.369 - 0.715)	1.320
B	26	0.163	0.445 (0.351 - 0.552)	1.370
C	11	0.090	0.470 (0.126 - 0.696)	0.720
D	42	0.079	0.512 (0.352 - 0.639)	1.426
F	517	0.043	0.342 (0.317 - 0.374)	1.841
E	388	0.016	0.078 (0.070 - 0.085)	0.403

In building #1 (rooms A – D and F) the ACR ranged from<0.1 h^{-1} when wind speed was lower than 5 km/h to1.2–1.4h^{-1}at a wind speed up to 30–40 km/h. The strong influence of the wind caused significant and uncontrollable heat losses through leaky windows which, in turn, caused unhealthy draught effects.

Surprisingly, the results obtained for room E in building #2 equipped with insulated windows and rubber seals, were significantly dependent on wind influences, as well, although the wind effects were much smaller than those of building #1. For room E,ACR were less or equal to 0.05h^{-1} when wind speed was lower than <5 km/h but increased up to 0.25h^{-1}with a wind speed of more than 40 km/h. The relations described here essentially confirm the results formerly obtained by Wegner (1983). Besides the wind speed, the flow direction of the wind had a significant impact on the amount of the ACR. This is evident from the graphical presentation in Figure12, grouping ACRs according to the wind direction prevailing in test periods.

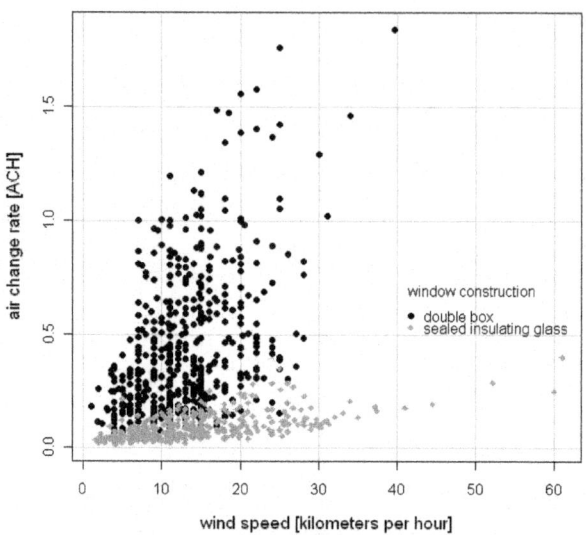

Figure 11. Influence of wind speed on ACR in office spaces equipped with windows of different construction. Windows and doors closed. Rooms with double box windows (A-D, F: n= 611), room with insulating glass window: n= 388).

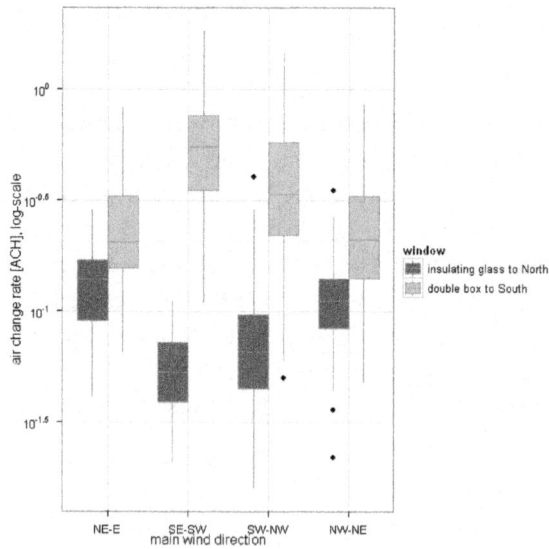

Figure 12. Wind direction and ACR in office spaces equipped with windows of different construction. Windows and doors closed. Rooms with double box windows (A-D, F: n= 601), room with insulating glass window: n= 388).

Adverse wind blowing against the window or in parallel to the window front are of higher impact than wind flow from the backside of the building, the latter causing lower ACR than the two former (see Table 8).

Table 8. Dependence of ACR on wind direction, raw and adjusted ACR quotients, Rooms A-D, F: windows to south, room E windows to north-east.

room	wind direction	sample size	ACR [ACH] Median (95% CI)	Quotient$_{raw}$ Median (95% CI)	Quotient$_{adj}$ Median (95% CI)
A - D, F	NE-E	122	0.202 (0.171 - 0.243)	0.92 (0.779 - 1.107)	1.06 (0.941 - 1.189)
	SE-SW	242	0.55 (0.505 - 0.596)	2.50 (2.295 - 2.709)	2.89 (2.601 - 3.207)
	SW-NW	163	0.329 (0.294 - 0.395)	1.50 (1.340 - 1.801)	1.38 (1.239 - 1.547)
	NW-NE	74	0.22 (0.179 - 0.261)	1 (Ref.)	1 (Ref.)
E	NE-E	83	0.138 (0.109 - 0.153)	2.09 (1.651 - 2.317)	2.21 (2.038 - 2.399)
	SE-SW	81	0.054 (0.047 - 0.063)	0.82 (0.714 - 0.957)	1.10 (1.013 - 1.198)
	SW-NW	171	0.066 (0.056 - 0.072)	1 (Ref.)	1 (Ref.)
	NW-NE	53	0.111 (0.095 - 0.135)	1.68 (1.438 - 2.043)	1.92 (1.742 - 2.115)

In case of the rooms with double box windows, this led to relative threefold increase of air change, expressed as ratio ac(luv) : ac(lee), for wind directions from south to west. In the room with double glazed windows, we measured an almost doubled increase of air change (70– 100%)for wind directions from north to east. The overall air change ratios depending upon the relative wind direction, respectively were not remarkably influenced by the wind speed (Quotient$_{adj}$ in Table8). Summarizing wind effects we could demonstrate that regardless of wind strength and the window type an airflow directed to the window front can yield a 1.5- to more than 2-fold increase of the ACR in comparison to wind from the backside.

Temperature Influences

The temperature difference between inside and outside is one of the major forces driving ventilation. We investigated the impact of the temperature gradient (room air temperature minus outside air temperature) on the ACR in room E and found a statistically significant effect on the extent of air change (Fig. 14a). Whereas the average air change rate λ, after correction for the wind influence, was only 0.06 h^{-1} at a temperature difference ΔT of 0K, it increased up to 0.1 h^{-1} when the temperature gradient rose toabout30 K. However, as compared to wind effects (Fig 14b), the exponential relationship between ACR and temperature gradient (Fig.14a), turned out to be much weaker, at least for room E with its tightly closing window.

Air Change Rate during Window Ventilation

Wind action and temperature gradient both proved to be significant factors influencing the amount of air change through joints and cracks in buildings. For indoor spaces in buildings with high leak tightness, window ventilation plays a crucial role in the removal of excess humidity, carbon dioxide, odors and other volatile compounds. The effectiveness of ventilation via opened windows is not only dependent on their number, size and arrangement in the room, but also on the modus and frequency of active window ventilation. Windows that are only opened partway (i.e. tilted) ventilate less effective than fully opened windows. Cross-ventilation can be achieved when windows on opposite sides of a room are opened simultaneously. Cross-ventilation removes moisture and contaminants most effectively from the indoor air to the outside. The effectiveness of the different modes of window ventilation expressed in terms of the ACR, was examined with a couple of tests in room E, the results of which are shown in Figure 13 in conjunction with the joint ventilation results.

ACR determinations were carried out by the decay method; duration of the measurements varied between 0.5h and3h with recording intervals of 12 s to 1 min. After doping the room with CO_2, one window in the room (dimensions: width 0.8m and length 1.2m) was either in tilted position(i.e. 10 cm wide upper opening)or widely open(i.e. opening angle80). To achieve cross-ventilation, an opposite window was also widely open. For both ventilation modes – tilted or full open – the following ACR were determined (Table 9a): With the tilted window a total of 239 measurements was performed; the average ACR (i.e. median) was 1.5 h^{-1} with day-to-day variations that ranged from 0.5 h^{-1}to3.4h^{-1}. In the case of full open window, the median ACR was10h^{-1}, with a minimum of 2.4h^{-1} and a maximum of 30h^{-1} (321 measurements). For cross ventilation a number of 179measurementswere performed. The ACR varied between 15 h^{-1} and 146h^{-1}(median: 40h^{-1}). On the basis of these data we calculated the ventilation times which are necessary to reduce an indoor CO_2 concentration of 1500 ppm by 98% (corresponding to a CO_2 concentration that is about 30 ppm higher than outside) and yielded 2 hours for the tilted window (range 1 – 7 hours), ≈0.3 hours for the wide open window (range 0.1 – 1.5 hours) and 0.1 hours for cross-ventilation (range 1 - 14 min; see also Table 9b).

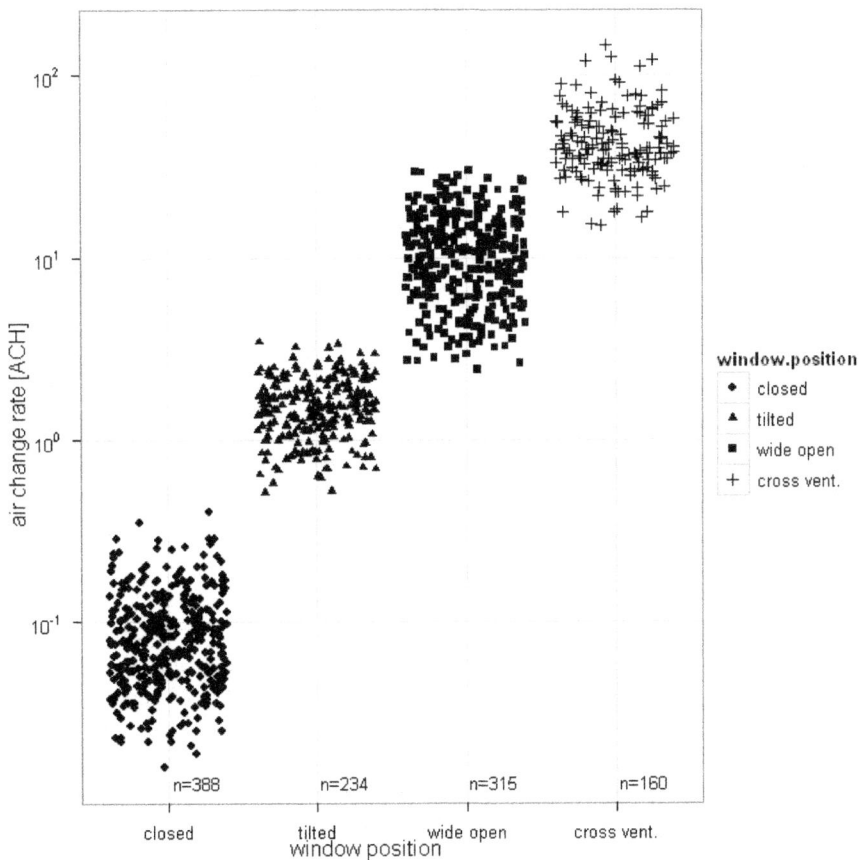

Figure 13. ACR and window position. Office room E.

Table 9A. Dependence of ACR on window construction and opening

| room | window construction | window position | Air change rate [ACH] | | | | | sample size |
			Min	05. P	Median	95. P	Max	
A – D, F	double box	closed	0.043	0.111	0.36	1.030	1.841	602
E	insulating glass	closed	0.016	0.029	0.078	0.200	0.403	388
		tilted	0.52	0.70	1.50	2.60	3.43	239
		wide open	2.4	3.5	10.4	23.4	29.7	321
		cross vent	15.0	22.5	40.1	104.7	146.4	179

Table 9B. Ventilation time needed to reduce CO_2 concentration of 1500 ppm by 98%

room	window construction	window position	Ventilation time [hours]				
			Max	95..P	Median	05. P	Min
A – D, F	double box	closed	84	32.4	10	3.5	2.0
E	insulating glass	closed	225	124	46	18	8.9
		tilted	6.9	5.1	2.4	1.4	1.0
		wide open	1.5	1.0	0.35	0.15	0.12
		cross vent.	0.24	0.16	0.09	0.03	0.02

These results clearly confirm the great importance of ventilation with fully open windows and, even more, of cross-ventilation for the efficient removal of indoor pollutants. The broad variation in ACR can be explained to a great deal by the effect of the temperature gradient between indoor and outside (Fig. 14a).

Compared to the temperature gradient, wind effects played only a minor role at least for the window openings "tilted" and "widely open" tested here (Fig. 14b). The dependence of the ACR on the temperature gradient could be estimated by non-linear exponential curve fitting for all tested window opening modes (Fig. 14a). From these findings we can conclude that in case of ventilation via tilted window, with a temperature gradient of 10K, an ACR of $0.92*exp(0.045 * 10) = 1.4h^{-1}$ can be expected for the room examined here (95% prediction interval: $1.0h^{-1}-2.0h^{-1}$). With a temperature gradient of 20K an average ACR of $2.3h^{-1}$ can be expected (95% prediction interval $1.6h^{-1} - 3.1 h^{-1}$). For ventilation with fully open window an ACR of $4.5 * exp (0.071 * 10) = 9.2 h^{-1}$ ($\Delta T = 10$ K) can be expected and 18.6 to h^{-1} ($\Delta T = 20$ K) with 95% prediction intervals $5.5 h^{-1}- 15.3 h^{-1}$ and $11.1 h^{-1}- 31.2 h^{-1}$, respectively. In the case of cross ventilation an average ACR would expected to be $39.0 h^{-1}$ and $50.2 h^{-1}$ for $\Delta T = 10$ K and 20 K, with 95% prediction intervals of $23.0 h^{-1}$ - $65.8 h^{-1}$ and $29.6 h^{-1} - 85.1 h^{-1}$, respectively.

Air Change Rates in Berlin's Housing Stock

The results presented so far do not allow general statements on the air change in buildings, since the number of rooms examined was far too small. As part of site inspections and indoor air investigations during 1999–2005, ACR were determined for a further number of different buildings and rooms applying the CO_2 concentration decay method. A total of 198 living and working spaces, situated in 152 buildings in the Berlin area and in the immediate surrounding were studied, comprising of 143 residential buildings of different age and storey numbers, and nine different office buildings.

None of the rooms had an additionally installed ventilation system. Tenants were asked not to enter the rooms during the tracer gas measurements and to keep windows and doors closed. The duration ofthetracergasmeasurementsvariedbetween2 hours and 24hours. To obtain decay curves with a high number of nodes we choose time intervals of 1 to 3 minutes between CO_2 measurements. Two Testo 400 devices with CO_2 probe(infrared absorption) were used for recording and data logging. The results show that under worst case conditions the vast majority of the ACR are below the range of 0.5 h^{-1} – 1 h^{-1} which was recommended by the former German Federal Health Office (Bundesgesundheitsamt, 1993) for common living quarters. The ACR of the studied rooms ranged from minimum0.02 $h^{-1}$to maximum1.98 h^{-1}. Of the198naturallyventilatedrooms the majority (167=84%) had ACR of 0.5h^{-1}orbelow. ACR of 0.8h^{-1}or higher could be determined for only 12of the rooms (6%). The50[th]percentile(median) of the ACRwas0.2h^{-1}witha95% confidence interval(CI) of 0.17 h^{-1} – 0.24h^{-1}; arithmetic mean was 0.31h^{-1} (95% CI 0.27 h^{-1} – 0.36h^{-1}) and geometric mean was 0.22h^{-1} (95% CI 0,20 h^{-1} – 0,25 h^{-1}), respectively (see Table 10). Thus, our studies essentially confirm the results obtained by Münzenberg (2004) and Salthammer et al. (1995), who conducted ACR measurements under similar methodological conditions in Germany, and received results consistent with those from surveys in Scandinavia by Andersen et al. (1997); Bornehag et al. (2005) and Harving et al. (1992).

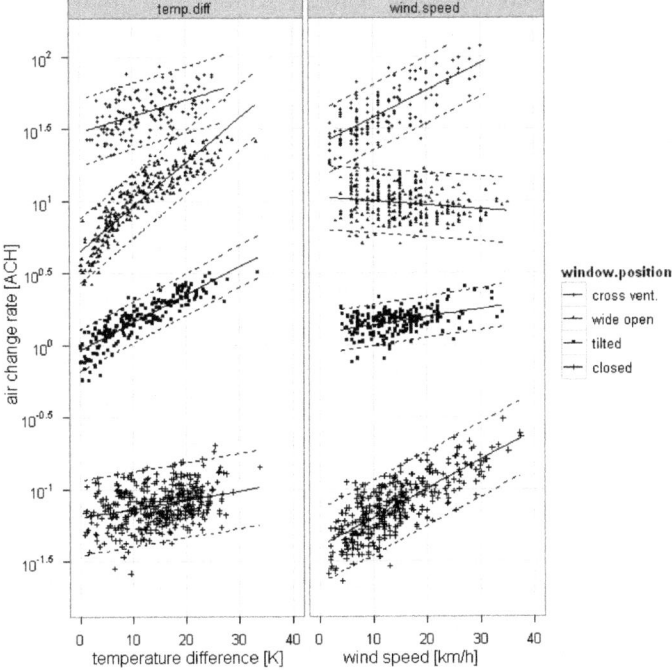

Figure 14. a. Influence of temperature gradient (indoor to outdoor) on ACR, corrected

for wind effects, b. Influence of wind speed on ACR, corrected for temperature effects, in an office space (room E) for different ventilation conditions: natural ventilation: "closed", "window tilted", "widely open" and "cross ventilation", (from bottom to top); room volume 40 m³, solid lines: fitted curve, dashed lines: 95% prediction intervals.

Measurements performed in the same rooms using diffusion tubes and hexafluorobenzene (C_6F_6) or PFT as a tracer gas yielded somewhat higher ACR compared to that obtained with the CO_2 concentration decay method. This can be explained by necessarily longer exposure times of collection tubes ranging from several days to weeks, which essentially means that there have been one ore more active ventilation phases during the measurement period influencing the air change (Table 10). However, even in these studies, the proportion of rooms with ACR below 0.5 h⁻¹ was remarkably high. Studies conducted in the USA also show a high proportion of buildings (50 percent and more) with ACR below 0.5 h⁻¹ (Persily et al., 2010), suggesting that a basically desirable ACR of at least 0.5h⁻¹ is difficult to achieve under normal conditions. Similar conclusions have been drawn by Erhorn and Gertis (1986) and Münzenberg (2004).

Table 10. ACR measurements in Germany and Scandinavia, results of surveys. TFE: 1.1.2-trichloro-1.2.2-trifluoroethane, [1] before reconstruction. [2] after reconstruction, [3] single family houses, [4] multi-family houses, AM: arithmetic mean,. GM: geometric mean, n.a. not available, < 0.5 h- 1: proportion (%) of air change rates < 0.5 h⁻¹

study	method	tracer gas	sample size	Min [ACH]	Median [ACH]	Max [ACH]	AM [ACH]	GM [ACH]	< 0.5 ACH (%)
Salthammer et al. (1995)	concentration decay	N₂O	150	< 0.1	0.25	1.7	0.36	n.a.	> 70
Münzenberg (2004)		CO₂	80	< 0.05	0.18	> 1.5	0.26	n.a.	90
own results		CO₂	198	0.02	0.21	1.98	0.31	0.22	84
Harving et al. (1992)		TFE	114	< 0.1	0.28	> 1.5	n.a.	n.a.	72
Bekö et al. (2010)	concentration decay and build-up	CO₂	300 cases		0.44		0.62	0.46	57
			200 bases		0.42		0.62	0.46	
Hirsch et al. (2000)		C₆F₆	78					0.73[1)	
								0.52[2)	
Andersen et al. (1997)		PFT	117	0.16	n.a.	0.96	0.37	0.34	82
Ruotsalainen et al. (1992)	constant-injection (diffusion tubes)	PFT	242	0.07	≈0.5	1.55	0.52	n.a.	48
Øie et al. (1997)		PFT	38	0.15	n.a.	1.4	n.a.	n.a.	58
Øie et al. (1998)		PFT	344	n.a.	n.a.	n.a.	n.a.	n.a.	36
Bornehag et al. (2005)		PFT	320[3)					0.36[3)	80
			43[4)					0.48[4)	60

Stratification for window types showed significantly higher ACR (factor ≈2) for rooms with double box windows or composite windows than for rooms with specially insulated windows (double glazing and continuous rubber seal; Table 11). However, also for rooms with double box windows or composite windows the majority of the ACR determinations yielded values <0.5 h⁻¹. The50ᵗʰ percentile(median) was 0.33h⁻¹, and the75ᵗʰ percentilewasabout0.56h⁻¹. In comparison, rooms with double glazed windows had a median ACR of 0.16 h⁻¹ and the 75ᵗʰ percentile was 0.25 h⁻¹.

Table 11. ACR in naturally ventilated rooms (windows and doors closed) equipped with different window types. Min: Minimum. Max: Maximum. 05. P: 5ᵗʰ percentile. 25. P: 25ᵗʰ percentile. 75.P: 75ᵗʰpercentile. 95. P: 95ᵗʰ percentile. 95% CI: 95% confidence interval. AM: arithmetic mean. GM: geometric mean

window construction	Sample size	Min	05. P	25. P	Median (95 % CI)	75. P	95. P	Max	AM (95 % CI)	GM (95 % CI)
Double box/ composite window	80	0.05	0.11	0.20	0.33 (0.27 – 0.44)	0.56	1.50	1.98	0.47 (0.37 – 0.57)	0.34 (0.28 – 0.41)
Insulating glass	118	0.02	0.05	0.10	0.16 (0.13 – 0.20)	0.25	0.54	1.00	0.22 (0.18 – 0.25)	0.16 (0.14 – 0.19)

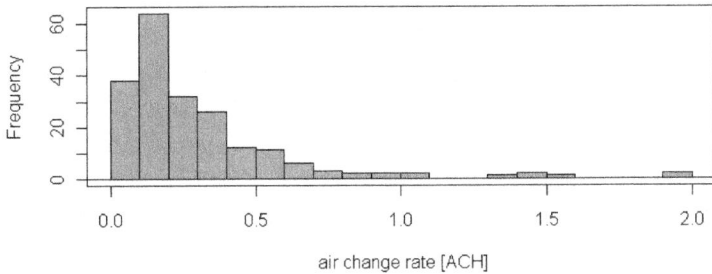

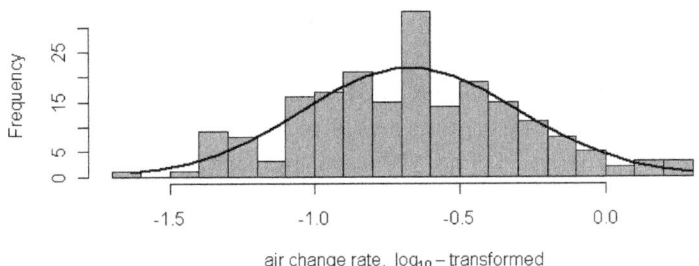

Figure 15. Empirical frequency distribution of ACR (natural ventilation) for 198 rooms in the City of Berlin; a, values as determined; b, log-transformed values. Solid line represents normal distribution.

The empirical frequency distribution of the ACR for the total sample had the typical shape of a log-normal distribution with a steep rise on the left and a long tail on the right side (Fig. 15a). A good fit to the normal distribution could be achieved by calculating the logarithms of the ACR (Fig. 15b).

This is consistent with the results of Scandinavian studies by Andersen et al. (1997); Bekö et al. (2010); Harving et al. (1992); Øie et al. (1998) and Ruotsalainen et al. (1992) who found the same form of empirical frequency distributions of ACR in Danish, Finnish and Norwegian residential buildings.

Ventilation and Pollutant Concentrations

Adequate ventilation is the cheapest and easiest way to effectively remove indoor pollutants. Previous studies have shown that insufficient air change is mainly due to renovation and other structural changes, namely in most cases the installation of new airtight windows (Wegner 1983, 1984). The results of our own investigations presented here have led to similar conclusions, apart from significantly lower ACR as compared to previous studies (Wegner 1983, 1984). Generally insufficient fresh air supply resulted in an considerable increase of indoor air pollutants. This is the case for formaldehyde (Li et al., 2002; Salthammer et al., 1995; Wegner, 1983) and other volatile compounds (Hodgson et al., 2003) as well as for the radioactive noble gas radon (Andersen et al., 1997; Chao et al., 1997; Sentikova, 1999). Our own studies also found a correlation between the concentration of volatile organic compounds and the ventilation rate in naturally ventilated rooms, as described below.

We have chosen some frequently found indoor air pollutants, which may be taken as typical representatives of different substance classes, namely formaldehyde, acetaldehyde, acetone, 2-ethyl-1-hexanol, hexanal, d-limonene, α-pinene, decamethylcyclopentasiloxane (D5-siloxane), and toluene. Air sampling to analyze these contaminants was done in the rooms under study (closed windows and doors), either after a CO_2 tracergasmeasurement was finished or directly before the start of a CO_2 decay recording with Air check samplers (model 224-PCXR8) SKC Inc., Eighty Four, PA, USA. For enrichment we used the adsorption media recommended for these pollutants. Air volumes varied from 90 l to 120 l and the sampling airflow was 1.5– 2 l/min. Formaldehyde was somewhat special since it was measured semi-quantitatively in situ in most rooms. An Inters can 4000 device (Ansyco, Lenzkirch, Germany) was used to carry out the measurements. Prior to the determination of the formaldehyde concentration, a zero balance was produced by pumping air through an upstream activated carbon filter for 10 minutes. Thereafter, the concentration value was recorded when the reading remained constant. Comparison of measurements for 24 different rooms yielded a good agreement of concentration levels obtained with the electrochemical formaldehyde method using the Inters can 4000 and the laboratory results obtained by analysis of the air samples that were collected with DNPH cartridges.

Statistics for the pollutants investigated here are listed in Table12; if the sample size was sufficient the95% confidence intervals were also included. Acetone, form aldehyde and acetaldehyde showed the highest concentrations, with median valuesof75µg/m³, 23 µg/m³and 31 µg/m³, respectively. Due to boiling pointsbelow60 C, these pollutants are classified as very volatile organic compounds (VVOC). Therefore, they are–with the exception of formaldehyde–not regularly recorded indoors.

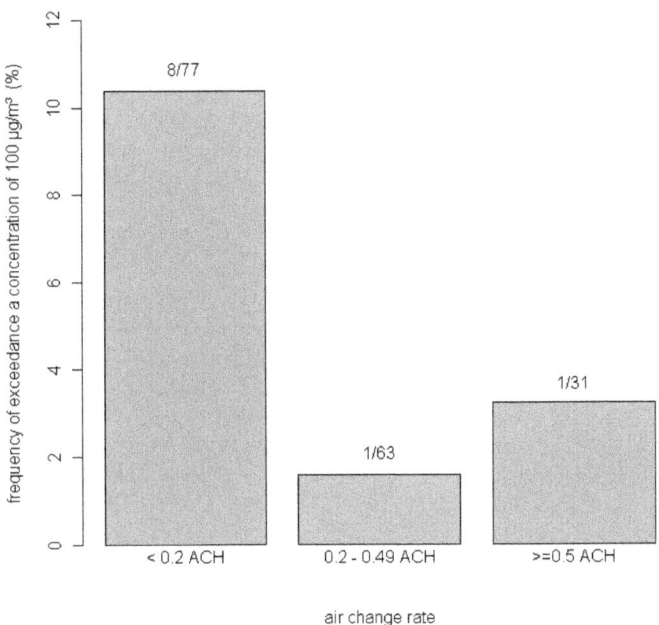

Figure 16. Frequency of formaldehyde concentrations >100 µg/m³ and ACR (natural ventilation) in rooms of the City of Berlin.

As Figure16shows, the frequency of formaldehyde concentrations >100µg/m³ seems to correlate with the ACR. In8of77 category-1 rooms with ACR<0.2 h⁻¹formaldehyde amounted to ≥100µg/m³, i.e. the frequency of exceeding values was 10.4% (95% CI 4.6% –19.5%). Among the category-2 und -3 rooms was only one each with exceeding formaldehyde (1.6% and3.2% relative frequency, 95% CI 0% –8.5% and0% –16.7%, respectively). According to X^2 testing these differences were close to be significant (p=0.060), but this seems unlikely due to strong overlaps of the very broad 95% CI. Our finding thus points to successful emission reduction measures through low-emission products by which pollutant concentrations can be accomplished that are within the range of the accepted limits, even in very airtight rooms.

Unlike formaldehyde, the concentrations of acetaldehyde and acetone showed a clear dependence of the air change rate λ, which could be described by the power function $(C(\lambda) = a*\lambda^{-k})$. The independent variable air change rate could explain about 50% ($R^2 = 0.5$) of the total variance of the acetaldehyde concentration values and about 34% ($R^2=0.34$) of the acetone concentration values. The results of a curve fitting are shown in Fig. 17 a and b.

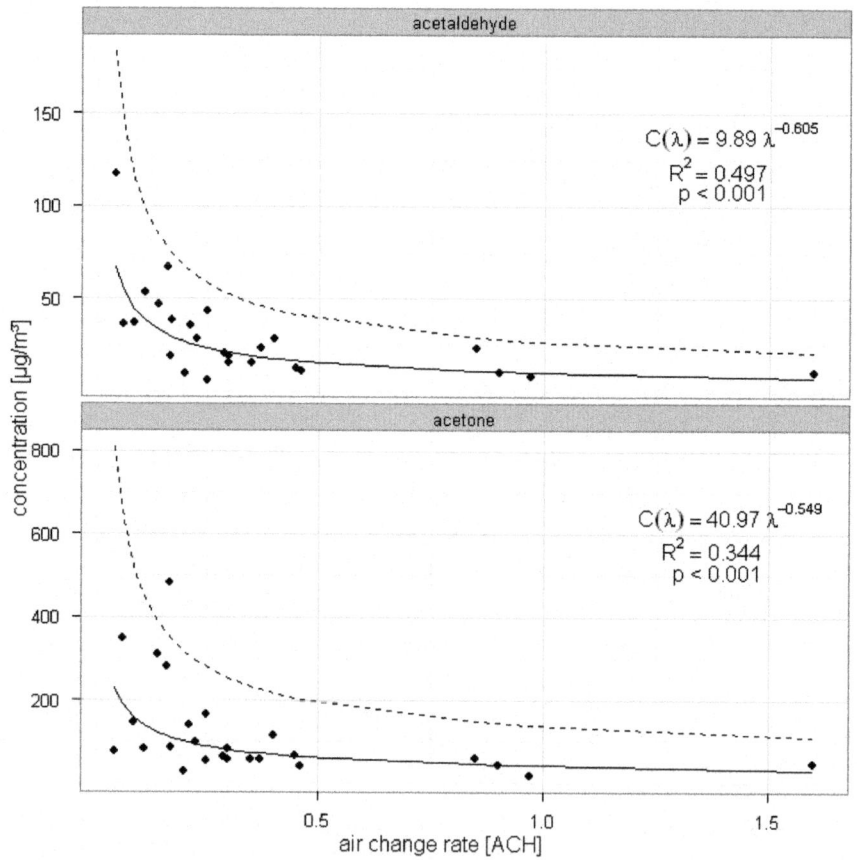

Figure 17. a. and b. Concentrations of acetaldehyde (a) and acetone (b) and corresponding air change rates in rooms of the City of Berlin. Sampling: worst case conditions, solid line: mean concentration, dotted line: 90% prediction limit.

According to the resulting function parameters it can be estimated that an increase of the ACR from $0.1h^{-1}$ to $0.5h^{-1}$ would yield a reduction of the indoor concentrations from $40\mu g/m^3$ to $15\mu g/m^3$ foracetaldehyde, and from $145\mu g/m^3$ to $60\mu g/m^3$ for acetone. This is a reduction factor of $2.4 - 2.7$.

Similar relations were found for other volatile organic compounds (Table 12). Again, the dependence of their indoor air concentrations on the ACR could be characterized by a power function (see Figures 17c to 17h). Regression analysis revealed that an increase of the ACR from $0.1h^{-1}$ to $0.5h^{-1}$ decreased the concentrations of 2-ethyl-1-hexanol and hexanal (or hexanaldehyde) by a factor of ≈ 2.3, since 2-ethyl-hexanol declined from $6.2\mu g/m^3$ to $2.6\mu g/m^3$, and hexanal declined from $21.2\mu g/m^3$ to $9.1\mu g/m^3$. For limonene and α-pinene the reductionfactorswere2.6and2.2, respectively. For these pollutants, the indoor air concentrations would be reduced from $24 \mu g/m^3$ and $30 \mu g/m^3$ to values of $9 \mu g/m^3$ and $14 \mu g/m^3$, respectively For decamethylcyclopentasiloxane(D5) and toluene the reduction factors amounted to2.7and2.4, corresponding to a decline $6.5\mu g/m^3$ to $2.4\mu g/m^3$ (D5) or $32\mu g/m^3$ to $13\mu g/m^3$ (toluene).

Table 12. Concentrations $[\mu g/m^3]$ and 95% confidence intervals of selected VVOC and VOC in the air of rooms in the City of Berlin; n.d.: 95% confidence interval not detected (small sample size), 10. P: 10th percentile, 90. P: 90th percentile, AM: arithmetic mean, GM: geometric mean.

adsorbent	compound	Sample size	Min	10. P (95 % CI)	Median (95 % CI)	90. P (95 % CI)	MAX	AM (95 % CI)	GM (95 % CI)
	very volatile								
In situ detection	formaldehyde	171	< 20	< 20 (< 20)	31 (25 – 38)	70 (63 – 96)	190	41 (36 – 45)	33 (30 – 37)
DNPH-Silica	acetaldehyde	25	6	9 (n.d.)	23 (11 – 37)	53 (n.d.)	117	29 (19 – 39)	22 (16 – 30)
DNPH-Silica	acetone	25	18	39 (n.d.)	75 (53 – 128)	308 (n.d.)	481	118 (71 – 166)	84 (60 – 117)
	volatile								
Anasorb 747 / Tenax	2-ethyl-1-hexanol	148	< 1	1 (1 – 1,5)	4 (3 – 6)	12 (10 – 19)	65	7 (5 – 8)	4 (4 – 5)
Anasorb 747 / DNPH	hexanal	93	< 2	4 (3 – 6)	17 (14 – 21)	37 (30 – 45)	75	20 (17 – 22)	14 (12 – 17)
charcoal (NIOSH)	limonene	133	< 1	2 (1 – 3)	10 (8 – 13)	66 (36 – 89)	256	24 (17 – 31)	10 (8 – 13)
charcoal (NIOSH)	α-pinene	133	< 1	2 (2 – 5)	15 (11 – 22)	102 (57 – 143)	623	45 (28 – 62)	16 (12 – 20)
charcoal (NIOSH)	d5-siloxane	99	< 1	< 1 (<1 - 1)	3 (2 – 4)	94 (39 – 321)	736	39 (17 – 61)	4 (3 – 6)
charcoal (NIOSH)	toluene	133	2	4 (3 – 5)	14 (10 – 21)	91 (46 – 144)	396	34 (25 – 43)	16 (13 – 20)

These results may allow to conclude that increasing the ACR to the fourfold will reduce indoor air concentration of VVOC and VOC to half. This means that the concentration is inversely proportional to the square root of the ACR, but seems to be contradictory to theoretical considerations, if one assumes a nearly constant emission rate of volatile compounds. In this case, the concentration would be inversely proportional to the ACR under equilibrium conditions, meaning that a fourfold increase in the ACR would reduce the concentration to a quarter.

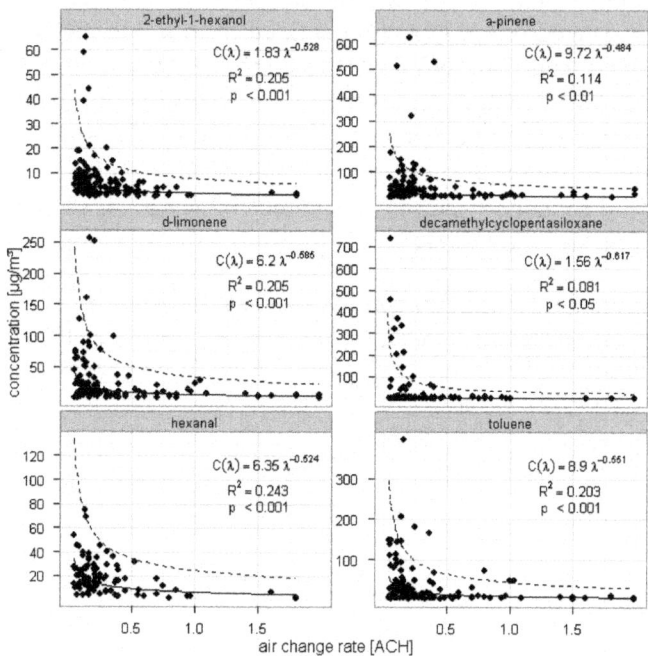

Figure 17. c-h. Concentrations of selected volatile organic compounds (VOC) and corresponding ACR in rooms of the City of Berlin. Sampling under worst case conditions, solid line: mean concentration, dotted line: 90% prediction limit.

One of the reasons for the deviating results is surely the fact that nowadays many buildings are constructed with such a high air-tightness that a kinetic equilibrium cannot be achieved under typical sampling conditions (i.e. time span of several hours between last window ventilation and start of the measurement; see also Figure 3b). That leads, firstly, to low ACR (<0.2 h⁻¹) and, secondly, to an underestimation of the VOC concentrations. The regression analysis yields a flatter curve and an exponent $k<1$ for the air change rate λ. Another reason is the different interior decoration (furniture, wallpapers, other equipment) of the spaces investigated here, and thus their differing pollution loads. Our VOC sample comprised rooms which differed greatly in respect to number and kind of possible sources. Therefore, we can assume that even in rooms with low air change either no VOC sources existed or only such sources with low emission rates. As a consequence, only low concentrations could have been measured in the air of these rooms. This fact would also contribute to the flattening of the air change-concentration curve.

The influence of ACR on the concentrations of VOC is also valid when we consider substance classes. The spectrum of individual compounds of these classes resembles that described by Schleibinger et al. (2001, 2002). As with formaldehyde, the analysis was performed with grouped data to reveal the dependency on the ACR, and categorization was done as above. For determining of frequencies of exceeding concentrations the individual measurement values were summed up. Subsequently, the sum values were dichotomized on the basis of the medians of the empirical distributions. The empirical medians agreed quite well with the limits proposed by Schleibinger et al. (2002) – especially when taking into account the 95% confidence intervals of the percentiles (Table 13). Exceptions here are the siloxanes, for which the 90th percentile was clearly above Schleibinger's recommendation (Schleibiger, 2002). The marked deviation from previous studies (Schleibinger et al., 2001; Scholz, 1998) may indicate an increased (and still increasing) use of siloxanes as constituents of coating materials, paints, cosmetics etc., thus contributing more than ever to indoor air pollution.

Table 13. Indoor air concentrations and 95% confidence intervals [$\mu g/m^3$] for selected classes of chemical compounds. Results of a survey performed in the city of Berlin. Published target and reference values by Schleibinger et al. (2002) are included. 90. P: 90th percentile, AM: arithmetic mean, GM: geometric mean.

Compound class	Sample size	Median (95% CI)	90. P (95% CI)	MAX	AM (95% CI)	GM (95% CI)	Target value	Reference value
aldehydes (C4 – C10)	93	49 (36 – 59)	88 (76 – 127)	203	51 (43 –58)	37 (30 – 45)	50	120
aromatics	133	39 (30 – 51)	206 (149 - 248)	1140	82 (60 – 103)	45 (37 – 54)	50	200
siloxanes (D3 – D5)	99	4 (2 – 6)	125 (51 - 338)	741	43 (21 – 65)	5 (4 – 8)	5	10
terpenes und sesquiterpenes	133	51 (36 – 61)	230 (150 – 297)	1008	103 (73 – 133)	47 (37 – 58)	40	150

The relationship between ambient air concentrations of selected substance classes and the ACR the following picture can summarized as follows (Fig. 18): Depending on the substance class, between 50% and 70% of rooms with ACR <0.2 h^{-1}(sample size of this subgroup varied from n= 41ton= 53) showed concentrations which were above the median of the respective substance class.

For ACR between 0.2 h⁻¹ and 0.5 h⁻¹ (sample size between n = 37 and n = 52) this was true for only 40 – 50% of the rooms. This proportion was further reduced to 10 – 20% when the ACR were equal to or greater than 0.5 h⁻¹ (n = 15-27). This trend was statistically significant for all substance classes studied here ($\chi 2$ test for trend: aldehydes $p < 0.01$, aromatics $p < 0.001$, terpenes $p < 0.001$), Even for siloxanes the trend was significant, although not very pronounced ($p < 0.05$).

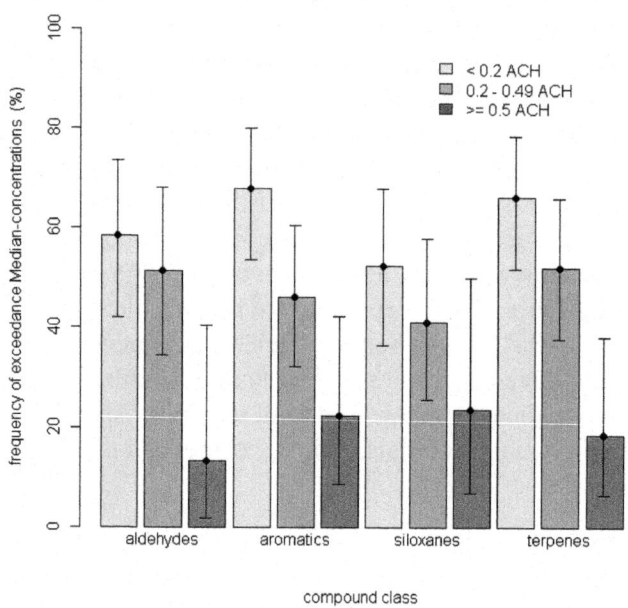

Figure 18. Frequency of concentrations above median (50th percentile), given for selected compound classes and stratified by ACR. Active air sampling under worst case conditions, error bars: 95% confidence intervals.

Overall, it can be deduced that the probability of exceeding pollutant concentrations not only depends on the number of indoor sources and their emitting power, but also on the airtightness of rooms. However, no exceeding concentrations can occur without the presence of indoor sources. For this reason, the use of low-emission products should given priority in order to avoid preventable future exposure to air pollutants. Whether a given indoor air concentration is above the usual burden and appropriate remedial measures are necessary, can be clearly evaluated by the ACR. It would therefore be desirable to define air-changerelated reference values for common and tolerable concentrations of volatile indoor air pollutants.

CONCLUSION

Tracer gas measurements facilitate the determination of air change between indoor and outdoor environment under different ventilation conditions. In this paper methods are described which can be used to determine the air change rate with tracer gas measurements. Special emphasis was given to carbon dioxide, since this tracer gas is often used to assess indoor air quality and air tightness of rooms. We have shown ways to determine air change rates with carbon dioxide and compared the results with those obtained by other tracer gases. The dependence of the natural air change from the prevailing weather conditions, such as the current wind and temperature conditions could be demonstrated by own investigations. Wheather effects in conjunction with different window types were leading to considerable day-to-day variability, thereby limiting the reproducibility of measurements to a large extent. However, focusing on the relationship between air change and window ventilation we could show that ventilation with wide open windows or cross-ventilation are by far the cheapest ways to remove indoor pollutants as quickly as possible from indoor air. The effectiveness of window ventilation strongly depends on the temperature difference between room air and outside air, whereas the influence of wind seems to be less important.

Studies on the air quality in Berlin's housing stock revealed a very high proportion of rooms with low air change rates. About 80% of the rooms examined by us showed air change rates below 0.5 h^{-1}, half of which were even below0.2 h^{-1}, when determined under so-called "worst case" conditions. Especially in rooms that had newly installed windows with insulated glass and peripheral sealing strip, very low air change rates were frequently determined. Of these, about 50% (median) had air change rates below 0.16 h^{-1}. This was in significant contrast to rooms in older buildings equipped with wooden framed double box windows or composite windows, which displayed mean air change ratesof0.47h^{-1}.

Low air change rates often contribute to elevated VOC concentrations, in rooms with insulated windows, unless emissions are not limited by appropriate low-emission products. This relationship could be confirmed by our investigations. Improvement of ventilation i.e. increasing the air change rate is an efficient measure to additionally reduce the pollutant load in indoor spaces. Therefore, to achieve limitation of unhealthy indoor-borne pollutants a combination of both measures can strongly be recommended.

ACKNOWLEDGEMENT

We are grateful to Dipl. Ing. P. Tappler, Innenraum Mess- und Beratungsservice,

Vienna, Federal Republic of Austria, for providing instrumentation and sulphur hexafluoride for the measurement of air change rates. We also thank Dr.-Ing. H. Schleibinger, Institute for Research in Construction, National Research Council of Canada, Ottawa, for his generous assistance in the acquisition of literature, and to the ALAB staff, Analytic Laboratory Berlin, for carefully conducting the sample analyses. We gratefully thank Dr.Liv Bode for her support in revising the manuscript.

REFERENCES

1. C. E. Andersen, N. C. Bergsøe, B. . Majborn, K. Ulbak, 1997 Radon and Natural Ventilation in Newer Danish Single-Family Houses.Indoor Air, 7 4 (December 1997), 278286 , 0905-6947

2. ASTM International 1998 ASTM Standard D 6245- 98 (Reapproved 2002): Standard guide for using Carbon dioxide concentrations to evaluate indoor air quality and ventilation. American Society for Testing and Materials. West Conshohocken, Pennsylvania

3. ASTM International 2001 ASTM Standard E 74100 . Standard Test Method for Determining Air Change in a Single Zone by Means of a Tracer Gas Dilution.American Society for Testing and Materials. West Conshohocken, Pennsylvania

4. BIA-Report   (2001).TracergasmesstechnikzurErmittlung   von Kenngrößen,   Berechnungsverfahren   und   Modellbildung   in   der Arbeitsbereichsanalys,. U. Eickmann (Ed.), 99114 99114 Hauptverband der   gewerblichenBerufsgenossenschaften   (HVBG),   3-88383-588-9 Augustin, Federal Republic Germany

5. P. Baránková, 2005 Měřeníintenzityvětránímetodouznačkovacíhoplynu CO2.Vytapeni, vetrani, instalace, 14 1 2, (únor 2005/duben 2005), 3134 /77-81, 1210-1389

6. G. Bekö, T. Lund, F. Nors, J. Toftum, G. Clausen, 2010 Ventilation rates in the bedrooms of 500 Danish children, Building and Environment, 45 10 (October 2010), 22892295 , 0360-1323

7. C. G. Bornehag, J. Sundell, L. . Hägerhed-Engman, T. Sigsgaard, 2005 Association between ventilation rates in 390 Swedish homes and allergic symptoms in children, Indoor Air, 15 4 (August 2005), 275280 , 0905-6947

8. Bundesgesundheitsamt: KommissionInnenraumlufthygiene des BGA 1993 Raumklimabedingungen in Schulen, Kindergärten und Wohnungen undihreBedeutungfür die Bestimmung von Formaldehydkonzentrationen, Bundesgesundheitsblatt, 36 2 (February 1993), 7678 , 0000-1436-9990

9. C. Y. H. Chao, T. C. W. Tung, J. Burnett, 1997 Influence of ventilation on indoor radon level. Building and Environment, 32 6 (November 1997), 527534 , 0360-1323

10. C. Y. Chao, M. P. Wan, A. K. Law, 2004 Ventilation performance measurement using constant concentration dosing strategy.Building and Environment, 39 11 (November 2004), 12771288 , 0360-1323

11. K. W. Cheong, S. B. Riffat, 1995 New approach for measuring airflows in buildings using a perfluorocarbon tracer. Applied Energy, 51 3 (July 1995), 223232 , 0306-2619

12. Y. K. Chuah, Y. M. Fu, C. C. Hung, P. C. Tseng, 1997 Concentration variations of pollutants in a work week period of an office.Building and Environment, 32 6 (November 1997), 535540 , 0360-1323

13. K. Ch. Chung, S.-P. Hsu, 2001 Effect of ventilation pattern on room air and contaminant distribution. Building and Environment, 36 9 (November 2001), 989998 , 0360-1323

14. R. N. Dietz, E. A. Cote, 1982 Air infiltration measurements in a home using a convenient perfluorocarbon tracer technique.Environment International, 8 1-6 , (July-December), 419433 , 0160-4120

15. R. N. Dietz, R. W. Goodrich, A. Cote, R. F. Wieser, 1986 Detailed description and performance of a passive perfluorocarbon tracer system for building ventilation and air change measurements. In: Measured Air Leakage of Buildings, Trechsel, H.R. & Lagus, P.M., 203264 , American Society for Testing and Materials, 978803149717

16. W. S. Dols, A. K. Persily, 1992 Study of Ventilation Measurement in an Office Building, In: Airflow Performance of Building Envelopes, Components, and Systems (ASTM STP 1255), Modera, M.P. &Persily, A.K., 2346 , American Society for Testing and Materials, 978-0-80312-023-5. West Conshohocken, Pennsylvania

17. P. J. Drivas, F. G. Simmonds, F. H. Shair, 1972 Experimental characterization of ventilation systems in buildings.Environmental Science and Technology, 6 7 (July 1972), 577666 , 0001-3936X

18. H. Erhorn, K. Gertis, 1986 Mindestwärmeschutzoder/und Mindestluftwechsel?giGesundheits-Ingenieur: Haustechnik- Bauphysik- Umwelttechnik, 107 1 (February 1986), 1214 & 71-76, 0932-6200

19. P. O. Fanger, 1988 Introduction of the olf and the decipol units to quantify air pollution perceived by humans indoors and outdoors.Energy and Buildings, 12 1 (April 1988), 16 , 0378-7788

20. N. L. Gregory, 1962 Detection of nanogram quantities of Sulphur

Hexafluoride by electron capture methods. Nature, 196 4859 (October 1962), 162 0028-0836

21. D. T. Grimsrud, M. H. Sherman, J. E. Janssen, A. N. Pearman, D. T. Harrje, 1980 An intercomparison of tracer gases used for air infiltration measurements. ASHRAE Transactions, 86 258267 , 0001-2505

22. L. Guo, O. J. Lewis, 2007 Carbon Dioxide Concentration and its Application on Estimating the Air Change Rate in Typical Irish Houses. International Journal of Ventilation, 6 3 (December 2007), 235245 , 1473-3315

23. H. Harving, R. Dahl, J. Korsgaard, S. A. Linde, 1992 The indoor environment in dwellings: a study of air-change, humidity and pollutants in 115 Danish residences. Indoor Air, 2 2 (June 1992), 121126 , 1600-0668

24. F. Heidt, D. , H. Werner, 1986 Microcomputer-aided measurement of air changerates. Energy and Buildings, 9 4 (December 1986), 313320 , 0378-7788

25. F.-D. Heidt, 1987 Fortschrittebei der Luftwechsel messungdurch Mikrocomputereinsatz. Heizung- Lüftung- Haustechnik, 38 8 (August 1987), 391395 , 0341-0285

26. M. Hill, R. Gehrig, V. Dorer, A. Weber, P. Hofer, 2000 Are measurements of air change rates with the PFT-method biassed by sink and temperature effects? Proceedings of Healthy Buildings 2000, 2 333338 , 9-52523-605-6 Finland, August 6-10

27. T. Hirsch, M. Hering, K. Burkner, D. Hirsch, W. Leupold, M. L. Kerkmann, E. Kuhlisch, L. Jatzwauk, 2000 House-dust-mite allergen concentrations (Der f 1) and mold spores in apartment bedrooms before and after installation of insulated windows and central heating systems. Allergy, 55 1 (January 2000), 7983 , 0105-4538

28. A. T. Hodgson, D. Faulkner, D. P. Sullivan, D. L. Di Bartolomeo, M. L. Russell, W. J. Fisk, 2003 Effect of outside air ventilation rate on volatile organic compound concentrations in a call center.Atmospheric Environment, 37 39-40 , (December 2003), 55175527 , 1352-2310

29. C. Howard-Reed, L. A. Wallace, W. R. Ott, 2002 The effect of opening windows on air change rates in two homes. Journal of the Air & Waste Management Association, 52 2 (February 2002), 147159 , 1047-3289

30. C. M. Hunt, D. M. Burch, 1975 Air infiltration measurements in a four bedroom townhouse using Sulphur Hexafluoride as a tracer gas.ASHRAE Transactions, 81 1 186201 , 0001-2505

31. R. Keller, J. Beckert, 1994 Untersuchungenzum Vorkommenorganischerflüchtiger Verbindungen in dem Musterraumeines Neubausunter der Berücksichtigung des „natürlichen" Luftwechsels. Zentralblattfür Hygiene und Umweltmedizin, 195 5-6 , (June 1994), 432443 , 0934-8859

32. J. Krooß, U. Siemers, P. Stolz, N. Weiss, K. Clausnitzer, D. , 1997 Luftwechselraten in Wohn- und Arbeitsräumen.Gefahrstoffe- Reinhaltung der Luft, 57 9 (September 1997), 357362 , 0949-8036

33. R. Kumar, A. D. Ireson, H. W. Orr, 1979 An automated air infiltration measuring system using SF6 tracer gas in constant concentration and decay methods.ASHRAE Transactions, 85 2 385395 , 0001-2505

34. P. L. Lagus, R. A. Grot, 1997 Control room envelope unfiltered air inleakage test protocols. In: Proceedings of the 24th DOE/NRC Nuclear Air Cleaning and Treatment Conference, First, M.W., 400427 , The Havard Air Cleaning Laboratory, NUREG/CP-0153, CONF-960715, Boston, MA

35. J. Lembrechts, M. Janssen, P. Stoop, 2001 Ventilation and radon transport in Dutch dwellings: computer modelling and field measurements. The Science of the Total Environment, 272 1-3 , (May 2001), 7378 , 0048-9697

36. Y.-Y. Li, P.-C. Wu, H.-J. Su, P.-C. Chou, C.-M. Chiang, 2002 Effects of HVAC ventilation efficiency on the concentrations of formaldehyde and total volatile organic compounds in office buildings.Proceedings of the 9th International Conference on Indoor Air Quality and Climate, 376381 , 0-97218-320-5 CA, June 30- July 5 2002

37. A. Maas, 1997 TracergasmeßtechnikenzurErmittlung des Luftwechsels. gi Gesundheits-Ingenieur: Haustechnik- Bauphysik- Umwelttechnik, 118 5 (October 1995), 256267 , 0932-6200

38. W. Mailahn, B. Seifert, D. Ullrich, H. Morikse, J. , 1989 The use of a passive sampler for the simultaneous determination of long-term ventilation rates and VOC concentrations. Environment International, 15 1-6 , (January-December 1989), 537544 , 0160-4120

39. R. Menzies, K. Schwartzman, V. Loo, J. Pasztor, 1995 Measuring ventilation of patient care areas in hospitals.Description of a new protocol.American Journal of Respiratory and Critical Care Medicine, 152 6 (December, 1995), 19921999 , 0107-3449X

40. U. Münzenberg, 2004 Der natürlicheLuftwechsel in Gebäuden und seine Bedeutungbei der Beurteilung von Schimmelpilzschäden. In: Umwelt, Gebäude&Gesundheit: Innenraumhygiene, Raumluftqualität

und   Energieeinsparung.   Ergebnisse   des   7.Fachkongresses
der   ArbeitsgemeinschaftÖkologischerForschungsinstitute
(AGÖF)   am   04.   und   05.   März   2004   in   München,   AGÖF-
ArbeitsgemeinschaftÖkologischerForschungsinstitute, (Ed.), 263271 ,
AGÖF, 3-93057-605-8-Eldagsen

41.   S. J. Nabinger, A. K. Persily, W. S. Dols, 1994 Study of Ventilation and
      Carbon Dioxide in an Office Building.ASHRAE Transactions, 100 2
      12641274 , 0001-2505

42.   R. Niemelä, A. Lefevre, J. P. Muller, G. Aubertin, 1991 Comparison of
      threetracer gases for determination ventilation effectiveness and capture
      efficiency.Annals Occupational Hygiene, 35 4 (August 1991), 405417 ,
      0003-4878

43.   L. Øie, P. Magnus, B. V. Johansen, 1997 Suspended particulate matter
      in Norwegian dwellings in relation to fleece and shelf factors, domestic
      smoking, air change rate, and presence of hot wire convection heaters,
      Environment International, 23 4 (July-August 1997), 465473 , 0160-4120

44.   L. Øie, H. Stymne, C. Boman, A. , V. Hellstrand, 1998 The ventilation
      rate of 344 Oslo residences. Indoor Air, 8 3 (September 1998), 190196 ,
      0905-6947

45.   M. D. Pandian, W. R. Ott, J. V. Behar, 1993 Residential air change rates
      for use in indoor air and exposure modelling studies. Journal of Exposure
      Analysis and Environmental Epidemiology, 3 4 (October 1993), 407416
      , 1053-4245

46.   G. B. Parker, 1986 Measured air change rates and indoor air quality in
      multifamily residences.Energy and Buildings, 9 4 (December 1986),
      293303 , 0378-7788

47.   J. N. Penman, 1980 An experimental determination of ventilation rate
      in occupied rooms using atmospheric carbon dioxide concentration.
      Building and Environment, 15 1 (April-March 1980), 4547 , 0360-1323

48.   J. N. Penman, A. A. M. Rashid, 1982 An experimental determination of
      air flow in a naturally ventilated room using metabolic carbon dioxide.
      Building and Environment, 17 4 253256 , (October-December 1982),
      0360-1323

49.   A. K. Persily, 1997 Evaluating building IAQ and ventilation with indoor
      Carbon Dioxide.ASHRAE Transactions, 103 2 193204 , 0001-2505

50.   A. Persily, A. Musser, S. J. Emmerich, 2010 Modeled infiltration rate
      distributions for U.S. housing. Indoor Air, 20 6 (December 2010), 473485
      , 1600-0668

51. M. Pettenkofer, 1858 Besprechungallgemeiner auf die Ventilation bezüglicherFragen. Über den Luftwechsel in Wohngebäuden,Cottaische Verlagsbuchhandlung, München, 69126

52. W. Raatschen, 1995 Tracergasmessungen in der Gebäudetechnik: Luftaustausch- Messung und Simulation. GI- Gesundheits-Ingenieur: Haustechnik, Bauphysik, Umwelttechnik, 116 2-3 , (April, June 1995), 7887 and 129-138, 0932-6200

53. C. A. Roulet, F. Foradini, 2002 Simple and cheap air change rate measurement using CO2 concentration decays. International Journal of Ventilation, 1 1 (June 2002), 3944 , 1473-3315

54. R. Ruotsalainen, R. Rönnberg, J. Säteri, A. Majanen, O. Seppänen, J. J. K. Jaakkola, 1992 Indoor climate and the performance of ventilation in Finnish residences.Indoor Air, 2 3 (September1992), 137145 , 0905-6947

55. J. Sakaguchi, S. Akabayashi, 2003 Field survey of indoor air quality in detached houses in Niigata Prefecture.Indoor Air, 13 S6 (January 2003), 4249 , 0905-6947

56. L. G. Salmon, G. R. Cass, K. Bruckman, J. Haber, 2000 Ozone exposure inside museums in the historic central district of Krakow, Poland. Atmospheric Environment, 34 22 (June 2000), 38233832 , 1352-2310

57. T. Salthammer, 1994 LuftverunreinigendeorganischeSubstanzen in Innenräumen.Chemie in unsererZeit, 28 6 (December 1994), 280290 , 0009-2851

58. T. Salthammer, F. Fuhrmann, S. Kaufhold, B. Meyer, A. Schwarz, 1995 Effects of Climatic Parameters on Formaldehyde Concentrations in Indoor Air.Indoor Air, 5 2 (April 1995), 120128 , 0905-6947

59. H. Schleibinger, U. Hott, D. Marchl, P. Braun, P. Plieninger, H. Rüden, 2001 VOC-Konzentrationen in Innenräumen des Großraums Berlin imZeitraum von 1988- 1999.Gefahrstoffe- Reinhaltung der Luft, 51 1 2, (January-February 2001), 2638 , 0949-8036

60. H. Schleibinger, U. Hott, D. Marchl, P. Braun, P. Plieninger, H. Rüden, 2002 Ziel- und RichtwertezurBewertung der VOC-Konzentrationen in der Innenraumluft- einDiskussionsbeitrag.Umweltmedizin in Forschung und Praxis, 7 3 (June 2002), 139147 , 1430-8681

61. H. Scholz, 1998 Vorkommenausgewählter VOC in Innenräumen und derenBewertung. In: Gebäudestandard 2000: Energie und Raumluftqualität, ArbeitsgemeinschaftökologischerForschungsinstitute (Ed.), 205214 , AGÖF, 3-93057-601-5Eldagsen

62. H. G. Schulze, G. Schuschke, 1990 Studieüber die Notwendigkeit und

Zuverlässigkeit von Luftwechsel und Luftvolumenstrommessungen. Gesundheits-Ingenieur, 111 1 (February 1990), 1216 , 0016-9277

63.   B. Seifert, T. Salthammer, 2003 Innenräume.Belastung der Umweltmedien. In: Handbuch der Umweltmedizin, H.-E. Wichmann, H.-W.Schlipköter& G. Fülgraff(Ed.), 26 Erg.Lfg. 4/03- 1-34, ecomedFachverlag, 978-3-60971-180-5 Landsberg/Lech

64.   S. C. Sekhar, 2004 Enhancement of ventilation performance of a residential split-system air-conditioning unit.Energy and Buildings, 36 3 , (March 2004), 273279 , 0378-7788

65.   I. Senitkova, 1999 Building redesign for Radon elimination.Proceedings of the 8th International Conference on Indoor Air Quality and Climate, 2 953958 , 860812953 Scotland, 8- 13 August 1999

66.   O. A. Seppänen, W. J. Fisk, 2004 Summary of human responses to ventilation.Indoor Air, 14 S7 (August 2004), 102118 , 0905-6947

67.   O. A. Seppänen, W. J. Fisk, M. J. Mendell, 1999 Association of ventilation rates and CO2 concentrations with health and other responses in commercial and institutional buildings. Indoor Air, 9 4 (December 1999), 226252 , 0905-6947

68.   C. Y. Shaw, 1984 The effect of tracer gas on the accuracy of air-change measurements in buildings.ASHRAE Transactions, 90 2816 =pt 1A), 212225 , 0001-2505

69.   M. H. Sherman, 1990 Tracer-gas techniques for measurement ventilation in a single zone.Building and Environment, 25 4 (April 1990), 365374 , 0360-1323

70.   P. N. Smith, 1988 Determination of ventilation rates in occupied buildings from metabolic CO2 concentrations and production rates. Building and Environment, 23 2 (January 1988), 95102 , 0360-1323

71.   P. Stavova (Barankova), A. K. Melikov, J. Sundell, F. Drkal, 2006 Validation of a CO2 Method for Ventilation Measurement in Dwellings-Laboratory Study.Proceedings of Healthy Buildings 2006, 8th International Conference on Indoor Air Quality and Health, 375 9-89950-671-0 Portugal, June 4-8, 2006

72.   TRGS 900 1999 TechnischeRegelnfürGefahrstoffe: Grenzwerte in der Luft am Arbeitsplatz- Luftgrenzwerte (TRGS 900), Bundesarbeitsblatt, 6 87 0007-5868

73.   VDI 2001 VDI 4300 - Messen von Innenraumluftverunreinigungen. Bestimmung der Luftwechselzahl in Innenräumen. In: VDI/DIN-HandbuchReinhaltung der Luft Band 5, KommissionReinhaltung der

Luft (KRdL) im VDI und DIN- Normenausschuß (Ed.), Beuth-Verlag, ICS: 13.040.01, Berlin

74. I. S. Walker, T. W. Forest, 1995 Field measurements of ventilation rates in attics.Building and Environment, 30 3 (July 1995), 333347 , 0360-1323

75. P. Wargocki, J. Sundell, W. Bischof, G. Brundrett, O. Fanger, F. Gyntelberg, S. O. Hanssen, P. Harrison, A. Pickering, A. Seppänen, P. Wouters, 2002 Ventilation and health in non-industrial indoor environments: report from a European Multidisciplinary Scientific Consensus Meeting (EUROVEN). Indoor Air, 12 2 (June 2002), 113128 , 0905-6947

76. J. Wegner, 1983 Untersuchungen des natürlichen Luftwechsels in ausgeführtenWohnungen, die mitsehrfugendichtenFensternausgestattetsind. gi- Gesundheitsingenieur, 104 1 (February 1983), 156 , 0932-6200

77. J. Wegner, 1984 Schadstoffanfall, Luftwechsel in Wohnungen, freieLüftung. Haustechnik- Bauphysik- Umwelttechnik- Gesundheits-Ingenieur, 105 3 (June 1984), 117123 , 0172-8199

78. A. L. Wilson, S. D. Colome, Y. Tian, E. W. Becker, P. E. Baker, D. W. Behrens, I. C. H. Billick, 1996 California residential air change rates and residence volumes. Journal of Exposure Analysis and Environmental Epidemiology, 6 3 (July-September 1996), 311326 , 1053-4245

Chapter 3

STATISTICAL CONSIDERATIONS FOR BIOAEROSOL HEALTH-RISK EXPOSURE ANALYSIS

M. D. Larrañaga[1], E. Karunasena[2], H. W. Holder[3], E. D. Althouse[4] and D. C. Straus[5]

[1] Oklahoma State University, USA

[2] Texas Tech University, USA

[3] 3SWK, LLC, USA

[4] Air Intellect, LLC, USA

[5] Texas Tech University Health Sciences Center, USA

INTRODUCTION

Air and surface sampling was conducted to confirm the types of microbiological contamination within a hospital facility in the southern United States, identify indicators of indoor microbiological contamination, and profile the aerobiological makeup of the inside air (ISA) for comparisons to outside air (OSA) and reference concentrations, where applicable. The investigation strategy recommended by the American Conference of Governmental Industrial Hygienists (ACGIH) was utilized to assess indoor environmental quality and conditions found within the hospital facility. The ACGIH methodology is guided by the text Bioaerosols: Assessment and Control. (Macher 1999) Biological contamination within a hospital environment is of great concern as bacteria and fungi are important causes of nosocomial infection (NI). It is estimated that the overall hospital-acquired infection rate in Europe and North America is between 5 and 10%. (Kalliokoski 2003) A substantial number of bacteria and fungi are capable of spreading via the airborne route in hospitals, and airborne transmission accounts for approximately 10% of all NI. (Eickhoff 1994; Kalliokoski 2003) Contaminated Heating, Ventilating, and Air Conditioning (HVAC) systems and infiltration of unfiltered outside air have been implicated in airborne outbreaks of NI via infective aerosols, dust, and contaminated filters. (Lentino, Rosenkranz et al. 1982; Rhame 1991; Eickhoff 1994) Certain underlying diseases, procedures, hospital services, and categories of age, sex, race, and urgency of admission have been shown to be significant risk factors for nosocomial infection. (Freeman and McGowan 1978) The day-specific incidence of nosocomial infection rises from near zero on the first hospital day to maximal during the fourth through seventh weeks of hospital stay. (Freeman and McGowan 1981) Nosocomial infections can affect patients in any location within a hospital. (Boss and Day 2003)

The Centers for Disease Control (CDC) estimates that 2 million patients develop hospital-acquired infections annually and as many as 88,000 die as a result. (CDC 1992) Hospitals typically maintain a patient population with increased susceptibility to infection and factors inherent to the healthcare environment contribute to the risk associated with acquiring an infection during a hospital admission. (Dulworth and Pyenson 2004) Environmental control and high efficiency filtration are critical to preventing person-to-person and environmentally related infections in hospitals. (Wenzel 1997; Boss and Day 2003) Current evidence indicates that excessive moisture indoors promotes microbial growth and is associated with an increased prevalence of symptoms due to irritation, allergy, and infection. It is widely accepted in various scientific communities that indoor microbiological contamination presents unacceptable conditions for the preservation of human health, and that removal and prevention of microbial contamination is necessary and prudent. (Pope, Patterson et al. 1993; Macher 1999; Agency 2001; ACOEM 2002; Fung and Hughson 2002; Redd 2002; CDC 2003; Fung and Hughson 2003)

The inherent variability of microbiological organisms in air presents a challenge for conducting air sampling that provides meaningful results for the evaluation of human exposure and health risk. In general, bioaerosol air sampling is highly variable and prone to error. Multiple and replicate samples over subsequent days are necessary to characterize exposure and multiple samples per sample location are required to evaluate human exposure in that particular location. An air and surface sampling plan was designed to address the inherent variability and error associated with air sampling and to evaluate the exposure and subsequent health risk to patients, visitors, and staff in a hospital facility. The objectives of this chapter are to highlight the necessity for multiple (replicate) air samples per sample location to conduct valid assessments of the airborne concentrations of bioaerosols.

MATERIALS AND METHODS

Sampling was conducted to provide a bioaerosol profile of the air within the spaces under the control of 9 separate HVAC systems over a two-year period. The spaces under the environmental control or each HVAC system are identified as air handling units (AHUs) 10, 11 (no final filters present), 13

(90% final filters), 15, 16, 17, 18, and 19 during the summer of 2005. In early 2006, a follow-up investigation was conducted within the spaces controlled by AHUs 17, 19, and 21. Except where indicated, AHU final filters had a filtration efficiency of 95%. Air and surface sampling data and analysis, observations, and the collective experience of a team experienced in moisture intrusion in hospital facilities along with input from professionals in medical microbiology, industrial hygiene, medicine, engineering, and public health were utilized to interpret data for hypothesis testing. The hypotheses were:

- Hypothesis A: The 90-95% final filters control particulate matter generated by the AHUs and preventing contamination downstream of the filters. Note: AHU 11 does not have final filters and therefore this hypothesis does not apply to AHU 11.

- Hypothesis B: The 90-95% final filters prevent microbial contamination downstream of the filters. Note: AHU 11 does not have final filters and therefore this hypothesis does not apply to AHU 11.

- Hypothesis D: Staff is being exposed to potentially harmful concentrations of biological contaminants.

- Hypothesis E: Patients are being exposed to harmful quantities of biological contaminants.

Air sampling was conducted to establish mean airborne concentrations of fungal and biological aerosols indoors and outdoors to characterize the fungal and bacterial aerobiological profiles of the areas controlled by each AHU. Culturable air samples were analyzed at the species level as information on species is crucial for determining whether indicator organisms are present. Indicator species of fungi whose presence indicates excessive indoor moisture or a health hazard were evaluated. Indicator organisms identified via air sampling in the hospital were *Aspergillus versicolor, A. flavus, A. fumigatus, Fusarium* species, yeasts, and species of *Penicillium*. (Macher 1999) In addition, the American Industrial Hygiene Association (AIHA) has consistently recommended urgent risk management decisions be made when the confirmed presence of these indicator organisms are identified indoors These indicator organisms include those listed above in addition to *Stachybotryschartarum*. The confirmed presence is defined as colonies in several samples, many colonies in any sample, or, where a single colony was found in a single sample, evidence of growth of these fungi on building materials by visual inspection or source sampling. (Macher 1999) As early as 1996, AIHA stated that urgent risk management decisions are required of the industrial hygienist in the following conditions: a) the confirmed presence of facultative pathogens (fungi capable of inducing pulmonary infections in humans) such as *A. versicolor and Fusarium moniliforme,* and b) The presence of fungi, such as *S. chartarum* and *F. moniliforme*, known to result in occupational diseases in part due to their potent toxins. AIHA recommended that these urgent risk management decisions be made promptly as opposed to weeks or months later. (Dillon, Heinsohn et al. 1996; Prezant, Weekes et al. 2008) Based on the literature and in consideration of the recommendations made elsewhere by governmental and nongovernmental entities and other professional societies, the 1996 and 2008 AIHA recommendations and the 1999 ACGIH recommendations (Macher 1999) continue to be appropriate risk management guidance for the industrial hygienist and indoor environmental professional.

The estimated cumulative sampling and analytical error for each air sample is defined as $E_c = (P^2 + T^2 + A^2 + O^2)^{1/2}$ where E_c is the cumulative error, P is the pump error (estimated at $\pm$ 5%), T is the time error (estimated at $\pm$ 2.5%), A is the analytical error (estimated to be $\pm$ 25%), and O is other error associated with calibration and technician variability (estimated to be $\pm$ 25%). (Macher 1999; Burton 2006) Solving for E_c, the cumulative sampling and analytical error for each sample is estimated to be approximately $\pm$ 36%. Therefore, data derived from individual samples should be viewed as qualitative. The inherent variability of the air concentrations of bioaerosols over time or within a space far outweighs any errors associated with measurement of airborne microbiological concentrations. Thus, the interpretation of a single sample is difficult without information on the variability of the concentrations of biological agents identified in the environment because the variability in the measurement is almost always large. (Macher 1999)

Duplicate, side-by-side air samples were taken at each location to address the error associated with individual samples. Duplicate side-by-side air sampling is considered adequate to define the mean and the random sampling error given the high temporal and spatial variability of bioaerosol concentrations in air. (Dillon, Heinsohn et al. 1996) Multiple samples from multiple random and non-random locations were taken on separate days to characterize exposures within the space under the environmental control of each air handling unit. (Macher 1999) Replicate samples within the space controlled by each air handling unit were taken to address sampling variability (Weber and Page 2001) and allow for the estimation of the sampling data's variances so that differences between two environments (e.g. ISA v. OSA concentrations) could be identified. (Macher 1999) A minimum of 6 replicate samples were taken at 10 indoor locations ($6 \leq n \leq 48$ samples per location) and 2 outdoor locations ($12 \leq n \leq 72$ total outdoor reference samples per indoor location). An ANOVA was utilized to compare indoor and outdoor concentrations of biological agents using SAS Statistical Software. The level of significance was prescribed as $\alpha = 0.05$. A significant difference identified by the ANOVA indicates that the difference is unlikely to have occurred by chance and that there is statistical evidence that there is a difference. At the $\alpha = 0.05$ (5%) level of significance, the result could have occurred by chance one time in 20. Statistical techniques evaluate an observed difference in view of its precision to determine with what probability it might have arisen by chance (the level of significance). Values with a low probability of occurring by chance are called statistically significant and are considered to represent a real effect (e.g. difference between means). (Conover 1999; Montgomery 2001)

Ideally, human respiratory exposure is measured using air samples taken near the breathing zone, or within 12 inches of the mouth. This was not feasible considering the large size and weight of bioaerosol air sampling equipment. Most bioaerosol sampling is done to characterize ambient aerosols and the ambient conditions are utilized as quantitative estimates of bioaerosol exposure. Although the characterization of the ambient environment is not the ideal exposure sampling scheme, when low flow rate suction impactors (e.g. Andersen type) are utilized, the error introduced is small. (Pope, Patterson et al. 1993) Low flow rate suction impactors were utilized to take bioaerosol samples.

A comparison of total fungal or bacterial concentrations may be used as a preliminary indicator of a difference in two environments, but not as evidence of similarity or dissimilarity. Indoor/outdoor comparisons are used to document the presence or infer the absence of indoor, biologically derived contamination. These comparisons cannot be made unless the genera and species found indoors and outdoors have been identified. (Macher 1999) Air sampling mean total counts were utilized as a preliminary indicator of a difference in two environments (e.g. indoor vs. outdoor air concentrations) to determine the effectiveness of the filters in removing aerobiological particulates from the air stream. Air sampling indicators of indoor microbiological contamination were identified from sampling results where both the genera and species are identified (culturable fungal and bacterial samples incubated at 25°C and 37°C). Culturable air samples were taken for incubation at two separate temperatures (25°C and 37°C) to enhance the detection of both environmental and pathogenic microorganisms in the air. (Dillon, Heinsohn et al. 1996; Macher 1999)

Air sampling indicators of indoor contamination were compared to both surface sampling results and outdoor air sampling concentrations of like-organisms for interpretation. Note that monitoring for allergens can help characterize environments with respect to specific allergens (e.g., fungi and/or bacteria), and measurements can be semi-quantitative (e.g., "presence or absence" or "low, medium, or high"). (Pope, Patterson et al. 1993) Airborne bioaerosol sampling was conducted so that comparisons and interpretations could be made between ISA and OSA culturable and non-culturable bioaerosols, mean concentrations and variability of concentrations, and species could be compared. Each set or type of sampling results should be viewed in consideration with the other sampling results and not independently. For example, one should not consider spore trap (total and non-culturable fungal samples) alone, as spore trap sampling does not identify fungi at the species level and may mask important differences in species present between test and reference locations. Analysis of spore traps alone or total fungal or bacterial concentrations could lead to incorrect conclusions. The sampling interpretation considered all sampling results for the development of interpretations and conclusions.

The American Society for Heating, Refrigerating, and Air Conditioning Engineers (ASHRAE) defines critical care areas as the following functional spaces within a hospital: 1) Surgery and Critical Care, 2) Nursing, 3) Ancillary, 4) Diagnostic and Treatment, and 5) Sterilization and Supply. Critical care areas include but are not limited to intensive care units, coronary care units, angiography laboratories, cardiac catheterization laboratories, delivery rooms, operating rooms, recovery rooms, emergency departments, and other special care units where enhanced engineering controls are required for the protection of patients and staff. Although patients spend most of their time within a specific area, they may be exposed to bioaerosols in non-critical locations of a hospital where engineering controls are not as stringent. Non-critical functional areas are administration and service locations within the Hospital. (ASHRAE 2003) Table 1 lists the critical and non critical locations within the areas investigated.

Table 1. Classification of Hospital areas by critical or non-critical area.

AHU #	Area Served	Critical Area
10	Endoscopy, EKG, EGG, PFT, Pharmacy, Sterile Preparation, Recovery	Yes
11	1st floor PBX, Waiting Room, HR, Medical Records, Security	No
13	Patient Rooms	Yes
15	Intermediate Nursery, Ambulatory Surgical Center, Endoscopy, Operating Rooms, Sterile Supply, Pre- and Post- Op areas	Yes

16	Patient Rooms, Wound Therapy, Lactation Center	Yes
17	Emergency Room, Radiology and CT-Scan	Yes
18	Radiology, Ultrasound, Mammography, Laboratory, Laboratory Biohazard, Histology, Pathology	Yes
19	1st Floor, Administration Area, Labor and Delivery, Recovery, C-Section Room, Pre-Op Area, Intensive Care, Rehabilitation	Yes
21	Radiology, Purchasing, Shipping and Receiving, Plant Operations	Yes

RESULTS

Total Outside Vs. Total Indoor Concentrations (2005 Data)

Figure 1 depicts the percent difference of bioaerosols from the OSA reference concentrations to the ISA concentrations for the space controlled by each AHU. The results show that the existing filters within the facility were removing bioaerosols from the air within the Hospital.

95% final filters were installed in the AHUs investigated, with the exception of AHUs 11 and 13. AHU 11 did not have final filters and AHU 13 had 90% final filters installed. Ninety-five percent final filters are rated to remove greater than 90% of particles between 1.0 and 10 micrometers in size and 85-95% of particles between 0.3 and 1.0 micrometers in size. Ninety percent final filters are rated to remove 90% of particulate matter between 1 and 10 micrometers in size and 75-85% of particles between 0.3 and 1.0 micrometers in size.

(ASHRAE 1992; ASHRAE 1999) Therefore, with the exception of AHU 11, the percent differences for fungal particulates between OSA and ISA should approach 90% for organic (e.g. fungi and bacteria) particulate matter ranging from 1.0 to 10 micrometers in aerodynamic diameter and 78-85% of organic matter for particles ranging from 0.3-1.0 micrometers in aerodynamic diameter.

Non-culturable fungal air sampling via spore trap measures the airborne fungal particle concentrations in spores/m³ of air. The minimum percent difference (reduction) of 79% (AHU 19) for total spore count (spore trap) results indicates that at least 79% of the non-culturable total spore concentrations from the outside air are being removed by the HVAC systems prior to entering the building. The minimum percent difference of all fungal air sampling results is 63% for the fungal samples incubated at 25°C in the space controlled by AHU 11. Excluding AHU 11 because it does not have final filters, the minimum percent difference of all fungal sampling results is 66% for AHU 13, which has 90% final filters. This indicates that at least 66% of the fungal bioaerosols are being removed by the final filters of the AHUs investigated, with the exception of AHU 11. The concentrations of total and culturable fungal bioaerosols within the spaces controlled by the AHUs with 95% final filters (all except AHUs 11 and 13) are at least 75% lower than outside air concentrations, indicating that the filters are performing and removing particulates from the air.

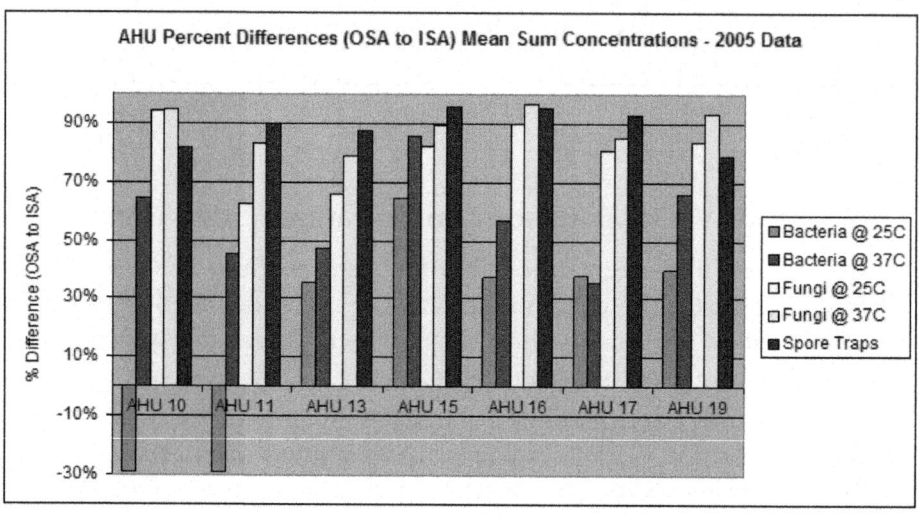

Figure 1. AHU percent differences from outside air to inside air. Note: A negative percent difference indicates that the indoor concentration was higher than outdoors.

The bacterial concentrations (incubated at 25°C) within the space controlled by AHUs 10 and 11 were higher indoors than outdoors. Unlike fungi, bacteria have natural reservoirs indoors (including humans), and total bacterial concentrations are often higher indoors than outdoors. (Macher 1999) The bacterial organisms identified (incubated at 25°C) as indicators of an indoor source were *Micrococcus* species, *Micrococcus luteus*, *Staphylococcus capitus*, and *Staphylococcus hyicus*, which are human-shed bacteria. (Wilson 2005) Because these organisms are human-shed and were not identified as indoor contaminants via surface sampling, it cannot be concluded that these higher concentrations of bacteria detected via air sampling were the result of building-related sources of bacterial contamination. (Macher 1999) Higher indoor concentrations of human-shed bacteria are an anticipated condition within a building. The bacteria *Bordetella bronchiseptica* was identified inside AHUs 10, 11, 13, 16, 18, 19 and was not identified indoors via air sampling. Viridans streptococci was identified inside AHU 18 and not identified in indoor air samples. *Staphylococcus aureus* was identified inside AHU 21 and not identified in indoor air samples. This is a strong indication that the filters may prevent the transmission of *Bordetella bronchiseptica*, Viridans streptococci, and *Staphylococcus aureus* through the filters and into the occupied space of the hospital.

With the exception of the bacterial (incubated at 25°C) air samples in AHUs 10 and 11, the percent differences for bacterial sample sets comparing OSA to ISA were at least 35%. This is an indication that the filters were removing bacteria from both the outside and re-circulated air of the Hospital.

Total Outside Vs. Total Indoor Concentrations (2006 Data)

The 2006 sampling data indicate that, in general, the total air concentrations are lower indoors than outdoors. However, indicators of indoor contamination were identified. The 2006 sampling strategy was to compare outdoor fungal and bacterial airborne concentrations with the concentrations identified 1) within each AHU before the filter, 2) within each AHU after the filter, and 3) within a room controlled by the AHU. AHUs 17, 19, and 21 were tested in early 2006.

AHU 17

AHU 17 serves the first floor emergency room and radiology, which are critical areas. Figure 2summarizes the 2006 data for AHU 17.

The AHU 17 data indicate fungal and bacterial percent differences from OSA to ISA greater than 74%, with the exception of total fungi (spore traps), which is shown as a difference of 1%. Percent differences between before filter and after filter concentrations also show minimum percent differences of 58%, which indicate bioaerosol removal from the air stream after it passes through the filter. The low percent differences for spore traps and culturable air sampling (25°C and 37°C samples) between OSA and ISA (room) indicate indoor contamination as shown by the negative percent difference shown for the after filter to room samples. This is not an indication that the filters are inadequately filtering particles, but rather an indication of an indoor source and/or infiltration due to negative pressure contributing to the indoor concentrations of bioaerosols.

Percent differences show an increase in airborne fungal and bacterial concentrations between the locations after the filter and within the room, indicating an increase in airborne concentrations in the air for both culturable bacteria and fungi and total fungi after it leaves the AHU. Negative percent differences are identified for spore trap samples from before the filter to room and after the filter to room. The increase in concentrations as the air moves from the AHU to the room indicates the presence of an indoor source of fungi and bacteria, specifically yeasts, fungal species of Cladosporium, Penicillium, Aspergillus, and Fusarium, and bacterial species of Staphylococcus, Micrococcus, and Sphingomonas paucimobilis.

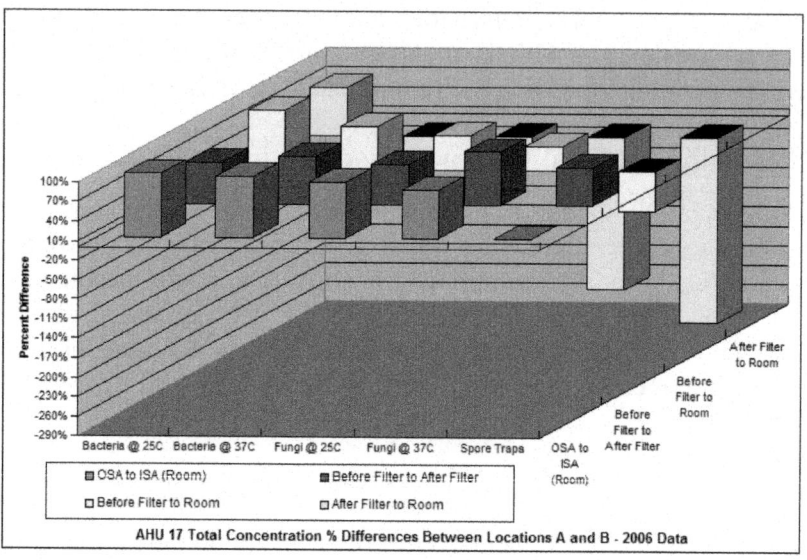

Figure 2. AHU 17 percent differences in mean total concentrations between spaces (2006 data).Note: A negative percent difference indicates that the indoor concentration was higher than outdoors for OSA to ISA comparisons.Note: A black top of a bar in the graph in the 0% plane indicates a negative percent difference indicating an increase in concentration from one space to another when a decrease is expected.

AHU 19

AHU 19 serves the 1st Floor Administration Area, Labor and Delivery, Recovery, the Cesarean Section room, Pre-Op Area, the Intensive Care Unit, and a Rehabilitation area. All areas supplied by AHU 19 are considered critical care areas except the Administration Area. The AHU 19 data indicate percent differences that approach or are greater than 90% for fungi and bacteria between OSA and ISA. This indicates that greater than 90% of the bioaerosols in the OSA are being removed from the air stream. The AHU 19 data indicate positive percent differences signifying decreasing concentrations from before the filter to after the filter, before the filter to room, and after the filter to room. Figure 3summarizes the 2006 data for AHU 19.

AHU 21

AHU 21 serves Radiology, Purchasing, Shipping and Receiving, and Plant Operations. Radiology is a critical area. The AHU 21 data indicate positive percent differences for airborne concentrations comparisons of OSA to ISA. The minimum percent difference is 54% (spore traps) for OSA to ISA comparisons and show more than 50% of the mean total concentrations of fungi and bacteria were being removed from OSA. Percent differences for bacteria incubated at 25°C were negative for the before filter to after filter, before filter to room, and after filter to room sample sets, indicating an indoor source of bacteria.

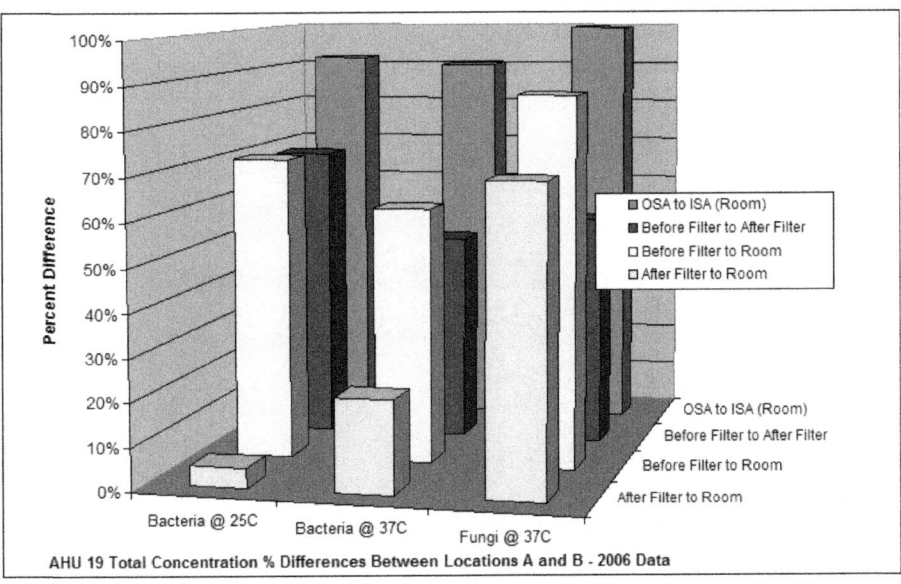

Figure 3. AHU 19 percent differences in mean total concentrations between spaces (2006 data).Note: A negative percent difference indicates that the indoor concentration was higher than outdoors for OSA to ISA comparisons.Note: A black top of a bar in the graph in the 0% plane indicates a negative percent difference indicating an increase in concentration from one space to another when a decrease was expected. Note: The 2006 AHU 19 laboratory results were not received from the analytical laboratory for spore trap and fungal samples incubated at 25°C.

Percent differences for bacteria incubated at 37°C were negative for the before filter to room and after filter to room sample sets, indicating an indoor source of bacteria. Percent differences were negative for all sampling sets comparing after filter to room, indicating indoor sources of bacteria and fungi. Percent differences show an increase in airborne fungal and bacterial concentrations between air concentrations after the filter and air concentrations within the room, indicating an increase in airborne concentrations in the air for both culturable bacteria and fungi and total fungi after the air leaves the AHU.

The low percent differences for spore traps and culturable air sampling (25°C and 37°C samples) between OSA and ISA (room) indicate indoor contamination as shown by the negative percent difference shown for the after filter to room samples. This indicates an indoor source or infiltration due to negative building pressure as opposed to inadequate particle filtration by the AHU filters. The increase in concentrations as the air moves from the AHU to the room indicates the presence of an indoor source of fungi and bacteria, specifically fungal species of *Rhodotorula, Penicillium, Aspergillus, Verticillium, Nigrospora, Paecilomyces, Cladosporium, Engyodontium, Rhizopus,* and*Scytalidium,* and bacterial species of *Acinetobacter, Chryseomonas, Pseudomonas, Tatumella,* and*Staphylococcus.* Table 4 summarizes the 2006 data for AHU 21.

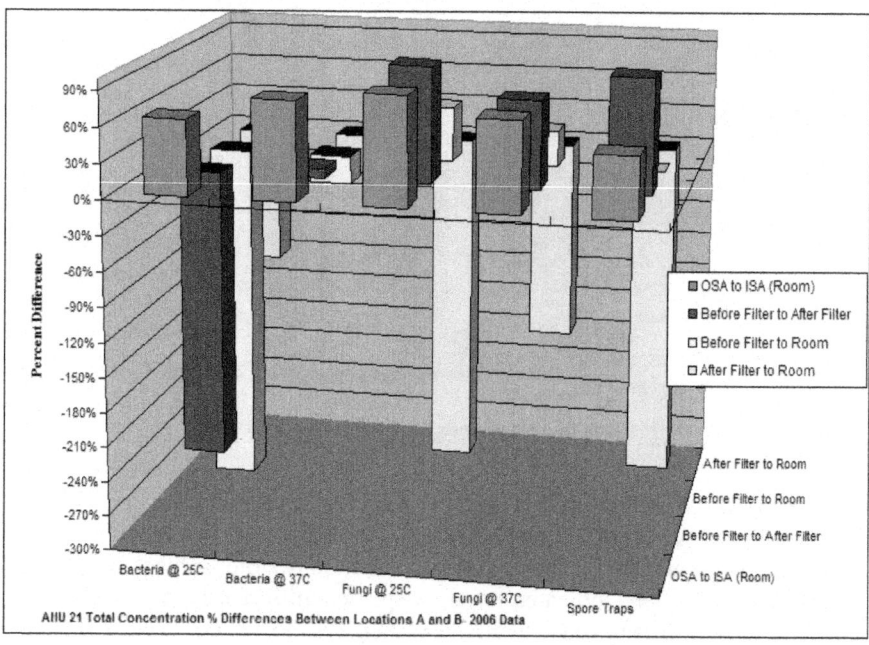

Figure 4. AHU 21 percent differences in mean total concentrations between spaces (2006 data).Note: Negative scale was truncated at -300% for simplicity in graphical presentation.Note: A negative percent difference indicates that the indoor concentration was higher than outdoors for OSA to ISA comparisons.Note: A black top of a bar in the graph in the 0% plane indicates a negative percent difference indicating an increase in concentration from one space to another when a decrease was expected.

Air Sampling Indicators of Indoor Contamination

Indoor vs. outdoor comparisons of fungi and bacteria are used to document the presence or infer the absence of indoor, biologically derived contamination. (Macher 1999) A substantial number of bacteria and fungi are capable of spreading via the airborne route in hospitals. (Eickhoff 1994) The presence of contamination in dust or on surfaces or water is often considered de facto evidence of human exposure to fungal aerosols. (Burge 2000)

When evaluated as total concentrations, the air sampling results generally indicate lower concentrations of fungi and bacteria indoors compared to outdoors. Unless the genera and species have been identified total counts merely indicate gross numbers and indoor vs. outdoor comparisons are not meaningful. (Macher 1999; Weber and Page 2001) An investigator cannot make meaningful indoor vs. outdoor comparisons unless the genera and species found indoors and outdoors have been identified. (Macher 1999) For this project, culturable air samples were analyzed at the species level so that meaningful comparisons could be made.

When evaluated according to genera, air sampling results indicate the potential presence of indoor contamination sources for both bacteria and fungi. Air sampling indicators of an indoor source were identified in the space controlled by each AHU. Indicators of indoor contamination were defined as:

1. Potential indoor biological source (indoor mean concentration > outdoor mean concentration [tested by the ANOVA, $\alpha = 0.05$, with differences identified by SNK Post Hoc Grouping] or the organism was identified in the indoor air samples but not identified in the outdoor reference samples ($n \geq 12$) for the space controlled by each AHU),

2. Meets the criteria of 1 above (a potential indoor biological source) and is a confirmed indoor contamination source via surface sampling in the space controlled by each AHU,

3. Meets the criteria of 1 above (a potential indoor biological source) and the organism was not identified in all outdoor air samples ($n \geq 48$), indicating an indoor source of air contamination, and

4. Meets the criteria of both 2 and 3 above, confirming an indoor source of air contamination.

Air sampling identified indicators of indoor contamination within each space investigated in both the 2005 and 2006 data. While indoor bacterial sources are expected within occupied buildings, the species of fungi found in indoor and outdoor air should be similar. (Weber and Page 2001) Outdoor air samples serve as the primary comparison to indoor bioaerosol samples and the types of fungi present indoors should not be significantly different from the outdoor environment. (Spicer and Gangloff 2000) To determine if the indoor and outdoor bioaerosol profiles were similar, the means of the sampling data were tested for biodiversity. The numbers of air sampling indicators identified per AHU from the 2005 air sampling data are shown in Figure 5.

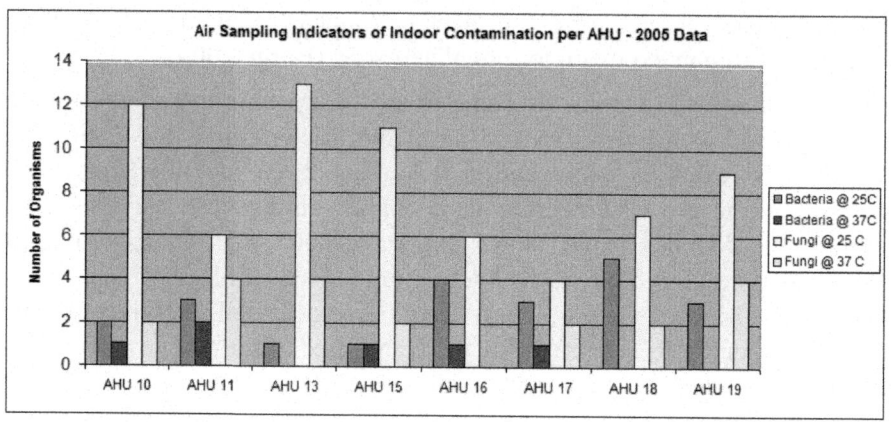

Figure 5. Number of Air Sampling Indicators of Indoor Contamination per AHU.

Air sampling indicators identified per AHU from the 2006 air sampling data are shown in Figure 6.

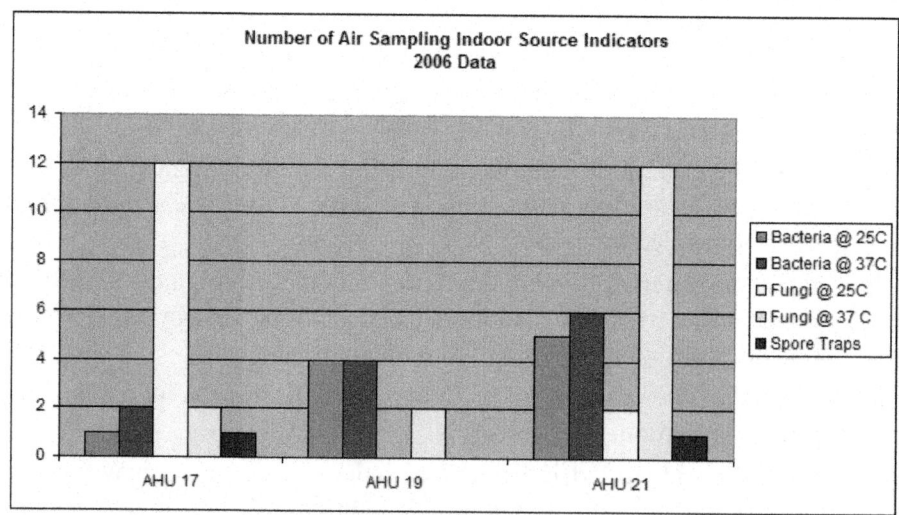

Figure 6. Number of air sampling indicators identified via air sampling (2006 data).

Figures 7 and 8 illustrate the percentage of air sampling indicators discussed in the Air Sampling Indicators of Indoor Contamination section that were confirmed via surface sampling or not detected in the outside air reference samples, indicating an indoor source of contamination.

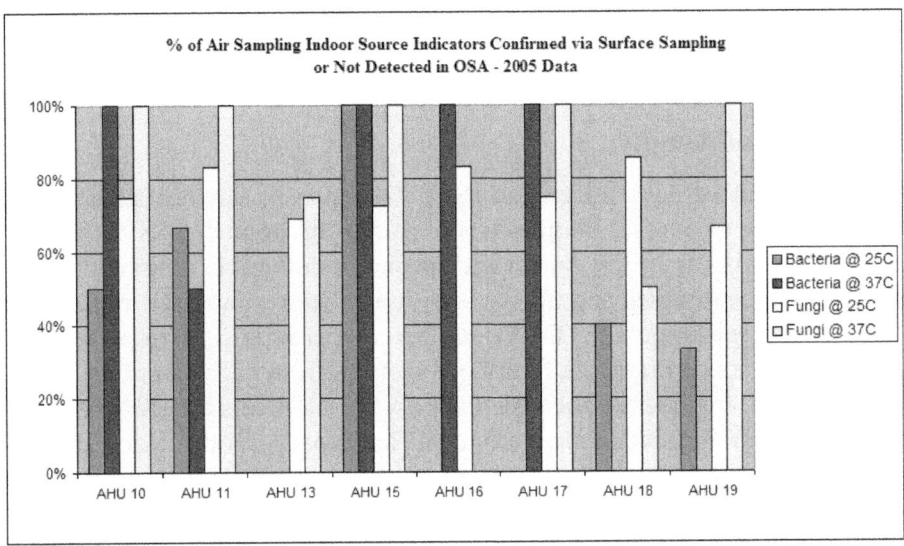

Figure 7. Percentage of air sampling indoor source indicators confirmed via surface sampling or not detected in the OSA.

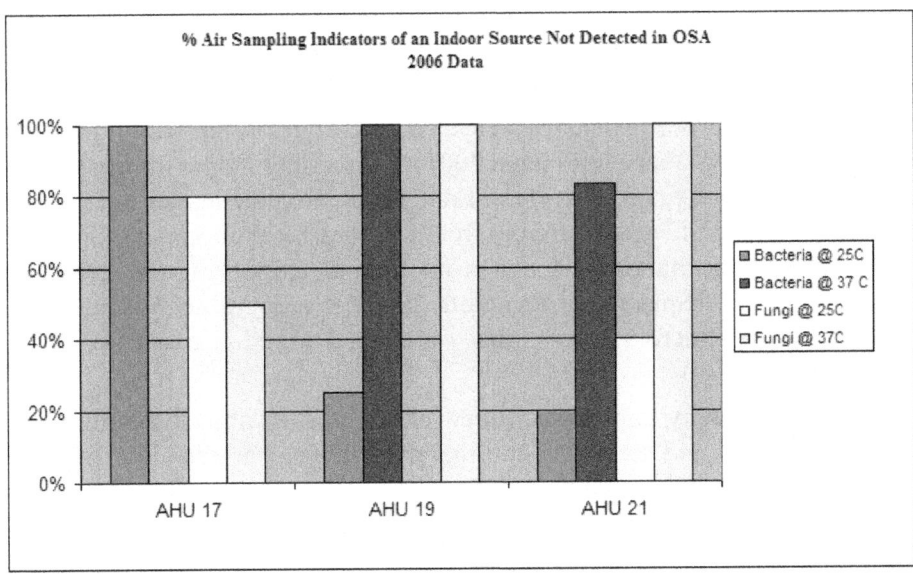

Figure 8. Percentage of air sampling indoor source indicators not detected in the OSA.

Indicators of indoor contamination that were not identified in outdoor air samples or were identified in indoor samples were found within the space controlled by each AHU investigated. Indicators of indoor contamination identified on surface sampling are associated with the building while indicators not identified in the outdoor air are likely to be associated with the building, building occupants, or another indoor source.

Tests for Biodiversity

The Spearman's Rank Correlation (SRC) is a non-parametric statistical test for comparing bioaerosol samples from separate environments. SRC is used to assess the similarity of the genera and species of culturable fungi and bacteria in air or source samples and the types of fungal spores in spore-trap samples. Data from a reference site (OSA) can be compared to data from a test site (ISA). Mean concentrations of multiple samples from each sampling location are preferred for use in the calculations for biodiversity. (Macher 1999) Indoor air sampling locations were identified broadly as the space under the environmental control of each AHU investigated. Several samples were taken from each sampling area (space under the environmental control of the AHU) so that mean concentrations could be analyzed statistically. The SRC test for biodiversity allows inferences to be made based upon whether the bioaerosol profile of one location is statistically similar to the bioaerosol profile of another. For fungi, indoor and outdoor profiles should be statistically similar. (Dillon, Heinsohn et al. 1996; Macher 1999; Spicer and Gangloff 2000) For bacteria, however, it is not unusual to have differences in biodiversity, as many bacteria are human shed. (Macher 1999)

The mix of airborne fungal species indoors should be similar to that found in the outdoor air. (Weber and Page 2001) In assessing indoor air quality with regards to airborne fungi, the types of fungi present in the indoor environment should not be significantly different from the outdoor environment. Similarity indicates that the building is not promoting or amplifying the growth of microorganisms. (Spicer and Gangloff 2000) A zero value was used as a replacement for microorganisms that were not detected in a sample. (Spicer and Gangloff 2000)

The biodiversity can be measured either as a combined aerobiological profile of fungi and bacteria, or separately. Here, bacteria and fungi are considered separately. (Macher 1999) Indoor source aerosols tend to be dominated by the readily released spores of *Aspergillus* and *Penicillium* species. (Burge 2000) The culturable air sampling results of *Aspergillus* and *Penicillium* species (at both 25°C and 37°C) were tested to determine if indoor vs. outdoor species identified were similar. Table 2summarizes the SRC test results for biodiversity. Tests for biodiversity were conducted with SPSS 14 Statistical Software. The level of significance was prescribed as $\alpha = 0.05$. A significant correlation (p-value $\le \alpha \le 0.5$) indicates that there is statistical evidence that the biodiversity of the indoor air is similar to the biodiversity of the OSA. Similarity indicates that the building is not promoting or amplifying the growth of microorganisms, while dissimilarity indicates that the biodiversity of indoor and OSA are independent and that the indoor environment is amplifying the growth of microorganisms. (Dillon, Heinsohn et al. 1996; Macher 1999; Spicer and Gangloff 2000) Similarities in the biodiversity of indoor and outdoor air are unlikely to have occurred by chance, and at the $\alpha = 0.05$ (5%) level of significance, the result could have occurred by chance one time in 20. Statistical techniques evaluate an observed difference in view of its precision to determine with what probability it might have arisen by chance (the level of significance). Values with a low probability of occurring by chance are called statistically significant and are considered to represent a real effect (e.g. difference or similarity in biodiversity). (Dillon, Heinsohn et al. 1996; Conover 1999; Macher 1999) The SRC applied to bioaerosol samples in the tests for differences between OSA and ISA biodiversity at the species level for culturable bacteria and fungi and at the genus level for spore traps.

The 2005 data show consistent dissimilarity between the bioaerosol profiles of ISA and OSA for species of *Penicillium* and *Aspergillus* at both 25°C and 37°C (86% dissimilar). For AHUs 16, 17, 18, and 19, bacterial and fungal biodiversities are similar to OSA. Yet, the biodiversity of *Penicillium*species and *Aspergillus* species analyzed independently are consistently dissimilar (86% dissimilar) indicating indoor amplification. Spore traps consistently showed 100% similarity because only non-culturable fungi were analyzed at the genus level, masking differences in species present in OSA and ISA. This is an expected occurrence with spore trap analysis because spore traps analysis is not sufficient for identification at the species level. For the 2005 data, the biodiversity of bacteria between OSA and ISA were consistently similar (75% similar). The tests for biodiversity indicate that the diversity of bacterial taxa identified indoors were consistently dissimilar to the bacterial taxa identified outdoors in the space controlled by AHUs 10, 11, 13, and 15.

The 2006 data indicate that the biodiversity between OSA and ISA was dissimilar for the space under the environmental control of AHUs 17, 19, and 21. For the 2006 data, the biodiversity of bacterial tests were consistently dissimilar (83% dissimilar). This may be due to differences in seasons that the tests were conducted. The 2005 samples were taken in summer and the 2006 samples were taken in winter. During winter, OSA bioaerosol concentrations may be lower due to less than ideal growth conditions. Conversely, indoor bioaerosol concentrations of bacteria may be similar year-round due to the presence of human shed sources of bacteria.

SAMPLING INTERPRETATION SUMMARY

Allergic reactions to indoor allergens [including fungi and bacteria] can produce inflammatory conditions of the eyes, nose, throat, and bronchi. A substantial number of bacteria and fungi are capable of spreading via the airborne route in hospitals. (Eickhoff 1994) Contaminated ventilation or air conditioning systems have been implicated in nosocomial outbreaks, via infective aerosols, dust, and colonized filters. (Eickhoff 1994) Biofilms or slime on HVAC system pan or coil surfaces is an indicator of microbiological amplification. (Pope, Patterson et al. 1993) The growth of microorganisms downstream from the cooling coils can be promoted by water droplets being blown off coil surfaces and into the air. (Pope, Patterson et al. 1993; CDC 2003) Poorly designed HVAC systems may provide for amplification of fungi and *Actinomycetes* in wet niches of the system. (Pope, Patterson et al. 1993; CDC 2003) In a hospital with high efficiency filters, airborne fungal spores reflect incomplete filtration, infiltration of outside air, and shedding of adherent spores from indoor growth. (Rhame 1991) Nosocomial aspergillosis occurs in direct proportion to the mean ambient hospital airborne spore content. (Rhame 1991) Mini-bursts of spores occur from disturbance of settled spores in dust, shedding spores from clothes or sources such as indoor growth. (Rhame 1991) Immunocompromised persons, children and the elderly are at risk of from exposure to infectious microorganisms. (Boss and Day 2003) In general, the filters show that total quantities of aerobiological contaminants are being removed from the air stream. See the filter discussion in the Total Outside Concentrations vs. Total Indoor Concentrations (2005 Data) section. However, both surface and air samples indicate the presence of indoor microbiological amplification, which is problematic in the indoor environment.

Table 2. Spearman's Rank Correlation Test for Biodiversity between outdoor and indoor sampling locations. (Significance level $\alpha \leq 0.05$)

Results for Spearman's Rank Correlation Testing for Biodiversity									
AHU	10	11	13	15	16	17	18	19	21
2005 Data									

Bacteria @ 25°C	D	D	S	D	S	S	S	S	Not tested in year 1.
Bacteria @ 37°C	S	S	D	S	S	S	S	S	
Fungi @ 25°C	D	S	D	D	S	S	S	S	
Fungi @ 37°C	D	S	D	D	S	S	S	S	
Spore Traps	S	S	S	S	S	S	S	S	
Aspergillus @ 25°C	D	D	D	D	D	D	D	D	
Penicillium @ 25°C	D	D	D	D	D	S	D	D	
Aspergillus @ 37°C	D	S	D	D	*	D	S	S	
Penicillium @ 37°C	*	D	D	D	*	D	D	D	

2006 Data

Bacteria @ 25°C	Not tested in year 2.					D	Not tested in year 2.	D	D
Bacteria @ 37°C	D								
Fungi @ 25°C	D								
Fungi @ 37°C	S								
Spore Traps	S								
Aspergillus @ 25°C	S								
Penicillium @ 25°C	**								
Aspergillus @ 37°C	S								
Penicillium @ 37°C	D								
	*								
	D								
	S								
	**								
	S								
	*								
	*								

	*				
	S				
	*				
	*				
	*				
	*				
	*				
	*				
	*				
	*				

S indicates that the aerobiological profiles between indoor and outdoor air are "similar" and not independent (i.e. the populations appear to be related or the samples could have been drawn from the same environment), ($\alpha \leq 0.05$, Spear

man's Rank Correlation). (Dillon, Heinsohn et al. 1996;Macher 1999)

D indicates that the aerobiological profiles between indoor and outdoor air are "dissimilar" and independent (ie. The populations appear to be unrelated and drawn from separate environments). (Macher 1999; Dillon, Heinsohn et al. 2005)

* indicates that the Spearman's Rank Correlation test could not be performed due to indoor concentration data that were constant (zero) or the number of species identified for each test was not sufficient to warrant testing via Spearman's Rank Correlation. (Spicer and Gangloff 2000)

** Note: Sample results were not received from the analytical laboratory.

Note: One tailed test for correlation. Significant correlation indicates that the indoor and outdoor biological profiles are similar. (Conover 1999; Macher 1999)

Fungal Samples

The mix of airborne fungal species indoors should be similar to that found in the outdoor air. Fungal organisms identified indoors, that are not present in the outdoor air or control locations, suggests the presence of an amplifier (growth site) for that species in the building. (Macher 1999; Weber and Page 2001) Most fungi can become opportunistic pathogens in a severely immunocompromised patient. (Weber and Page 2001) Airborne fungi within the hospital setting are especially dangerous because antifungal therapy is still rather ineffective. (Kalliokoski 2003) Although average air concentrations of total culturable fungi indoors were consistently lower than those found outdoors, many types of fungi identified indoors were not found in outdoor reference samples; especially within the species of *Aspergillus* and *Penicillium*. (Weber and Page 2001) The biodiversity of species of *Penicillium* and *Aspergillus* between indoor and outdoor air was consistently dissimilar for sampling results within the sampling space controlled by each air handling unit. Species of fungi identified within the hospital in air and on surfaces that are associated with airborne transmission or NI are *Penicillium, Aspergillus,Rhizopus, Acremonium*, and *Fusarium*. (CDC 2003)

Indicator organisms identified via air sampling in the hospital are *Aspergillus versicolor, A. flavus, A. fumigatus*, species of *Fusarium* and *Penicillium*, and yeasts. (Macher 1999) Some indicator organisms identified on indoor surfaces within the Hospital are species of *Aspergillus, Chaetomium, Stachybotrys, Penicillium, Fusarium, Acremonium, Trichoderma*, and yeasts. The presence of indicator organism contamination on surfaces indicates long-term or severe moisture problems. (Boss and Day 2003) Indicator organisms were identified on surfaces within the space controlled by each AHU investigated in 2005.

Exposure to fungi actively growing indoors may present unusual health risks even when total fungal concentrations are higher outdoors. (Macher 1999) Exposure to damp indoor environments and the presence of molds in damp indoor environments are associated with asthma symptoms in sensitized asthmatic persons. (Institute of Medicine Committee on Damp Indoor Spaces and Health 2004) Serious respiratory infections resulting from exposure to *Aspergillus* species and *Fusarium* species are common in persons who are immunocompromised. It is likely that many of these fungal infections are contracted through contact with fungi in indoor environments, because poor health conditions limit people with severely impaired immune systems to spend most of their time indoors. (Institute of Medicine Committee on Damp Indoor Spaces and Health 2004) The lungs of persons with chronic pulmonary disorders such as cystic fibrosis, asthma, and chronic obstructive pulmonary disorder may become colonized and potentially infected with *Aspergillus* species. (Institute of Medicine Committee on Damp Indoor Spaces and Health 2004) Healthy persons exposed to damp or moldy indoor environments report that they are more prone to respiratory infections, including the common cold, sinusitis, tonsillitis, otitis, and bronchitis. (Institute of Medicine Committee on Damp Indoor Spaces and Health 2004) Fungi have become the largest cause of occupational diseases among healthcare workers in Finland. (Kalliokoski

2003). Indoor source aerosols tend to be dominated by the readily released spores of *Aspergillus* and *Penicillium* species (Burge 2000), which produce large numbers of spores that are easily released into the air. (Institute of Medicine Committee on Damp Indoor Spaces and Health 2004) Indoor exposure to *Aspergillus* and *Penicillium* species spores has been shown to be associated with an increased risk of allergic sensitization in children (Wilson, Holder et al. 2004) and are chiefly involved in the genesis of asthma and allergic alveolitis (pulmonitis due to hypersensitivity). (Perdelli, Christina et al. 2006) Bronchial asthma is frequently provoked by airborne fungal spores belonging to the genera *Aspergillus*and *Penicillium*. (Smith 1990) Species of *Aspergillus* and *Penicillium* are considered indicator organisms that may signal unwanted moisture intrusion and/or a potential for health problems. (Macher 1999) As such, the biodiversity of both *Penicillium* and *Aspergillus* species was tested to determine whether the outdoor air was the primary source for the fungi identified in the indoor air. (Macher 1999)

Aspergillus and Penicillium species are two of the most ubiquitous fungi known. Large quantities of fungal spores are produced when these fungi are actively growing. During germination, large quantities of spores are produced, and when sporulation occurs, several thousand spores may be disseminated per cubic meter of air. (Wenzel 1997) *Penicillium* and *Aspergillus* spores are sphere-like, measuring from approximately 2-5 micrometers in diameter and can be suspended very easily in the air. (Wenzel 1997; Straus 2004) Spores of *Penicillium* species often cannot be distinguished via microscopic examination from spores of *Aspergillus* species and vice versa. (Stetzenbach and Yates 2003) Once suspended, they may remain suspended for prolonged periods, and when those spores settle, they can contaminate any surface in contact with air. (Wenzel 1997) Once inhaled, these spores travel through the airways into the lower regions of the lungs, leading to the potential development of respiratory symptoms. (Straus 2004)

Aspergillus Species

The presence of *Aspergillus* species in health-care environments is a substantial extrinsic risk factor for opportunistic invasive aspergillosis (invasive aspergillosis being the most serious form of the aspergillosis). (CDC 2003) The presence of *Aspergillus* contamination and/or growth within a health care facility is of particular concern due to the presence of immunocompromised persons. Causative agents of aspergillosis are *Aspergillus flavus, Aspergillus fumigatus, Aspergillus niger, Aspergillus terreus*, and *Aspergillus nidulans*. (CDC 2003) Airborne concentrations of *Aspergillus* species at or below 0.1 cfu/m^3 have been recommended for the prevention of nosocomial aspergillosis.

(Weber and Page 2001) Low concentrations of *A. fumigatus* and *A. flavus* have been associated with nosocomial aspergillosis in immunocompromised patients. (Arnow, Sadigh et al. 1991) Among immunosuppressed patients in general, invasive aspergillosis remains a serious complication and may be lethal. (Perdelli, Christina et al. 2006)

The genus *Aspergillus* is one of the most ubiquitous fungi known. Large quantities of fungal spores are produced when actively growing. The *Aspergillus* spores are inhaled easily because of their small aerodynamic size and can easily reach and colonize the upper respiratory tract, including paranasal sinuses and terminal airways. (Wenzel 1997) Nosocomial pulmonary and disseminated aspergillosis arises from inhalation of fungal spores. (Rhame, Streifel et al. 1984) The use of powerful new chemotherapy protocols for malignancies and certain immunologic disorders and the increasing use of organ transplantation are risk factors for nosocomial aspergillosis. Patients with acute or chronic myelogenous leukemia and AIDS are particularly susceptible to nosocomial aspergillosis. In the transplant population and in patients with aplastic anemia, *Aspergillus* has emerged as a major cause of death. (Wenzel 1997)

Nosocomial aspergillosis is primarily established when an immunocompromised host inhales fungal spores present in the air. (Rhame, Streifel et al. 1984; Wenzel 1997) Any dust-generating activity, such as maintenance of ventilation systems, cleaning, vacuuming, and dry mopping, can render *Aspergillus* spores airborne and potentially cause outbreaks of nosocomial aspergillosis. The achievement of a spore-free air within an area or ward of a hospital may not be sufficient to eradicate nosocomial aspergillosis because of "non-ward" sources of *Aspergillus* within the hospital, such as radiology, radiation therapy units, and other areas visited by patients where engineering and environmental controls may not be as stringent. (Wenzel 1997) Fungi actively growing indoors compounds the problem associated with the prevention of nosocomial aspergillosis.

Species of *Aspergillus* found indoors should be similar to species identified outdoors. With the exception of AHU 11, all AHUs have filters of 90% efficiency or greater. 90% and 95% filters are considered high efficiency (Boss and Day 2003) and rated to remove 90% of particles from 1-10 micrometers and almost all particles greater than 10 micrometers in size. (ASHRAE 1992; ASHRAE 1999) Due to the presence of high efficiency filters in the AHUs investigated (except AHU 11), indoor concentrations should be statistically lower than outdoor concentrations and organisms not detected outdoors should not be detected indoors (Weber and Page 2001) (with the exception of indoor-source bacteria (Macher 1999)). However, many species of *Aspergillus* were detected indoors and not outdoors or indoor concentrations were greater than or not statistically different than outdoor concentrations, indicating an indoor source or infiltration.

The average concentration of *A. fumigatus* for culturable fungal air samples at 25°C was 0.50 cfu/m3 (n=14 indoors, n=38 outdoors) in the space controlled by AHU 15. The biodiversity of species of *Aspergillus* between indoor and outdoor air was consistently dissimilar for sampling results (incubated at 25°C) within the sampling space controlled by each air handling unit, indicating the presence of indoor fungal amplifiers; the building appears to be promoting or amplifying the growth of species of *Aspergillus*. (Macher 1999; Spicer and Gangloff 2000; Weber and Page 2001)

Of particular concern in a hospital setting are the presence of thermotolerant fungi, including *A. fumigatus* and *A. flavus*. Thermotolerant fungi are of primary concern in healthcare facilities, since they can cause infection in at-risk patients, even when concentrations are very low. (Page and Trout 2001) Concentrations of thermotolerant *A. fumigatus* (incubated at 37°C) were 1.17 (n=12 outdoors, n=6 indoors) and 3.50 cfu/m3 (n=12 indoors, n=22 outdoors) in the spaces controlled by AHUs 13 and 15, respectively. Indoor concentrations of *A. flavus* were not statistically different than outdoor concentrations in the spaces controlled by AHUs 11 and 17 for samples incubated at 25°C and AHUs 11, 17, and 18, for samples incubated at 37°C. This is an expected condition within the space controlled by AHU 11 due to the absence of final filters. This suggests the presence of an indoor source of *A. flavus* or outdoor air infiltration within the spaces controlled by AHUs 17 and 18. Indoor *A. flavus* concentrations detected in the space controlled by AHUs 10 and 19 were lower than OSA concentrations (statistically significant difference at the $\alpha = 0.05$ level of significance). The biodiversity of species of thermotolerant *Aspergillus* between indoor and outdoor air was consistently dissimilar for sampling results within the sampling space controlled by AHUs 10, 13, 15, and 17, indicating the presence of indoor fungal amplifiers; the building appears to be promoting or amplifying the growth of species of *Aspergillus*. (Macher 1999; Spicer and Gangloff 2000; Weber and Page 2001)

Indoor concentrations for culturable *Aspergillus* species incubated at 25°C (*A. flavus, A. sydowii, A. nidulans, A. niger, A. fumigatus, A. flavipes, A. sclerotiorum, A. terreus, A. versicolor*) that exceeded indoor concentrations, were not statistically different to outdoor concentrations, or not detected in the outdoor reference samples were identified in the space controlled by AHUs 11, 13, 15, 16, 17, 18, and 19. Indoor concentrations of thermotolerant species of *Aspergillus* (*A. flavus, A. sydowii, A. niger, A. fumigatus, A. terreus)* that exceeded indoor concentrations, were not statistically different to outdoor concentrations, or not detected in the outdoor reference samples were identified in the space controlled by AHUs 10, 11, 13, 15, 17, and 18.

Potentially hazardous concentrations of *A. fumigatus* were identified in the spaces controlled by AHUs 13 and 15. (Weber and Page 2001) Moisture intrusion via infiltration, leaks, inadequate HVAC control, etc. has provided a chronically moist indoor environment ideal for fungal growth. Air infiltration due to negative building pressurization allows the introduction of unfiltered air into the hospital. Air samples indicated the presence of indoor fungal reservoirs/amplifiers within the spaces controlled by all AHUs investigated. Air and surface sampling indicated microbial growth, dissemination and, hence, occupant exposure, from the indoor microbial reservoirs. (Weber and Page 2001) An indoor environment has been created in which immunosuppressed or allergic patients within the Hospital are not fully protected against the risk of infection and the allergenic effects of *Aspergillus* species (Perdelli, Christina et al. 2006) and otherwise healthy persons may suffer exacerbation of allergies and be prone to increased incidences of respiratory infections, including the common cold, sinusitis, tonsillitis, otitis, and bronchitis. (Institute of Medicine Committee on Damp Indoor Spaces and Health 2004)

General controls for prevention of nosocomial aspergillosis are: air filtration, positive pressurization, avoidance of dust-generating activities, attention to non-filtered air infiltration, protection of immunocompromised patients who enter areas without highly filtered air, and isolation of hospital construction. (Wenzel 1997) Highly filtered air is essential to preventing person-to-person and environmentally related infections. (Boss and Day 2003) However, it is important to note that the use of highly filtered air conditioning systems does not provide complete protection against fungi actively growing or infiltrating into a facility. (Perdelli, Christina et al. 2006) Nosocomial aspergillosis has been associated with poorly maintained and/or malfunctioning HVAC systems. (CDC 2003)

Penicillium Species

As with *Aspergillus* species, the genus *Penicillium* is one of the most ubiquitous fungi known and one of the most commonly isolated molds from contaminated buildings. With *Penicillium* species the main cause for concern is allergic disease, as infections due to *Penicillium* species are rare. Indoor concentrations of *Penicillium* species greater than outdoor concentrations are associated with negative health effects in humans. *Penicillium* species has been correlated with allergic asthma, allergic alveolitis, atopy, increased lower respiratory infections in children during the first year of life, and wheezing. (Straus 2004)

Inhalation of *Penicillium* species spores has been shown to provoke immediate and delayed-type asthma in individuals already sensitized to

Penicillium. (Straus 2004) Infants with high risk for the development of asthma (e.g. due to premature birth and/or ethnicity) may be at significant risk for persistent cough and wheeze when exposed to *Penicillium* species spores. *Penicillium* species have been shown to cause allergic alveolitis due to exposure from a faulty installation of a HVAC system. (Straus 2004) *Penicillium* is a large group of fungi valued as producers of antibiotics. *Penicillium* may cause allergic reactions, exacerbate asthma, and cause other adverse health effects when dispersed through air. (Boss and Day 2003; Institute of Medicine Committee on Damp Indoor Spaces and Health 2004) The blue-green molds of *Penicillium* are common contaminants of indoor environments. Inhalation of spores is the major route of entry. *Penicillium* species have been associated with asthma and hypersensitivity pneumonitis (Weber and Page 2001) and can cause NI in the immunocompromised host. (Fox, Chamberlin et al. 1990; Walsh and Groll 1999; CDC 2003) *Penicillium* species spores have been shown to be associated with an increased risk of allergic sensitization in children (Straus 2004) and are chiefly involved in the genesis of asthma and allergic alveolitis (pulmonitis due to hypersensitivity). (Perdelli, Christina et al. 2006) One 2.5 cm diameter colony of *Penicillium* species can produce 400,000,000 spores that can become airborne and generate increased concentrations of airborne spores. (Hitchcock, Mair et al. 2006)

The biodiversity of *Penicillium* species between ISA and OSA was consistently dissimilar for sampling results (incubated at 25°C) within the sampling space controlled by each AHU (except AHU 17), indicating the presence of indoor fungal amplifiers; the building appears to be promoting the growth of species of *Penicillium*. (Spicer and Gangloff 2000) Thermotolerant fungi are of primary concern in healthcare facilities, since they can cause infection in at-risk patients, even when concentrations are very low. (Weber and Page 2001) The biodiversity of species of thermotolerant *Penicillium* between indoor and outdoor air was consistently dissimilar for sampling results within the sampling space controlled by each AHU (except AHUs 10 and 16 because tests could not be performed), indicating the presence of indoor fungal amplifiers; the building appears to be promoting or amplifying the growth of species of *Penicillium*. (Macher 1999; Spicer and Gangloff 2000; Weber and Page 2001)

Species of *Penicillium* found indoors should be similar to species identified outdoors. With the exception of AHU 11, all AHUs have filters of 90% efficiency or greater and should remove greater than 90% of particles between 1 and 10 micrometers in size. Due to the presence of high efficiency filters in the AHUs investigated (except AHU 11), indoor concentrations should be statistically lower than outdoor concentrations and organisms not detected outdoors should not be detected indoors (with the exception of indoor-source bacteria). (Weber and Page 2001) However, many species of*Penicillium* were detected indoors and not outdoors or indoor concentrations were greater than or not statistically different than outdoor concentrations. Indicators of indoor contamination of *Penicillium*species were identified in the space controlled by each AHU.

Indoor airborne concentrations of culturable Penicillium species incubated at 25°C (P. citrinum, P. chrysogenum, P. corylophilum, P. decumbens, P. duclauxii, P. funiculosum, P. glabrum, P. implicatum, P. janthinellum, P. oxalicum, P. pinophilum, P. purporogenum, P. sclerotiorum, P. variabile, P. waksmani) that exceeded or were similar to outdoor concentrations or not detected in the outdoor reference samples were identified in the space controlled by each AHU. Indoor airborne concentrations of thermotolerant (incubated at 37°C) species of Penicillium (P. citrinum, P. chrysogenum, P. decumbens, P. funiculosum, P. janthinellum, P. oxalicum, P. pinophilum, P. simplicissimum) that exceeded or were similar to outdoor concentrations or not detected in the outdoor reference samples were identified in the space controlled by each AHU, with the exception of AHU 10 where no thermotolerant species of Penicillium were identified.

Yeasts

Yeasts are found in a variety of natural habitats or organic substrates such as plant leaves, flowers, soil, and salt water. Some yeasts are part of the normal human flora. Although yeasts may be part of the normal human flora, they were detected on indoor surfaces. Therefore, it can be concluded that the yeasts identified as indicators of indoor contamination were most likely from an indoor contamination source. (Macher 1999) The presence of yeasts actively growing indoors is of concern, as yeasts are considered an indicator organism and can cause infections in the immunocompromised host. Some yeasts are reported to be allergenic, and may cause problems in individuals with previous exposure and developed hypersensitivities. Yeast infections are among the most common fungal infections in humans. Their form ranges from localized cutaneous or mucocutaneous lesions, to fungemia or disseminated systemic mycoses. (AerotechP&K 2006) Indoor concentrations of yeasts exceeded outdoor concentrations or were detected indoors and not outdoors in the spaces controlled by AHUs 10, 13, 15, and 19. See indicators of indoor contamination for both 2005 and 2006 data. Yeasts were identified via surface sampling within the spaces controlled by AHUs 10, 11, 13, 15, 16, 17, 18, and 19.

Bacterial Samples

Like fungi, bacteria actively growing indoors can release spores into the air (Institute of Medicine Committee on Damp Indoor Spaces and Health 2004). Additionally, bacteria secrete enzymes that can act as allergens. Enzymes and spores from Gram-positive bacilli and thermophilic *Actinomycetes* have been implicated in epidemics of hypersensitivity pneumonitis and work-related asthma. Concentrations of bacteria associated with sensitization or provoking human allergic reactions are unknown. (Pope, Patterson et al. 1993) Bacteria are known to cause diseases either as pathogens or as opportunistic pathogens in the immunocompromised host. (Boss and Day 2003) Environmental bacteria also grow in all wet spaces and are found in most cases where there is fungal growth. (Institute of Medicine Committee on Damp Indoor Spaces and Health 2004) Some bacteria that are common in outdoor air may penetrate to building interiors and may also grow indoors. (Macher 1999) Unlike fungi, bacteria have natural reservoirs indoors (primarily humans), and total bacterial concentrations are often higher indoors than outdoors. (Macher 1999)

The bacterial organisms identified (incubated at 25°C) as indicators of an indoor source were Acinetobacter lwoffi, Gram (+) cocci, Micrococcus luteus, Micrococcus species, Staphylococcus species (S. auricularis, S. capitus, S. epidermis, S. hominis, S. hyicus, S. warneri, and S. xylosus) which are human-shed bacteria. (Wilson 2005) Because these organisms are human-shed and were not identified as indoor contaminants via surface sampling, it cannot be concluded that these higher concentrations of indoor-source bacteria detected via air sampling were the result of building-related sources of bacterial contamination. (Macher 1999)

The bacterial organisms identified (incubated at 37°C) as indicators of an indoor source were Gram (-) cocci, *Pseudomonas aeruginosa, Pseudomonas stutzeri, Rhizobium radiobacter, and Tatumella ptyseos*. Gram (-) bacteria are usually not present on the skin (with the exception of *Acinetobacter* species). *Pseudomonas aeruginosa* can be found on skin, but is also considered an environmental organism and may be associated with building contamination and/or infiltration. (Wilson 2005) *Pseudomonas stutzeri and Rhizobium radiobacter* are considered environmental source organisms. The presence of airborne concentrations of *Pseudomonas stutzeri* (identified via surface sampling on a ceiling tile in the space controlled by AHU 19), and *Rhizobium radiobacter* indicates the presence of an indoor environmental source (amplification).

Acinetobacter species may cause NI and death in infants during periods of airborne dissemination. Environmental conditions leading to an increase in air conditioner condensate in HVAC systems may increase the risk of nosocomial infection with *Acinetobacter* species. (McDonald, Walker et al. 1998)*Acinetobacter* species have been cultured from air conditioners in a hospital nursery. (CDC 2003)*Acinetobacter lwoffi* species was identified as an indicator of an indoor source via air sampling in the space controlled by AHUs 13 and 21. *Acinetobacter* species are widely distributed in the environment and are frequently found on human skin. (Wilson 2005) *Acinetobacter* sp. are a main causative agent of pneumonia, which is a leading cause of morbidity and mortality and is the sixth most common cause of death in the United Kingdom and the Untied States. (Wilson 2005) The presence of *Acinetobacter*sp. in the air is of concern in the hospital environment. There is no indication that the indoor source of*Acinetobacter* species was associated with building contamination, but is likely associated with infiltration.

Pseudomonas species are common in the outdoor air, but rarely occur in indoor air. (Macher 1999) *P. aeruginosa* was identified as an indicator of indoor contamination in the space controlled by AHU 11. This may be due to the absence of final filters in AHU 11. Airborne infections by *P. aeruginosa* have been reported. (Kalliokoski 2003) Transmission of *P. aeruginosa* may occur through direct patient-to-patient contact, environmental contamination, or via the hands of health care workers. (Kerr, Moore et al. 1995; Beggs and Kerr 2000) Airborne *P. aeruginosa* was detected within the hospital in 2005 during a pilot study of proposed sampling equipment in the basement of the Hospital. These sampling results must be viewed as a qualitative indication of an indoor source of *P. aeruginosa*, as the equipment utilized in the 2005 pilot study was determined to be inaccurate for the determination of airborne concentrations but sufficient for qualitative identification of the bacterium at high concentrations within the space tested. *P. aeruginosa* is an environmental organism typically found on skin of people and has been isolated environmentally from soil, manure, canal water, and straw. *P. aeruginosa* is a ubiquitous soil organism that proliferates in standing water and wet and warm materials such as leaking hot water pipe insulation and showers. (Boss and Day 2003; CDC 2003; Stetzenbach and Yates 2003) *P. aeruginosa* is an opportunistic pathogen and can be especially problematic for those with cystic fibrosis and burn victims. (Boss and Day 2003; CDC 2003; Gaynes and Edwards 2005) Nosocomial infections due to *P. aeruginosa* have been associated with poorly maintained and/or malfunctioning HVAC systems. (CDC 2003) *P. aeruginosa* causes urinary tract and skin infections, septicemia (blood infections), and meningitis. (Boss and Day 2003; Wilson 2005) *P. aeruginosa* is a main causative agent of pneumonia, which is a leading cause of morbidity and mortality and is the sixth most common cause of death in the United Kingdom and the Untied States. (Wilson 2005) The presence of *P. aeruginosa* in the air is of concern in the hospital environment.

Pseudomonas oryzihabitans was identified as an indicator of indoor contamination within the space controlled by AHU 21 (2006 data). *P. oryzihabitans* is an opportunistic pathogen and is a common soil bacterium. (Freney, Hansen et al. 1988; Bendig, Mayes et al. 1989; Munro, Buckland et al. 1990;Podbielski, Mertens et al. 1990; Reina, Odgardd et al. 1990; Esteban, Valero-Moratalla et al. 1993;Lam, Isenberg et al. 1994; Lucas, Kiehn et al. 1994; Rahav, Simhon et al. 1995; Romanyk, Gonzalez-Palacios et al. 1995; Liu, Shi et al. 1996; Anzai, Kudo et al. 1997; Kiris, Over et al. 1997; Lin, Hsueh et al. 1997; Marin, Garcia de Viedma et al. 2000)

Pseudomonas stutzeri was identified as an indicator of an indoor source in the space controlled by AHU 15 (2005 data) and AHU 19 (2006 Data). It is considered an environmental organism and has been isolated from soil, manure, canal water, and straw. It has been isolated from the respiratory tract, wounds, blood, urogenital tract, spinal and joint fluid of humans and is associated with NI. (Palleroni, Doudoroff et al. 1970; Reisler and Blumberg 1999; Taneja, Meharwal et al. 2004; Lalucat, Bennasar et al. 2006; Yee-Guardino, Danziger-Isakov et al. 2006) *P. stutzeri* surface contamination was confirmed from a ceiling tile within the space controlled by AHU 19. The presence of indoor contamination of *P. stutzeri* is of concern in the hospital environment.

Pseudomonas species are one of the most antibiotic resistant bacteria and resistant to antiseptics such as quaternary ammonium compounds. (Georgiev 1998; Boss and Day 2003) This property allows them to survive environmental conditions which are lethal to many other bacteria. (Georgiev 1998)*Pseudomonas* species in general are among the most clinically relevant healthcare associated pathogens. (CDC 2003) In general, the presence of *Pseudomonas* in a hospital setting is problematic. (Boss and Day 2003) *P. fluorscens* was identified as a contaminant in the drain pan water of AHU 19 and *P. Stutzeri* was identified as surface contamination in the space controlled by AHU 19. The presence of*Pseudomonas* species actively growing in the indoor healthcare environment is especially problematic. (Boss and Day 2003)

Rhizobium species are environmental source fungi typically found in the roots of plants and in soils. (Stetzenbach, Buttner et al. 2004) *Rhizobium radiobacter* was identified as an indicator of indoor contamination in the space controlled by AHU 11, and was not detected in the outdoor air (n = 118 outdoor air samples). This indicates the presence of an indoor source. *Rhizobium* sp. is recognized as an opportunistic human pathogen associated with NI in the immunocompromised host. (Lai, Teng et al. 2004) It is likely that the absence of final filters in the space controlled by AHU 11 prevented indoor concentrations of *Rhizobium* sp. from being removed from the air stream. It is likely that the *R. radiobacter* contamination originated from within the Hospital from an environmental source. *Staphylococcus aureus* was identified in the condensate water of AHU 21 (2006 data). The presence of *S. aureus* actively growing indoors should be considered a health risk in the hospital environment. *S. aureus* produces toxins and can infect surgical wounds, develop resistance to antibiotics, and is the agent of toxic shock syndrome. *S. aureus* also produces toxins that cause food poisoning. (Boss and Day 2003) *S. aureus* was not identified in the indoor air via air sampling.

Tatumella sp. is a member of the family *Enterobacteriaceae*, which are widely distributed in soil, water, plants, and animals. (Hollis, Hickman et al. 1981; Georgiev 1998) *Enterobacteriaceae* are responsible for over half of the NI in the United States. (Hollis, Hickman et al. 1981) *Tatumella ptyseos* was identified as an indicator of an indoor source in the space controlled by AHUs 10 (2005 data) and 21 (2006 data). *T. ptyseos* is associated with NI, but there is no indication that the indoor source of *Tatumella* species was associated with building contamination.

Actinomycetes

Actinomycetes were once considered fungi because of their resemblance to fungi. However, these organisms are not fungi, but are bacteria. (Georgiev 1998; AerotechP&K 2006) The presence of *Actinomycetes* is rare in buildings and outdoors. The presence of *Actinomycetes* indoors may be considered an indication of an indoor environmental source (Macher 1999) and may add to the complexity of the environmental problem. (Straus 2004) The *Actinomycetes* have the potential to become opportunistic, especially in immunocompromised hosts. (McNeil and Brown 1994; Georgiev 1998)

Thermophilic *Actinomycetes* are usually found in closed barns, silos, grain mills, and bagasse (sugar cane waste). *Actinomycetes* have been found in problematic or poorly maintained air conditioning ducts. (AerotechP&K 2006) Allergic respiratory disease caused by the *Actinomycetes* is referred to as farmer's lung, a hypersensitivity reaction from repeated exposure to antigens produced by the *Actinomycetes*. *Actinomycetes* may also cause other diseases such as ocular infections, periodontal disease, and abscess formations, which can infect humans. (Georgiev 1998) Several reports indicate that infections by these bacteria are not rare (especially from the *Actinomycete* genus *Nocardia*), are frequently misdiagnosed, or are under diagnosed, and that the incidence of infection is increasing. The spectrum of disease caused by *Nocardia* is broad and varies from a self-limited, asymptomatic infection to an aggressive, destructive disease resulting in death. *Nocardial* infections are commonly diagnosed in previously healthy adults with no predisposing factors. (McNeil and Brown 1994) The *Nocardiae* are frequently being recognized as emerging opportunistic pathogens; the most common underlying predispositions include organ transplantation, malignancies, use of corticosteroids, alcohol abuse, diabetes, or other debilitating factors. (McNeil and Brown 1994; AerotechP&K 2006)

Nocardioform bacilli and/or presumptive *Nocardioforms* were identified as indicators of an indoor microbiological source in the space controlled by AHUs 17, 18, and 19. *Nocardioforms* include the genus *Nocardia* and the *Nocardioform Actinomycetes*. (Georgiev 1998; Boss and Day 2003; Gibson, Gilleron et al. 2003; Stetzenbach and Yates 2003) *Nocardia* species are the *Nocardioform* most often isolated from NI. (Georgiev 1998) *Nocardia* are found in soil around the world, and the indoor concentrations of *Nocardioforms* were not identified via surface sampling and could not be associated with building contamination. (Macher 1999) *Nocardia* morphologically resembles *Actinomyces* species, and both are bacteria that are often pathogenic and opportunistic. (McNeil and Brown 1994; Georgiev 1998; Boss and Day 2003; AerotechP&K 2006)

Actinomycetes were detected indoors via air sampling and not outdoors (n=94 outdoor air samples) within the space controlled by AHU 10 and AHU 17 for the fungal samples incubated at 25°C.*Actinomycetes* were detected indoors and not outdoors (n=48 outdoor air samples) via air sampling within the space controlled by AHU 19 for the fungal samples incubated at 37°C. *Actinomycetes* were confirmed via surface sampling within the spaces controlled by AHUs 13, 15 (*Actinomyces* sp.), 16, and 17 (*Actinomycetes*-like). Therefore, it can be concluded that the airborne concentrations of*Actinomycetes* identified within the Hospital were due to an indoor source of surface contamination. (Macher 1999) The presence of *Actinomycetes* indoors is of concern in the indoor hospital environment.

SURFACE SAMPLING INTERPRETATION AND HEALTH RISK MODEL

Allergic reactions to indoor allergens can produce inflammatory diseases of the eyes, nose, throat, and bronchi, which are medical problems that come under the headings of allergic conjunctivitis, allergic rhinitis, allergic asthma, and hypersensitivity pneumonitis (extrinsic allergic alveolitis) respectively. (Pope, Patterson et al. 1993) The Health Risk Model (HRM) considers the type of microbial contamination and the type of person expected to be within a specific Hospital location. Critical care areas are areas of the Hospital where it is expected that immunocompromised persons will be present and therefore contamination within a critical care area is given a higher weight in the overall determination of health risk.

Risk assessment is a process designed to evaluate the potential relationship that may exist between exposure to aeroallergens and a particular effect (e.g. toxic effect, allergic sensitization, infection, allergic disease). (Pope, Patterson et al. 1993) A HRM was utilized to semi-quantitatively identify the health risk associated with fungal and bacterial surface contamination within the hospital. Monitoring for allergens can help characterize environments with respect to specific allergens (e.g., fungi and/or bacteria). Both fungi and bacteria secrete enzymes that act as allergens. (Pope, Patterson et al. 1993) Source or reservoir samples have been used as indicators of exposure to indoor allergens and measurement interpretations can be semi-quantitative (e.g., "presence or absence" or "low, medium, or high). (Pope, Patterson et al. 1993) Environmental bacteria also grow in all wet spaces and are found in most cases where there is mold growth. (Institute of Medicine Committee on Damp Indoor Spaces and Health 2004)

The American Industrial Hygiene Association's consensus document *A Strategy for Assessing and Managing Occupational Exposures* (Mulhausen and Damiano 1998) served as the basis for the HRM. The HRM utilized criteria and recommendations of the Centers for Disease Control and Prevention (CDC 2003), US Environmental Protection Agency (USEPA 2001), American Conference of Governmental Industrial Hygienists (ACGIH 1999), Institute of Medicine (Pope, Patterson et al. 1993), the New York City Department of Health and Mental Hygiene (NYCDHMH 2006), the American Society of Heating, Refrigerating, and Air Conditioning Engineers (ASHRAE 2003) and the Vanderbilt University Medical Center (VUMC 2006) in establishing the risk factors for the model. A literature search was conducted to determine if the organisms identified via surface sampling within the Hospital were allergenic, pathogenic or opportunistic, and capable of producing fungal or bacterial toxin. The HRM resulted in a Health Risk classification of the space controlled by each AHU.

Health Risk was classified as High, Medium, Low, and de Minimis. The risk classifications were determined with input from experts in medical microbiology, industrial hygiene, public health, engineering controls, infection control, and medicine. A de Minimis risk score means that no indoor environmental contamination was found. A low risk score means the environmental conditions present do not indicate extensive biological contamination and/or the risk associated with adverse health affects to building occupants is low. A medium risk score indicates that environmental conditions present an increased risk for adverse health effects to building occupants due to environmental contamination and remediation is necessary. A high risk score indicates that conditions exist for adverse health effects due to exposure to biological contaminants and immediate intervention is necessary. Figure 9 below displays the HRM scores for the indoor space controlled by each AHU.

Indoor surface fungal and bacterial surface contamination was identified in every area of the hospital investigated. Air sampling confirmed the presence of indicators of indoor contamination in each of the spaces investigated. See Section 4. Sampling Interpretation Summary above. The spaces under the control of every AHU placed within the medium risk category.

The environmental conditions are present such that immunocompromised or allergic patients are not fully protected against the risk of NI due to environmental bioaerosols. (Perdelli, Christina et al. 2006) Healthy hospital workers are not protected against allergic reactions to indoor bioaerosols growing within the facility and are at an increased risk of respiratory infections, including the common cold, sinusitis, tonsillitis, otitis, and bronchitis. (Institute of Medicine Committee on Damp Indoor Spaces and Health 2004) Hospital workers who are immunocompromised (e.g., diabetics, asthmatics, those undergoing cancer therapy or who have recent invasive surgery) are more susceptible to allergic reactions and the risk of work-related infections. The results of the HRM indicate that patients and staff are being exposed to microorganisms that are actively growing within the hospital which present a risk higher than what is expected in a hospital without water damage, microbial contamination, moisture infiltration, and OSA infiltration.

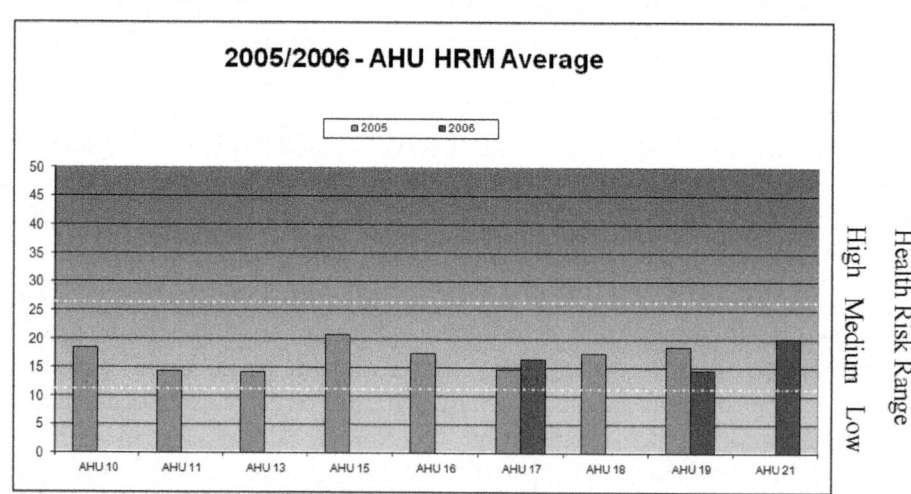

Figure 9. Health Risk Model Scores for the space controlled by each AHU.

Periods of maintenance and non-routine operation of HVAC systems within the hospital can result in filter bypass, dissemination of biological contamination, and the infiltration of unfiltered OSA into the hospital, placing the hospital within the High Risk category due the creation of an exposure pathway during these times. Hence, times during and immediately after maintenance and non-routine operation of the HVAC systems present a high risk for health effects due to bioaerosols in the indoor environment. (CDC 2003)

DISCUSSION AND HYPOTHESIS TESTING

Indoor microbial contaminants and infectious agents are closely related to water and moisture-related conditions. (Bartley 2000) The scenario that has emerged within the hospital is one in which immunosuppressed or allergic patients are not fully protected against the risk of NI or allergies due to environmental bioaerosols. (Perdelli, Christina et al. 2006) Where an indoor environment is exhibiting growth or airborne suspension of bioaerosols, risk may be determined to exist even if levels are less than outdoor baseline and control levels. (Boss and Day 2003) While indoor bioaerosol concentrations were consistently lower than OSA, the biodiversity was consistently different. Air sampling indicators of an indoor source and microbial surface contamination were identified in each of the areas investigated.

The objectives of environmental control in buildings are to prevent or minimize occupant exposures that can be deleterious to human health and to provide for the comfort and well-being of the occupants. (Pope, Patterson et al. 1993) The inability of the hospital's environmental systems in the areas investigated to manage moisture has resulted in a situation where patients, visitors, and staff are exposed to airborne microorganisms from indoor surfaces and OSA infiltration that normally would not be present within the building. This has resulted in an indoor environment that is not hygienic from the perspective of environmental contamination associated with the building and building systems.

Infection prevention in the hospital environment is one of the goals of healthcare workers and facilities. (Boss and Day 2003) Visitors, volunteers, and staff can both be infected and infectious. Highly filtered air is essential to preventing infection in healthcare facilities. (Boss and Day 2003) With the exception of AHU 11, all areas investigated were under the environmental control of an air conditioning system with >90% efficiency filters. The presence of the high efficiency final filters was largely responsible for the relatively low concentrations of total bioaerosols identified in the hospital ISA. Of particular concern is that a breach in the integrity of the final filters or dissemination of contamination during maintenance activities or non-routine operation could place the areas of the hospital which now fall into the medium risk category into the high risk category by creating an exposure pathway from contamination within the air handlers, OSA, and re-circulated air of the hospital to the occupant. The high efficiency filters within the AHUs are the hospital's main defense against nosocomial infection associated with indoor environmental contamination.

Although HVAC systems equipped with high efficiency filters can significantly reduce indoor concentrations of bioaerosols (Perdelli, Christina et al. 2006), the areas of the hospital investigated were under negative pressure, allowing the infiltration of unfiltered OSA, and had visible and confirmed microbial contamination on indoor surfaces and within each AHU. The presence of unfiltered air entering the Hospital via infiltration due to negative building pressurization is of concern (CDC 2003) and is significant risk factor for NI. (Streifel, Lauer et al. 1983; Rhame, Streifel et al. 1984; Rhame 1991; Nolard 1994) Biofilms (communities of bacteria in a matrix) and standing water were present in each of the AHUs investigated in 2005 and indicate the presence of excess moisture and poorly draining drain pans. Biofilms are conglomerates of microorganisms and indicate microbiological amplification within the AHUs. Visible surface contamination was identified within all the areas investigated.

Note that a breach in filter media or during times of HVAC system maintenance such as changing filters or starting and stopping the units, will cause an increase in airborne microorganisms or infiltration of unwanted OSA into the ISA of the Hospital. A failure or malfunction of any HVAC system component may subject patients to discomfort and exposure to airborne contaminants. Accumulation of dust and moisture in HVAC systems increases the risk for spread of health-care-associated environmental fungi and bacteria. If moisture is present in the HVAC system, periods of stagnation should be avoided. Bursts of organisms can be released upon system start-up, increasing the risk of airborne infection. If the ventilation system is out of service, rendering indoor air stagnant, sufficient time must be allowed to clean the air and re-establish the appropriate number of air changes once the HVAC system begins to function again. Reactivation of HVAC systems after shutdown can dislodge substantial amounts of dust and create a transient airborne increase of fungal spores. (CDC 2003) Therefore, during and after non-routine operation of the HVAC systems, the hospital is at high risk for nosocomial infection from environmental microorganisms.

The hypotheses testing results are:

Hypothesis A: The 90-95% final filters are controlling particulate matter generated by the AHUs and preventing contamination downstream of the filters. Note: AHU 11 does not have final filters.

Test: OSA versus ISA comparisons of total bioaerosols and spore traps.

Method: Investigator observations, interpretations, and literature review.

Test Result: Accept Hypothesis A—The final filters are controlling the dissemination of particulate matter and preventing the dissemination of particles downstream of the filters during routine operation. Statistical comparisons showed that indoor concentrations of non-culturable (spore traps) fungal bioaerosols were significantly lower than OSA concentrations. At least 79% (2005 data) of the non-culturable fungal bioaerosols are being removed by the final filters of the AHUs investigated, with the exception of AHU 11, which does not have final filters. At least 66% (2005 data) of the culturable fungal bioaerosols are being removed by the final filters of the AHUs investigated, with the exception of AHU 11, which does not have filters. With the exception of the bacterial (25°C) air samples in AHUs 10 and 11, the percent differences from outside air to indoor air were at least 35%. This is an indication that the filters were removing bacteria from both the outside and re-circulated air of the Hospital. The 2006 data showed a minimum percent reduction of 54% of all fungal samples (culturable and non-culturable) for the before filter to after filter comparisons. Therefore, particulate matter of similar aerodynamic diameters to fungi and bacteria are being removed from the air stream by the final filters.

Hypothesis B: The 90-95% final filters are preventing microbial contamination downstream of the filters.

Test: OSA versus ISA comparisons of bioaerosols.

Method: Investigator observations, interpretations, and literature review.

Test Result: Accept Hypothesis B—The final filters are preventing fungal particulate dissemination downstream of the filters by filtering the particles from the airstreams during routine operation. See Test Result for Hypothesis A. Indoor concentrations of non-culturable fungal bioaerosols were consistently lower and spore trap sampling results indicated that indoor concentrations of fungal bioaerosols were significantly lower than indoor concentrations. At least 66% (2005 data) of the culturable fungal bioaerosols are being removed by the final filters of the AHUs investigated, with the exception of AHU 11, which does not have filters. The 2006 data showed a minimum percent reduction of 54% of all fungal samples (culturable and non-culturable) for the before filter to after filter comparisons. The final filters are preventing the dissemination of bacterial contamination downstream of the filters during routine operation. Species of bacteria were identified within the AHUs and not identified post-filter, indicating that the filters were preventing the dissemination of contamination downstream. Indoor bacterial concentrations were consistently lower than outdoor concentrations. With the exception of the bacterial (25°C) air samples in AHUs 10 and 11, the percent differences from outside air to indoor air were at least 35% for the 2005 data. For the 2006 data, the minimum percent difference from OSA to ISA for bacteria was 64%. This is an indication that the filters were removing bacteria from the air streams. Note that indoor concentrations of bacteria are expected to be higher than outdoors but were consistently lower in the Hospital.

Hypothesis C: Staff is being exposed to potentially harmful concentrations of biological contaminants.

Test: OSA versus ISA comparisons of bioaerosols.

Method: Air samples, indoor surface sampling, HRM, Investigator observations, interpretations, and literature review.

Test Result: Accept Hypothesis D—Both the 2005 and 2006 data show indicators of indoor contamination and confirmed microbial surface contamination in each area investigated. Analysis for biodiversity showed that the bioaerosol profiles of ISA and OSA were consistently different. The HRM places all of the areas investigated in the medium risk category, indicating that staff are at risk for adverse health effects due to environmental contamination.

Hypothesis D: Patients are being exposed to harmful quantities of biological contaminants.

Test: OSA versus ISA comparisons of bioaerosols.

Method: Air samples, indoor surface sampling, HRM, Investigator observations, interpretations, and literature review.

Test Result: Accept Hypothesis E—Both the 2005 and 2006 data show indicators of indoor contamination and confirmed microbial surface contamination in each area investigated. Analysis for biodiversity showed that the bioaerosol profiles of ISA and OSA were consistently different, especially for species of *Penicillium* and *Aspergillus*. The HRM places all of the areas investigated in the medium risk category, indicating that patients are at risk for adverse health effects due to environmental contamination.

CONCLUSION

This 2-year investigation of bioaerosol concentrations in a hospital facility demonstrates that consideration of error, sampling variability, identification of microorganisms at the species level, and at least 6 replicate samples per location are necessary to detect significant differences in bioaerosol concentrations. Furthermore, spore trap samples were not sufficient to detect these differences regardless of the number of samples taken. Environmental investigators must consider this in their investigation strategy. Of particular importance is the failure of spore trap sampling to detect the same differences in airborne fungal bioaerosol species identified by culturable air sampling. Spore trap sampling should be considered a tool to determine a gross estimation of the quality of the indoor environment with regards to fungal bioaerosols and should not be used to make estimations on health or bioaerosol exposure. In all cases, spore trap sampling failed to detect differences in the aerobiological profile, masking significant concentrations of airborne fungal bioaerosols, including species of *Aspergillus* and *Penicillium*.

Statistical validity, analytical/sampling error, and species diversity should be considered for any bioaerol sampling intended to make comparisons on airborne concentrations of bioaersols between two or more environments. Identification of surface contamination should be considered de facto evidence of exposure to potentially harmful biological compounds. This study shows that high efficiency filtration can result in the environmental control of airborne bioaerosols. This is especially important in the hospital setting. PCR sampling shows promise because longer air sampling times can be used, surface samples can detect the presence of organisms on surfaces for an indication of a building's microbial burden, and because of relatively simple PCR sampling methods.

REFERENCES

1. ACGIH 1999 Bioaerosols: Assessment and Control. Cincinnati, OH, American Conference of Governmental Industrial Hygienists.

2. ACOEM 2002 Adverse Human Health Effects Associated with Molds in the Indoor Environment, American College of Occupational and Environmental Medicine: 110 .

3. AerotechP&K. 2006 "Microbial and IAQ Glossaries." Retrieved August, 2006, from www.aerotechpk.com.

4. U. S. E. P. Agency, 2001 "Mold remediation in schools and commercial buildings." USEPA EPA 402 -K-01-001.

5. Y. Anzai, Y. Kudo, et al. 1997 "The phylogeny of the genera Chryseomonas, Flavimonas, and Pseudomonas supports synonymy of these three genera." Int J Syst Bacteriol 47 2 249251 .

6. P. M. Arnow, M. Sadigh, et al. 1991 "Endemic and epidemic aspergillosis associated with in-hospital replication of Aspergillus organisms." J Infect Dis 164 5 9981002 .

7. ASHRAE 1992 Gravimetric and Dust-Spot Procedures for Testing Air-Cleaning Devices Used in General Ventilation for Removing Particulate

Matter Atlanta, GA, American Society of Heating, Refrigerating, and Air-Conditioning Engineers: 32.

8. ASHRAE 1999 Method of Testing General Ventilation Air-Cleaning Devices for Removal Efficiency by Particle Size Atlanta, GA, ASHRAE: 41.

9. ASHRAE 2003 2003 ASHRAE handbook : heating, ventilating, and air-conditioning applications. Atlanta, Ga., ASHRAE.

10. J. M. Bartley, 2000 "APIC state-of-the-Art report: the role of infection control during construction in health care facilities." Am J Infect Control 28 2 156169 .

11. C. Beggs, K. Kerr, 2000 "The Threat Posed by Airborne Micro-Organisms." Indoor Built Environmnet 9 241245 .

12. J. W. Bendig, P. J. Mayes, et al. 1989 "Flavimonas oryzihabitans (Pseudomonas oryzihabitans; CDC group Ve-2): an emerging pathogen in peritonitis related to continuous ambulatory peritoneal dialysis?" J Clin Microbiol 27 1 217218 .

13. M. J. Boss, D. W. Day, 2003 Biological risk engineering handbook : infection control and decontamination. Boca Raton, Fla., Lewis Pub.

14. H. A. Burge, 2000 The Fungi. Indoor Air Quality Handbook. J. D. Spengler, J. M. Samet and J. McCarthy, McGraw-Hill: 45.4145 .33.

15. D. J. Burton, 2006 Industrial Hygiene Workbook: the Occupational Health Sciences. Bountiful, UT, IVE, Inc.

16. CDC 1992 "Public health focus: surveillance, prevention, and control of nosocomial infections." Morb Mort Wkly Rep 42 42 783787 .

17. CDC 2003 Guidelines for environmental infection control in health-care facilities: Recommendations of CDC and the HealthCare Infection Control Practices Advisory Committee. Atlanta, GA.

18. W. J. Conover, 1999 Practical nonparametric statistics. New York, Wiley.

19. H. K. Dillon, P. A. Heinsohn, et al. 2005 Field guide for the determination of biological contaminants in environmental samples. Fairfax, VA, American Industrial Hygiene Association.

20. H. K. Dillon, P. A. Heinsohn, et al. 1996 Field guide for the determination of biological contaminants in environmental samples. Fairfax, VA, American Industrial Hygiene Association.

21. S. Dulworth, B. Pyenson, 2004 "Healthcare-associated infections and length of hospital stay in the medicare population." American Journal of Medical Quality 19 3 121127 .

22. T. C. Eickhoff, 1994 "Airborne nosocomial infection: a contemporary

perspective." Infection Control and Hospital Epidemiology 15 10 663672
.

23. J. Esteban, M. L. Valero-Moratalla, et al. 1993 "Infections due to
     Flavimonas oryzihabitans: case report and literature review." Eur J Clin
     Microbiol Infect Dis 12 10 797800 .

24. B. C. Fox, L. Chamberlin, et al. 1990 "Heavy contamination of operating
     room air by Penicillium species: identification of the source and attempts
     at decontamination." Am J Infect Control 18 5 300306 .

25. J. Freeman, J. E. Mc Gowan, 1978 "Risk factors for nosocomial
     infection." J Infect Dis 138 6 811819 .

26. J. Freeman, J. E. Mc Gowan, 1981 "Differential risk of nosocomial
     infection." Am J Med 70 4 915918 .

27. J. Freney, W. Hansen, et al. 1988 "Postoperative infant septicemia
     caused by Pseudomonas luteola (CDC group Ve-1) and Pseudomonas
     oryzihabitans (CDC group Ve-2)." J Clin Microbiol 26 6 12411243 .

28. F. Fung, W. G. Hughson, 2002 Health effects of indoor bioaerosol
     exposure. Indoor Air.

29. F. Fung, W. G. Hughson, 2003 "Health effects of indoor fungal bioaerosol
     exposure." Appl Occup Environ Hyg 18 7 535544 .

30. R. Gaynes, J. R. Edwards, 2005 "Overview of nosocomial infections
     caused by gram-negative bacilli." Clin Infect Dis 41 6 848854 .

31. V. S. Georgiev, 1998 Infectious diseases in immunocompromised hosts.
     Boca Raton, CRC Press.

32. K. J. C. Gibson, M. Gilleron, et al. 2003 "Structural and functional
     features of Rhodococcus ruber lipoarabinomannan." Microbiology 149
     14371445 .

33. P. J. Hitchcock, M. Mair, et al. 2006 "Improving performance of HVAC
     systems to reduce exposure to aerosolized infectious agents in buildings;
     recommendations to reduce risks posed by biological attacks." Biosecur
     Bioterror 4 1 4154 .

34. D. G. Hollis, F. W. Hickman, et al. 1981 "Tatumella ptyseos gen. nov.,
     sp. nov., a member of the family Enterobacteriaceae found in clinical
     specimens." J Clin Microbiol 14 1 7988 .

35. Institute of Medicine Committee on Damp Indoor Spaces and Health 2004
     Damp indoor spaces and health. Washington, DC, National Academies
     Press.

36. P. Kalliokoski, 2003 "Risks caused by airborne microbes in hospitals-
     source control is important." Indoor and Built Environment 12 4146 .

37. J. R. Kerr, J. E. Moore, et al. 1995 "Investigation of a nosocomial outbreak of Pseudomonas aeruginosa pneumonia in an intensive care unit by random amplification of polymorphic DNA assay." Journal of Hospital Infection 30 125131 .

38. S. Kiris, U. Over, et al. 1997 "Disseminated Flavimonas oryzihabitans infection in a diabetic patient who presented with suspected multiple splenic abscesses." Clin Infect Dis 25 2 324325 .

39. C. C. Lai, L. J. Teng, et al. 2004 "Clinical and microbiological characteristics of Rhizobium radiobacter infections." Clin Infect Dis 38 1 149153 .

40. J. Lalucat, A. Bennasar, et al. 2006 "Biology of Pseudomonas stutzeri." Microbiol Mol Biol Rev 70 2 510547 .

41. S. Lam, H. D. Isenberg, et al. 1994 "Community-acquired soft-tissue infections caused by Flavimonas oryzihabitans." Clin Infect Dis 18 5 808809 .

42. J. R. Lentino, M. A. Rosenkranz, et al. 1982 "Nosocomial aspergillosis: a retrospective review of airborne disease secondary to road construction and contaminated air conditioners." Am J Epidemiol 116 3 430437 .

43. R. D. Lin, P. R. Hsueh, et al. 1997 "Flavimonas oryzihabitans bacteremia: clinical features and microbiological characteristics of isolates." Clin Infect Dis 24 5 867873 .

44. P. Y. Liu, Z. Y. Shi, et al. 1996 "Epidemiological typing of Flavimonas oryzihabitans by PCR and pulsed-field gel electrophoresis." J Clin Microbiol 34 1 6870 .

45. K. G. Lucas, T. E. Kiehn, et al. 1994 "Sepsis caused by Flavimonas oryzihabitans." Medicine (Baltimore) 73 4 209214 .

46. J. Macher, 1999 Bioaerosols: assessment and control. Cincinnati, Ohio, American Conference of Governmental Industrial Hygienists.

47. M. Marin, D. Garcia de Viedma, et al. 2000 "Infection of hickman catheter by Pseudomonas (formerly flavimonas) oryzihabitans traced to a synthetic bath sponge." J Clin Microbiol 38 12 45774579 .

48. L. C. Mc Donald, M. Walker, et al. 1998 "Outbreak of Acinetobacter spp. bloodstream infections in a nursery associated with contaminated aerosols and air conditioners." Pediatr Infect Dis J 17 8 716722 .

49. M. M. Mc Neil, J. M. Brown, 1994 "The medically important aerobic actinomycetes: epidemiology and microbiology." Clinical Microbiology Reviews: 357417 .

50. D. C. Montgomery, 2001 Design and analysis of experiments, John Wiley

and Sons, Inc.

51. J. R. Mulhausen, J. Damiano, 1998 A strategy for assessing and managing occupational exposures. Fairfax, VA, AIHA Press.

52. R. Munro, G. Buckland, et al. 1990 "Flavimonas oryzihabitans infection of a surgical wound." Pathology 22 4 230231 .

53. N. Nolard, 1994 "[Links between risks of aspergillosis and environmental contamination. Review of the literature]." Pathol Biol (Paris) 42 7 706710 .

54. NYCDHMH 2006 "Guidelines on Assessment and Remediation of Fungi in Indoor Environments." The New York City Department of Health and Mental Hygiene.

55. E. H. Page, D. B. Trout, 2001 "The role of Stachybotrys mycotoxins in building related illness." AIHA J 62 644648 .

56. N. J. Palleroni, M. Doudoroff, et al. 1970 "Taxonomy of the aerobic pseudomonads: the properties of the Pseudomonas stutzeri group." J Gen Microbiol 60 2 215231 .

57. F. Perdelli, M. L. Christina, et al. 2006 "Fungal contamination in hospital environments." Infection Control and Hospital Epidemiology 27 1 4447 .

58. A. Podbielski, R. Mertens, et al. 1990 "Flavimonas oryzihabitans septicaemia in a T-cell leukaemic child: a case report and review of the literature." J Infect 20 2 135141 .

59. A. M. Pope, R. Patterson, et al. 1993 Indoor allergens: assessing and controlling adverse health effects. Washington, D.C., National Academy Press.

60. A. M. Pope, R. Patterson, H. Burge, 1993 Indoor Allergens. Assessing and controlling adverse health effects. Washington, D.C., National Academy Press.

61. B. Prezant, D. M. Weekes, et al. Eds, 2008 Recognition, Evaluation, and Control of Indoor Mold, American Industrial Hygiene Association.

62. G. Rahav, A. Simhon, et al. 1995 "Infections with Chryseomonas luteola (CDC group Ve-1) and flavimonas oryzihabitans (CDC group Ve-2)." Medicine (Baltimore) 74 2 8388 .

63. S. C. Redd, 2002 "State of the Science on Molds and Human Health." Center for Disease Control and Prevention, U.S. Department of Health and Human Services.: 111 .

64. J. Reina, J. Odgardd, et al. 1990 "Flavimonas oryzihabitans (formerly CDC group Ve-2) bacteremia in a pediatric patient on assisted ventilation." Eur J Clin Microbiol Infect Dis 9 10 786788 .

65. R. B. Reisler, H. Blumberg, 1999 "Community-acquired Pseudomonas stutzeri vertebral osteomyelitis in a previously healthy patient: case report and review." Clin Infect Dis 29 3 667669 .

66. F. S. Rhame, 1991 "Prevention of nosocomial aspergillosis." J Hosp Infect 18 Suppl A: 466472 .

67. F. S. Rhame, A. J. Streifel, et al. 1984 "Extrinsic risk factors for pneumonia in the patient at high risk of infection." Am J Med 76(5A): 4252 .

68. J. Romanyk, R. Gonzalez-Palacios, et al. 1995 "A new case of bacteraemia due to Flavimonas oryzihabitans." J Hosp Infect 29 3 236237 .

69. G. E. Smith, 1990 Sampling and identifying allergenic pollens and molds. San Antonio, TX, Blewstone Press.

70. R. C. Spicer, H. J. Gangloff, 2000 "Limitations in Application of Spearman's Rank Correlation to Bioaerosol Sampling Data." AIHA J 61(May/June): 362366 .

71. L. D. Stetzenbach, M. P. Buttner, et al. 2004 "Detection and enumeration of airborne biocontaminants." Curr Opin Biotechnol 15 3 170174 .

72. L. D. Stetzenbach, M. V. Yates, 2003 The dictionary of environmental microbiology. San Diego, Calif., Academic Press.

73. D. C. Straus, 2004 Sick Building Syndrome. Advances in Applied Microbiology. London, Elsevier Academic Press.

74. A. J. Streifel, J. L. Lauer, et al. 1983 "Aspergillus fumigatus and other thermotolerant fungi generated by hospital building demolition." Appl Environ Microbiol 46 2 375378 .

75. N. Taneja, S. K. Meharwal, et al. 2004 "Significance and characterisation of pseudomonads from urinary tract specimens." J Commun Dis 36 1 2734 .

76. USEPA 2001 Mold remediation in schools and commercial buildings. United States Environmental Protection Agency. EPA 402 -K-01-001.

77. VUMC 2006 Infection Control Interventions During Construction in Patient Care Areas. Procedure 1010 .17. Nashville, TN, Vanderbilt University Medical Center. 10-10.17.

78. T. J. Walsh, A. H. Groll, 1999 "Emerging fungal pathogens: evolving challenges to immunocompromised patients for the twenty-first century." Transpl Infect Dis 1 4 247261 .

79. A. Weber, E. Page, 2001 "Renovation of contaminated building materials at a facility serving pediatric cancer outpatients." Applied Occupational and Environmental Hygiene 16 1 231 .

80. R. P. Wenzel, 1997 Prevention and control of nosocomial infections.

Baltimore, William & Wilkins.

81. M. Wilson, 2005 Microbial inhabitants of humans : their ecology and role in health and disease. New York, Cambridge University Press.

82. S. C. Wilson, W. H. Holder, et al. 2004 "Identification, remediation, and monitoring processes used in a mold-contaminated high school." Adv Appl Microbiol 55 409423 .

83. S. Yee-Guardino, L. Danziger-Isakov, et al. 2006 "Nosocomially acquired Pseudomonas stutzeri brain abscess in a child: case report and review." Infect Control Hosp Epidemiol 27 6 630632 .

Chapter 4

IMPROVING THE QUALITY OF THE INDOOR ENVIRONMENT UTILIZING DESICCANT-ASSISTED HEATING, VENTILATING, AND AIR CONDITIONING SYSTEMS

M.D. Larrañaga[1], E. Karunasena[2], H.W. Holder[3], M.G. Beruvides[2] and D.C. Straus[4]

[1] Oklahoma State University, USA

[2] Texas Tech University, USA

[3] SWK, LLC, USA

[4] Texas Tech University Health Sciences Center, USA

INTRODUCTION

Throughout history, diverse cultures and societies have appreciated the importance of a clean and healthy environment (Bardana, Montanaro et al. 1988). In western societies today, most people spend over 90% of their time indoors (Teichman 1995; EPA 2003). Reports involving buildings with indoor air-related problems have appeared increasingly in the medical and scientific literature, although this problem has been with humans for centuries. Sick building syndrome (SBS), a common term for symptoms that result from individuals' exposure to poor IAQ (IAQ), was first recognized as an important problem that affects occupants in certain buildings in 1982. The first official study of SBS to examine more than one structure was published in 1984 (Finningan, Pickering et al. 1984). SBS has been difficult to define, and no single cause of the malady has been identified (Hodgson 1992). Early studies showed that many of the reported causes of SBS included undesirably high levels of known respiratory irritants such as nitrogen and sulfur dioxides, hydrocarbons, and particulates (National Academy of Sciences 1981), known or suspected carcinogens such as asbestos, radon, formaldehyde, and tobacco smoke (Sterling 1984), or chemicals released by new building materials. Although fungal spores are ubiquitous in the indoor and outdoor environments, they are now generally accepted as important causes of respiratory allergies (Solomon 1975; Bernstein 1983). Although no single cause for the symptoms induced by poor IAQ likely exists, the presence of fungi, spores, and fungal growth in sick buildings has become consistently associated with this problem (Miller 1992; Mishra 1992;Cooley 1998).

Since Biblical times, indoor fungal contamination and IAQ have been a cause for concern in society. Leviticus 14:33-45, in the Old Testament, states that a house needs to be cleaned and ridded of mildew to be clean and free of mold (Heller, Heller et al. 2003). If the house cannot be cleaned, then it is to be torn down. This is similar to the manner in which IAQ issues associated with fungal contamination are addressed today. Air pollution became an important environmental issue during the industrial revolution, and several instances of deadly environmental air pollution are cited in the literature. In the early 1900s, building design and construction underwent radical change in the United States; the introduction of air conditioning with cooling coils and forced ventilation, new construction materials, new lighting and heating standards, insulation, and similar advances changed the way buildings were designed and constructed. The new construction designs and methods allowed architects and engineers to build quality structures at lower costs. Schools of engineering and architecture in the United States then began to teach the new building designs and construction methods to students. Thus, thousands of years of collective architectural and engineering design experience, specifically natural ventilation design, were erased from the educational system in the United States.

World War II was a pivotal point in IAQ. At approximately the same time as building design and construction was experiencing radical change, American society also experienced radical change. America began to transform from an agricultural economy wherein the majority of time was spent outdoors, to a service and manufacturing economy, where the majority of time was spent indoors. Since 1945, a phenomenal transformation has occurred: Americans and people in other industrialized nations now spend more than 90 percent of their time indoors (Teichman 1995; EPA 2003). The Arab Oil Embargo of 1973-74 and the resulting energy supply shortages dramatically changed the public's attitude toward environmental control, because building energy usage had a monetary impact on every American. The energy crisis resulted in a tighter building design that produced more energy-efficient homes by closing the windows, redesigning the buildings, circulating re-heated and/or re-cooled air, and reducing the amount of fresh outside air brought into buildings. This "tight" building design resulted in an improvement of buildings and HVAC system energy efficiencies. This improvement in efficiency, however, was paralleled by an increase in SBS complaints. SBS is a term used to describe situations in which building occupants experience health and comfort effects that appear to be linked to time spent in a building, but for which no specific illness or cause can be identified. Indicators of SBS include headache; eye, nose, or throat irritation; dry cough; dry or itchy skin; dizziness and nausea; difficulty in concentrating; fatigue; sensitivity to odors; and other symptoms. Another important indicator was the relief of symptoms when complainants leave the building for extended periods of time (EPA 1991).

The first study to compare indoor versus outdoor fungal morphology was conducted in 1904 by Saito in Japan (Saito 1904); this work was followed by Rostrup in Denmark (Rostrup 1908) and Peyronel in Italy (Peyronel 1919). Since the 1930s, medical specialists in the allergy and immunology community recognized molds as being allergenic and capable of both exacerbating asthma and sensitizing patients (Bernton 1930; Flood 1931; Credille 1933; Conant 1936). With increasing awareness that poor IAQ may generate a variety of deleterious effects on human health, IAQ has become a serious public health concern (Samet 1990; Mishra 1992; Passon 1996; Hodgson, Morey et al. 1998; Sudakin 1998;Hagmann 2000; King and Auger 2002; Karunasena 2005; Karunasena E, Larrañaga MD et al. 2010). It has been estimated that more than 30% of the buildings in developed countries suffer from poor IAQ (World Health Organization 1983; Smith 1990). The Occupational Safety and Health Administration (OSHA) estimated that 30 to 70 million U.S. workers are affected by SBS (Bureau of National Affairs 1992).

Moisture problems have been encountered with increasing frequency both in family housing and in the workplace in the U.S. and Europe (Reijula 1996). Persistent water leaks and moist building materials inevitably lead to the growth of fungi and bacteria in these buildings. Several epidemiological studies suggested that dampness and fungal problems are present in 20% to 50% of modern homes (Dales 1991; Brunekreef 1992; Jaakkola 1993; Spengler 1994; Verhoeff 1995). Not only are dampness and fungi risk factors in the association between indoor dampness and respiratory symptoms (Hodgson, Morey et al. 1985; Dales 1991; Dales 1991b; National Academy of Sciences 1993; Flannigan 1994;Spengler 1994; Hodgson, Morey et al. 1998; Husman, Meklin et al. 2002), but damp homes tend to have higher levels of fungi than non-damp homes (Platt 1989; Verhoeff 1992). In addition, poorly maintained heating, ventilation, and air conditioning (HVAC) systems have been recognized as sources of microorganisms, including fungi.

Fungi are well known allergens that cause allergic rhinitis, allergic asthma, and hypersensitivity pneumonitis when inhaled (Salvaggio 1981; Tarlo 1988; Burge 1989; Flannigan 1994). Fungi also produce volatile organic compounds (VOCs) including alcohols, aldehydes, and ketones, which often produce moldy odors and can cause symptoms such as headaches, eye, nose and throat irritation, and fatigue (Tobin 1987; Flannigan 1991). Fungi also produce toxic metabolites called mycotoxins. Mycotoxins have been identified indoors or on materials indoors by several authors (Croft 1986;Hodgson, Morey et al. 1998; Nielsen, Hansen et al. 1998; Richard, Platnerr et al. 1999; Croft, Jastromski et al. 2002). Mycotoxins produced by *Stachybotrys chartarum* have been implicated in producing non-allergic respiratory symptoms in humans (Croft 1986; Johanning 1996; Andersson 1997; Croft, Jastromski et al. 2002) and have been shown to cause significant damage to cells of the neurological system in concentrations found indoors (Karunasena E, Larrañaga MD et al. 2010).

At present, no single environmental factor or group of factors has been established as the cause of SBS. Although fungal contamination in indoor environments has been shown to produce allergies in building occupants (Lehrer 1983; Licorish 1985; Verhoeff 1995), the role of fungi in SBS has become increasingly controversial. Numerous theories have been put forward (Mendell 1993). Along with the VOC theory, a heightened neurogenic inflammatory response to low-level chemical exposures has been suggested (Meggs 1993; Karunasena E, Larrañaga MD et al. 2010), while other theories have focused on particulates (Salvaggio 1994a) and physical factors (Levin 1995). Inadequate ventilation is a factor in all of these theories. Investigators who are more skeptical have emphasized the roles of psychosocial factors, stress, and gender (Stenberg 1994; Salvaggio 1994b; Bachmann 1995).

Today, children spend most of the day indoors, and because dampness in buildings has increased over the last decade, relationships have been identified between an increase in children's health symptoms, dampness, and fungal spores (Dill 1996; Li 1997; Garrett, Rayment et al. 1998). Children who attend school in buildings with dampness and fungal contamination have been shown to suffer higher rates of respiratory infections (Koskinen 1995). The literature suggests a strong association between the presence of fungal growth indoors and SBS (Cooley 1998). The Centers for Disease Control states that the inhalation of fungal spores inside buildings can cause allergic rhinitis, hypersensitivity pneumonitis, and exacerbate asthma (Redd 2002). Although associations have been made, far more research related to SBS, building-related illness (BRI), and IAQ is needed. Hodgson, et al., maintain that undesirable moisture levels indoors represent a public health concern inadequately addressed by building, health, or housing codes (Hodgson, Morey et al. 1998).

In warm and humid climates, conventional HVAC systems are incapable of adequately controlling humidity and simultaneously meeting the minimum fresh air requirements specified by the American Society for Heating, Refrigerating, and Air Conditioning Engineers' (ASHRAE) Standard 62, Ventilation for Acceptable IAQ. The bulk of a building's moisture load is carried by the incoming ventilation air, and simply drying the ventilation air will provide excellent humidity control at minimal cost (Harriman and Kittler 2001; Larrañaga, Beruvides et al. 2008). Drying the ventilation air via desiccant based cooling is a cost effective method of humidity control. A separate dedicated outdoor air system with humidity control is the simplest method and may be the only reliable method of meeting Standard 62 (Mumma 2001; Larrañaga, Beruvides et al. 2008). Controlling humidity is crucial for human comfort, minimizing adverse health effects associated with high humidity, and maximizing the structural integrity of buildings. The use of a desiccant pre-conditioner apparatus can improve the humidity control capabilities of HVAC systems arising from inherent evaporator coil limitations and can accommodate the minimum outdoor air ventilation rates specified by Standard 62 (Meckler 1994; Larrañaga, Beruvides et al. 2008).

Since the mid to late 1980s, desiccant based cooling systems have found increased applications as humidity control devices for non-industrial structures like schools, homes, hospitals, and commercial buildings (Hines 1992c; Harriman III, Witte et al. 1999; Larrañaga, Beruvides et al. 2008). Several authors have stated that the use of active desiccants enhances the quality of the indoor air by helping to maintain comfort criteria (temperature, humidity and ventilation) (Meckler 1994; Kovak 1997; Fischer and Bayer 2003), removing particulates and bioaerosols from the air (Hines 1992a; Kovak 1997), and removing chemical pollutants from the air (Hines 1992c; Popescu and Ghosh 1999). Several investigators explored the ability of solid and liquid desiccant materials to remove environmental tobacco smoke, particulates, radon, organic vapors, carbon dioxide, and several microorganisms responsible for the majority of nosocomial infections, including several bacteria and *Aspergillus niger,* from the airstream. Popescu and Ghosh used a fixed bed adsorber to simulate the operation of a rotary desiccant wheel, and showed that carbon dioxide and organic vapors were successfully removed by the desiccant materials (Popescu and Ghosh 1999). Hines, et al., used both fixed bed adsorbers in a column configuration for solid desiccants (Hines 1992a) and a packed bed absorber-stripper system for liquid desiccants (Hines 1992b) in their studies. Hines, et al., studied the desiccant removal capabilities of these column-type adsorbers and absorbers on carbon dioxide, volatile organic compounds (Hines 1992b), airborne particulates, environmental tobacco smoke, several bacteria, and *Aspergillus niger*(Hines 1992a). Kovak, et al., conducted a laboratory and field study of the capabilities of a solid-desiccant dehumidifier in removing seven microorganisms responsible for nosocomial infections (Kovak 1997).

This study quantified the removal capabilities of a rotary wheel solid-desiccant dehumidifier at removing selected IAQ-related fungal organisms from the airstream. While the above-mentioned authors studied the removal capabilities of solid and liquid desiccants and one rotary wheel solid desiccant dehumidifier, none explored the ability of a rotary wheel solid desiccant dehumidifier to remove IAQ-related fungal species from the air. Rotary wheel solid desiccant dehumidifiers in the honeycomb configuration are the most appropriate dehumidifier configuration for air-conditioning applications (Pesaran 1994). The use of active desiccation in warm and humid climates would result in energy savings from a reduction in latent cooling and an increase in sensible cooling, offsetting initial purchase costs (Dolan 1989; Pesaran 1994; Larrañaga, Beruvides et al. 2008).

Indoor environmental quality has been an issue throughout history. The interactions of a number of technological discoveries and historical events have resulted in the construction of buildings that make people sick. Every industrialized nation in the world has experienced an increase in asthma in the last 30 years. Asthma is an affliction of the industrialized world, and is not a prominent illness in third-world countries (Vogel 1997). The term sick-building syndrome was first used in the 1980s, when illnesses associated with buildings began surfacing throughout the industrialized world. SBS is a societal and public health issue that has become prevalent in today's world with a negative impact on the world's economy.

THE ECONOMIC IMPACT OF POOR IAQ

The incidences of illnesses related to building occupancy have brought about a number of financial penalties, ranging from lower worker productivity to expensive lawsuits (Addington 2000). Poor IAQ has a negative effect on productivity, worker health, and morale. Direct evidence established in the literature stated that characteristics of buildings and indoor environments can directly affect worker health and productivity. Estimates (in 1996 dollars) of annual savings and productivity gains included $6 to $14 billion from reduced respiratory disease; $2 to $4 billion from reduced allergies and asthma; $15 to $40 billion from reduced symptoms of sick building syndrome; and $20 to $200 billion from direct improvements in worker performance. Building characteristics and indoor environments have been linked to SBS experienced by building occupants. The most common reported sufferers of SBS are office workers and teachers, who make up approximately 50 percent of the total workforce (64 million workers). SBS symptoms include irritation of the eyes, nose, skin and upper respiratory tract, increased airway infections, dizziness, nausea, headache, fatigue, bleeding from the nose, and lethargy. Cognitive impairment, memory loss, permanent lung damage, and other physiological effects have been correlated to SBS. Psychosocial factors are also known to influence the symptoms of SBS. Building factors such as the amount of fresh air ventilation, indoor lighting levels including sunlight, levels of chemical and microbial contamination, and indoor temperature and humidity have been shown to influence SBS symptoms (Gordon, Johanning et al. 1999; Fisk 2000).

SBS symptoms hinder work. Symptoms also can cause not only absences from work, but also visits to doctors and costly emergency medical services. The investigations, maintenance, relocation, legal fees, and insurance costs associated with SBS quickly rise to astronomical levels, and impose a societal burden. The quantification of SBS costs is extremely difficult, due to economic and psychosocial factors, decreases in productivity, legal costs, and other influences. However, several attempts have been made at quantification of the costs of SBS based on the Gross Domestic Product (GDP) associated with office-type-work. The GDP of the United States in 1996 was $7.6 trillion. On the basis of an estimated two percent decrease in productivity due to SBS symptoms for office-type-work, which has a GDP of $3.8 trillion, the annual nationwide cost of SBS symptoms is $76 billion. The potential financial benefits of improving U.S. indoor environments exceed costs by a factor of 18 to 47 (Bayer 2000). While poor IAQ places a heavy economic burden on the workplace, poor IAQ can also have a negative impact on school-aged children, and teachers (Handal, Leiner et al. 2004).

Asthma as Epidemic

Mold and moisture indoors is a significant risk factor for asthma and the US Environmental Protection Agency identifies mold and moisture indoors as asthma triggers (Environmental Protection Agency 2010). Since 1970, a three-fold increase occurred in the incidence of asthma in the United States: 7 million cases in 1970 vs. 20 million cases in 2000. Persons with asthma collectively have more than 100 million days of restricted activity and 470,000 hospitalizations annually (Weiss 1992). Asthma is the most common chronic childhood disease, affecting 6.3 million children. 1 in 10 school-aged children has asthma (Environmental Protection Agency 2010), and the rate is rising more rapidly in children of preschool age than in any other age group. In 2000, there were nearly 2 million emergency room visits and approximately 500,000 hospitalizations due to asthma. Asthma symptoms that are not severe enough to require an emergency room physician visit can still be severe enough to prevent a child with asthma from living a fully active life. Asthma is the leading cause of school absenteeism due to chronic illness. During the past 20 years, the number of school absence days due to asthma has more than doubled. The CDC estimates 14 million school days were missed due to asthma in 2000

(EPA 2003). The economic impact of asthma is staggering; During 1994, total US costs of asthma were $10.7 billion (Weiss, Sullivan et al. 2000), yearly treatment costs alone approach $6 billion (Cookson 1999), and asthma costs 5,000 deaths yearly with no signs of leveling off (Vogel 1997). The roots of asthma may be traced to heating of the bedrooms in homes, tight building design, and air conditioning. Forced air ventilation allowed homes to be heated or cooled throughout at all times of the year. Prior to the industrial revolution, homes were designed with a central room with heating capabilities (i.e. kitchen and living room) and plumbing. The surrounding rooms, most commonly bedrooms, were not heated, and would freeze through the cold months of the year. This yearly cycle of freezes in bedrooms allowed beddings to freeze, killing dust mites. Few indoor environments routinely freeze today. Thus, dust mites survive throughout the year in carpets and beddings, and are able to affect those with asthma. Heating and air conditioning system design has re-circulated the indoor air to maintain efficiencies, decreasing the amount of fresh air available in buildings.

The Economics of Poor IAQ And Schools

Teachers and school-age children also suffer from SBS. One in five U.S. schools exhibits IAQ problems, and studies linking specific environmental conditions to student performance indicate impaired performance of students. A study of 627 Swedish secondary school pupils reported "impaired performances were more common in schools with lower air exchange rates, higher relative humidities, and higher concentrations of respirable dust, formaldehyde, VOCs, and total bacteria or molds. A relationship was demonstrated between subjective reports of impaired mental performance, measured indoor air pollutants, and low air exchange rate" (Bayer 2000). A similar study of 12 schools within the Denver, CO metropolitan area indicated an increased prevalence of nasal congestion, sore throat, headache, dustiness complaints, and red and watery eyes in schools with certain ventilation system types (Kinshella, Van Dyke et al. 2001). In a similar study of 85 schools in Canada, it was shown that children with allergies who displayed allergic-type symptoms during school were disproportionately in the below-average category for academic achievement (Landrus 1990). Several other studies demonstrated a link between student absenteeism and IAQ factors. Additionally, there seemed to be a link between unsatisfactory IAQ and the proportion of a school's students from low-income households.

IAQ and SBS appeared especially important in schools because (1) children are developing physically and affected by pollutants to a greater degree than adults, (2) the number of children with asthma has risen approximately 49 percent since 1982, (3) children below the age of 10 have three times as many colds as adults, (4) poor IAQ can lead to drowsiness, headaches, lack of concentration, and other symptoms, and (5) children have a higher rate of metabolism than adults and may ingest or inhale more air and surface contaminants than adults (Bayer 2000). Schools face separate epidemics: an epidemic of deteriorating facilities and an epidemic of asthma among children (Bascom 1997). Asthma is the principal cause of school absences, accounting for 20 percent of lost school days in elementary and high school (Richards 1986). Richards also states that allergic disease (nasal allergy, asthma, and other allergies) is the number one chronic childhood illness. It has been clearly established that SBS and poor IAQ affect productivity in a negative way. The factors associated with IAQ interact in a very complex relationship that sometimes requires extensive and diverse knowledge, experience, and diversity to solve. Ventilation, however, plays a key role as the underlying factor for SBS in the modern, sealed building with a controlled indoor environment (Cooley 1998).

Economic Incentives for Improving IAQ

In some cases, improving IAQ to acceptable levels can be quite expensive and uneconomical. In rare instances, structures are demolished rather than repaired or remediated because of poor IAQ. Potential savings from changes in building factors, that produced a 10 to 30 percent reduction in symptoms and associated costs, projected an annual savings of $2 to $4 billion, in addition to reducing the psychosocial and societal costs. Three general approaches for reducing allergy and asthma by changes in buildings are currently recognized: (1) control the indoor sources of the allergens and chemical compounds that cause symptoms or initial sensitization, (2) use cleaning systems to decrease the indoor concentrations of the relevant pollutants, and (3) modify buildings and IAQ in a manner that reduces viral respiratory infections among occupants (Fisk 2000).

Potential savings from changes in building factors creating a 20 to 50 percent reduction in symptoms and associated costs in office buildings projected an annual productivity increase on the order of $15 to $38 billion (Fisk 2000). Strong evidence existed that good IAQ can effectively increase health and productivity and the cost benefits associated with improving IAQ exceeded the cost of improving IAQ by a factor of 18 to 47. "For the United States, the estimated potential annual savings plus productivity gains, in 1996 dollars, are approximately $40 billion to $250 billion" (Fisk 2000). This evidence justified changes in (1) the components of building codes affecting IAQ and (2) company and institutional policies related to IAQ, (3) building design, operation, and maintenance to incorporate maintaining and promoting a desirable IAQ. Fischer concluded that the payback period associated with a desirable indoor environmental quality is probably very short (Fischer 1996). He indicated that the many benefits listed would be recognized year after year, whereas the costs associated with providing the desirable indoor environmental quality are a one-time expense with minimal maintenance costs. The expected benefits—which included reductions in absenteeism and health care costs, positive impacts on productivity, alertness, drowsiness, allergies, and illness, avoidance of property damage and remediation, and reduced maintenance costs—quickly exceeded the initial expense associated with an improved indoor environment (Bayer 2000).

CAUSES OF POOR IAQ AND DESICCATION AS AN IAQ CONTROL STRATEGY

Indoor dampness, water damage, and fungi have been associated with respiratory complaints (Martin 1987; Andrae 1988; Brunekreef 1989; Dales 1991; Dales 1991b; Brunekreef 1992; Summerbell 1992;Jaakkola 1993; Joki 1993; Husman, Meklin et al. 2002), and with both allergic and non-allergic respiratory disease (Salvaggio 1981; Lehrer 1983; Licorish 1985; Flannigan 1994; Spengler 1994;Verhoeff 1995; Jarvis and Morey 2001) in several industrialized nations. Asthma was associated with 'damp houses and fenny countries' three centuries ago by Sir John Floyer (Sakula 1984). In a study of 48 U.S. schools, Cooley, et al., associated the presence of propagules of *Penicillium* and *Stachybotrys*on building surfaces with SBS (Cooley 1998). In a survey of 59 homes selected on the basis of previously measured mold levels in 400 houses, White correlated measurements of mold growth and immunological reactions of occupants, noting that lymphocytes from children are chronically activated, and immuno-regulation may be altered in households with mold growth (White 1995). The presence of moisture damage in schools was identified as a significant risk factor for respiratory symptoms in children based on data from microbial IAQ studies in 24 schools (Meklin, Husman et al. 2002) and in the home (Hyvärinen, Pekkanen et al. 2002), and airborne *Penicillium* and *Aspergillus*species was identified as a risk factor for asthma, atopy, and respiratory symptoms in children (Garrett, Rayment et al. 1998). In a population-based incident case-control study in South Finland of 521 adults with newly diagnosed asthma and 932 controls, Jaakkola, et al., found that 35.1% (95% confidence interval, 1.0%-56.9%) of the adult-onset asthma cases were related to workplace mold exposure indoors and that indoor mold problems constitute an important occupational health hazard (Jaakkola, Nordman et al. 2002). It is unlikely that the number of associations in several different countries is a result of chance. It is more likely these associations represent a combination of factors leading to poor ventilation, moisture damage, fungal contamination, and poor IAQ because of systemic and synergistic effects of contaminants within modern built environments (Passon 1996). One important building parameter in controlling indoor moisture and mold growth is the HVAC system, which if not properly designed, maintained, or operated, can cause poor IAQ.

HVAC Systems as a Cause of Poor IAQ and Desiccation as an IAQ Control Strategy

HVAC systems are essential to modern life and can provide healthy and comfortable indoor environments when properly installed, operated, and maintained (Batterman 1995). Sietz categorized the primary factors leading to building-related illness (BRI), and found that 53% of 529 IAQ evaluations conducted by NIOSH from 1971 through 1988 were associated with inadequate ventilation (Seitz 1990). NIOSH attributed more than half (52%) of the SBS cases to unsuitable facility ventilation systems (Bayer 2000). Additionally, conventional HVAC systems cannot adequately dehumidify air in warm and humid climates (Bayer 1992a; Fischer 1996; Davanagere 1997; Larrañaga, Beruvides et al. 2008) and it is not economically feasible to use only materials that are not susceptible to moisture damage. A systemic relationship between the HVAC system, outdoor air, and indoor environment exists when indoor relative humidities are high. A properly designed, functioning, and operating HVAC system can have a significant positive impact on reducing the number of SBS symptoms experienced within buildings.

The application of a control strategy that aids in removing organic materials and microorganisms from the air, while introducing fresh air into a building, can improve the IAQ of a building and eliminate many problems associated with ventilation and lack of fresh air in buildings. Bayer suggested that IAQ improves when using active humidity control and continuous ventilation in schools (Bayer 2000). In a study of 10 schools in Georgia, Bayer states that of the five schools having conventional HVAC systems, none supplied outside air at the ASHRAE recommended 15 cfm/person. The schools having desiccant systems were delivering as much as three times more outside air, while maintaining equal or better control of the indoor relative humidity than the conventional systems. The average total volatile organic compound (TVOC) concentrations tended to be lower in schools having desiccant-based systems. The school showing the highest air exchange rate utilized a rotary desiccant system, and had the lowest carbon dioxide, TVOC, and airborne microbial concentrations, and the lowest average indoor relative humidity (Bayer 2000). In Phase II of the same project, Fischer and Bayer stated that increasing the air ventilation rate from 5 to only 8 cfm/student challenged the ability of the conventional systems to maintain the space relative humidity below the ASHRAE and ACGIH recommended 60% level. Increasing the ventilation rate of the conventional systems to the required 15 cfm/student allowed the space relative humidities routinely to exceed 70%. These data explained why all of the conventional HVAC system schools were designed and/or operated with only 6 cfm/student of outdoor air or less. The decreased ventilation rates were in direct response to the performance limitations of the conventional cooling equipment and contributed to the poor IAQ within the schools. Furthermore, the schools served by the conventional HVAC systems experienced absenteeism at a 9% greater rate than those served by the desiccant systems (Fischer and Bayer 2003). Kumar and Fisk proposed that the energy cost of providing additional ventilation may be more than offset by the savings that result from reduced employee sick leave, and that increasing ventilation rates above the minimum rates specified in ANSI/ASHRAE Standard 62, Ventilation for Acceptable IAQ, can yield substantial benefits, including the reductions of the incidence of allergy and asthma in building occupants (Kumar and Fisk 2002).

The general approaches for reducing allergy and asthma by changes in buildings are: (1) control the indoor sources of the allergens and chemical compounds that cause symptoms or initial sensitization, (2) use cleaning systems to decrease the indoor concentrations of the relevant pollutants, and (3) modify buildings and IAQ in a manner that reduces viral respiratory infections among occupants (Fisk 2000). Utilizing desiccant treatment to pretreat fresh air and maintain the desired airflows created benefits for building occupants by providing a means to meet all three general approaches for reducing allergy and asthma in buildings. The use of desiccation as an IAQ control strategy provided other benefits including an increase in sensible cooling and a decrease in latent cooling at the cooling coils.

IAQ studies found that molds, or bioaerosols, were primary link to building-related illness, infections, toxic syndromes, and hypersensitivity diseases (Kovak 1997). Outdoor air parameters can be controlled actively by pre-treating outdoor air prior to its entering a building. Studies have shown that forced desiccant treatment of air has been effective in reducing airborne levels of tobacco smoke and volatile organics (Hines 1992c; Kovak 1997; Popescu and Ghosh 1999). The objective of this research was to determine the effectiveness of a titanium dioxide/silica gel catalytic dehumidification system in removing IAQ-related bioaerosols from the air. The research consisted of subjecting a laboratory setup of the desiccation system to airborne concentrations of IAQ-related bioaerosols. Its success proved that adapting dehumidification technology for use in HVAC systems will allow for moisture control and removal of IAQ-related organisms from the air stream, and offered a viable control strategy for preventing moisture damage and mold growth in buildings. Furthermore, utilization of active desiccation in humid climates results in energy savings from a reduction in latent cooling and an increase in sensible cooling, offsetting initial purchase costs while providing an economic benefit.

Hines, et al., (1992), Kovak, et al., (1997), and Popescu and Ghosh (1999) showed that forced desiccant treatment of air effectively reduced airborne levels of bacteria, fungi, particulates from environmental tobacco smoke, and VOCs. Hines showed that a packed bed adsorber can act as a filter for airborne particulates by removing between 22% and 73% of particulate matter associated with environmental tobacco smoke (Hines 1992a). Kovak, et al., showed median reductions after exposure to desiccant based air conditioning (DBAC) systems in three field studies of 39%-64% for bacteria and 32%-72% for fungi (Kovak 1997). In laboratory tests, Kovak et al. showed an average reduction of 38% of seven organisms associated with nosocomial infections after exposure to a DBAC system (Kovak 1997). Popescu and Ghosh showed removal efficiencies of a desiccant bed to be 35% for formaldehyde, 70% for toluene, 29% for carbon dioxide, and 54% for 1,1,1-tricchloroethane using a packed bed adsorber configuration.

The above authors investigated removal capabilities of desiccant treatments in other configurations than this research utilized. For example, Hines et al. (1992) studied the desiccant setups in a packed bed filter-like configuration. Popescu and Ghosh (1999) simulated a rotating desiccant wheel using an experimental system consisting of a fixed bed adsorber, which acted as a filter. Kovak, et al., (1997) exposed a desiccant wheel to biological organisms that were both associated with nosocomial (hospital-induced) infections and inherently heat sensitive.

Fig. 1 shows a desiccant unit installed in a school in south Texas for the purposes of (1) drying the school, (2) preventing the structure from becoming wet, (3) reducing the amount of latent cooling by the cooling coils, (4) creating a dew point of 45 F to prevent condensation at the cooling coils, within the HVAC unit, and within the building itself, and (5) providing fresh pre-conditioned air for the interior space. Most school facilities utilize packaged HVAC equipment designed for inexpensive, efficient cooling. This type of equipment is not designed to handle the continuous supply of outdoor air necessary to comply with ASHRAE-62, Ventilation for Acceptable IAQ. As a result, these schools are likely to experience IAQ problems (Fischer 1996; Larrañaga, Beruvides et al. 2008).

The desiccant setup in Fig. 1 is depicted in the flow diagram in Fig. 2. This unit is designed to provide the school with 100% outdoor make-up air while maintaining the indoor dew point below 45° F (7° C.). The use of 100% fresh make-up air helps maintain the building at positive pressure to minimize unplanned moisture infiltrations into the building. The desiccant wheel was constantly rotating and adsorbing moisture from the air stream on the process side, while moisture was removed from the wheel on the regeneration side. This provided a constant adsorption medium with no phase change. Heat is necessary to release the moisture from the desiccant wheel, which results in heating of the airstream and an energy penalty. However, the use of active desiccation saves energy costs by: 1) providing an enhanced occupant comfort at a lower cost, 2) improved humidity control resulting in sensible versus latent cooling, 3) equipment expenditures by allowing the downsizing of the evaporator coil and condensing units for comparable design loads, 4) allowing independent temperature and humidity controls, and 5) allowing higher temperature set points (Meckler 1994).

Figure 1. Desiccant unit installed in a south Texas school to provide pre-conditioned fresh air for ventilation.

In units used for 100% outside air for ventilation (Fig. 2), or continuous ventilation, the outside air (process) is first filtered, cooled by the pre-cooling coil, dehumidified by the desiccant wheel, cooled again by the post-cooling coil, and delivered to the building's interior. This configuration does not allow for air within the building to be re-circulated, providing continuous fresh air to the system. Although the 100% make-up air configuration conditions more air, using more energy, than the typical desiccant setup depicted in Fig. 3, it is preferred in very hot and humid climates for the protection of the building and building systems from moisture infiltration, condensation, and latent heat loads of occupants.

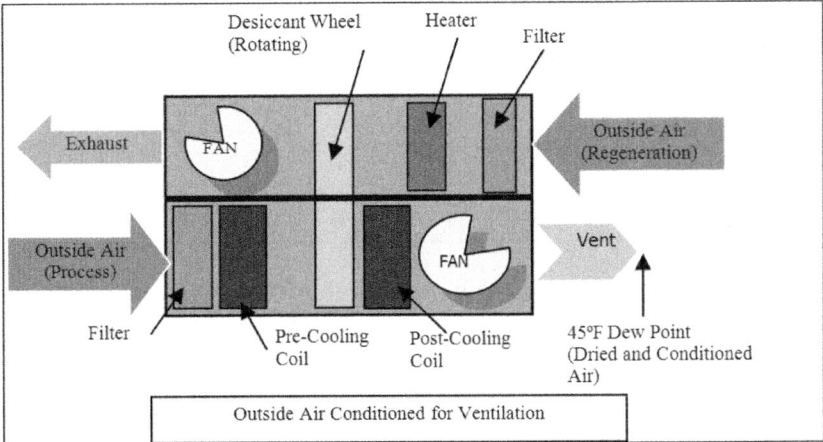

Figure 2. Depiction of desiccant unit for treatment of 100% outside air with no treatment of return air from the building. This setup is typical of large commercial buildings or schools.

MATERIALS AND METHODS

There are five typical configurations for desiccant dehumidifiers: liquid spray tower, solid packed tower, rotating horizontal bed, multiple vertical beds, and rotating honeycomb (Harriman III 2002c). The most typical method of presenting solid desiccants to a high volume air stream is to impregnate the material into a lightweight honeycomb-shaped matrix that is formed into a wheel (Harriman III, Witte et al. 1999).

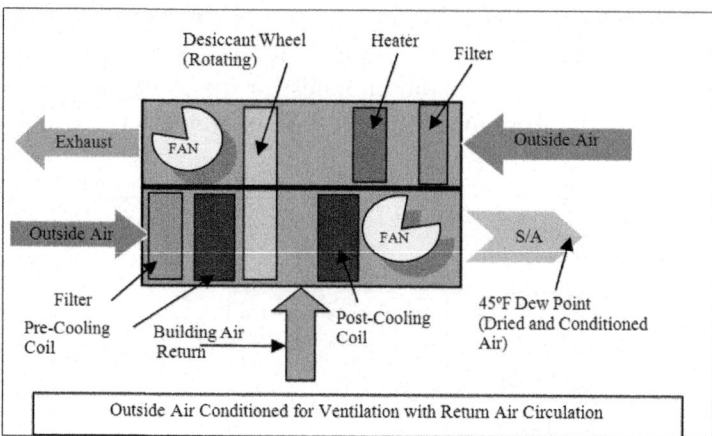

Outside Air Conditioned for Ventilation with Return Air Circulation

Figure 3. Typical desiccant setup with return air from the building treated by the Post-Cooling coil. This setup is typical of small commercial buildings, schools, and residential buildings.

The rotating honeycomb wheel is a finely divided desiccant impregnated into a semi-ceramic structure, maximizing the surface area of the desiccant material. The appearance of the honeycomb wheel resembles corrugated cardboard that has been rolled up into the shape of a wheel. The air passes through the flutes formed by the corrugations, and the wheel rotates through the process and reactivation airstreams. The flutes served as individual desiccant-lined air ducts, which maximizes the surface area of the desiccant presented to the air stream. The rotating honeycomb wheel design has several advantages. The structure is lightweight and very porous. Different types of desiccants can be arranged into a honeycomb wheel configuration for different applications. The design allowed for laminar flow within the individual flutes, reducing air pressure resistance compared to packed beds. This allowed the honeycomb wheel to operate efficiently for low dew point and high capacity applications. The honeycomb wheels are very light, and their rotating mass is very low compared to their high moisture removal capacity. The result is an energy efficient unit (Harriman III 2002c). The design is simple, reliable, and easy to maintain. The design is the most widely installed of all desiccant dehumidifiers in ambient pressure applications (Harriman III 2002c). Additionally, the honeycomb design is the most appropriate dehumidifier configuration for air-conditioning applications (Pesaran 1994). A working desiccant honeycomb unit with environmental chambers at the outlets of both the regeneration side and process side of the unit was utilized to determine removal efficiencies. See Fig. 4. The desiccant unit was modified with an additional heater (factory installed by Munters) and temperature controller (Omega CN132) to allow the operating temperature of the regeneration cycle (cycle that removes moisture from the wheel) to be varied between 100° (38° C) and 360° F (182° C). This allowed testing of the wheel at various increments of temperature. Spores were introduced using a 6-jet Collison nebulizer, a world standard for aerobiology research, that allowed for the aerosolization of fungal conidia for the purposes of testing the removal capabilities of the unit. The Collison nebulizer was operated at 40 psig, which generated a liquid generation rate of 16.5 ml/hr, allowing the experimenter to introduce a known concentration of fungi per volume of water as an aerosol into the test mechanism. For example, a fungal concentration of 1.25×10^5 spores/ml provided an aerosol generation rate of 3.44×10^4 spores per minute, allowing the experimenter to generate a known and constant airborne fungal concentration. Pressure was applied to the nebulizer with a Husky (model DK710700AV) Air Compressor.

The air was filtered with a HEPA Capsule prior to entering the Collison nebulizer. All microorganisms were concentrated, cultured, and speciated into a liquid solution. Lyophilization, a freeze-drying method under a vacuum, was used to concentrate spores (Virtis Freezemobile 6, Gardiner, NY). The organisms were then re-suspended in a known quantity of sterile water, and the concentration of spores determined through direct microscopic examination using a hemocytometer. The concentration was then manipulated by adding more water or spores to achieve the desired concentrations. The resulting fungal spore concentrates were frozen at -80 C.

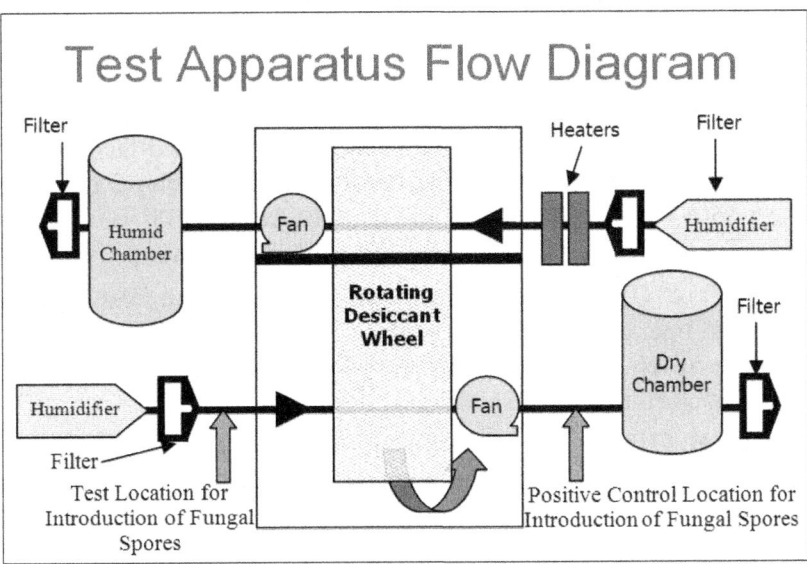

Figure 4. Munters Dew-150 Desiccant Wheel Test Setup

The test apparatus consisted of the Munters Dew 150 Desiccant Dehumidifier, four DeLonghi Dap-130 air purifiers with true high efficiency particulate air filters capable of capturing 99.97% of particles as small as 0.3 microns, and two environmental chambers (See Fig. 4). The environmental chambers were made of two 5-foot pieces of 24" agricultural high-pressure water pipe sealed at both ends with Plexiglas. Access doors were cut into each chamber, shelves installed, and a four plug-115V electrical outlet was installed to provide power for the sampling instruments. Each environmental chamber was sprayed twice with a Staticide ESD Clear Permanent Static Dissipative Coating to minimize the potential for electrostatic interference during the experiment. The environmental chambers were connected through metal duct to both the regeneration outlet (humid) and process outlet (dry) to permit the air exiting the desiccant unit to be sampled. All air supplied to the desiccant unit was pre-filtered to remove microorganisms and other particulates from the air for quality control purposes. The air was again filtered upon leaving the environmental chamber so that microorganisms were not unnecessarily introduced into the laboratory. The air filters were modified to accept ducting so that the air could be supplied directly to the desiccant unit and filters and vice versa. Filters were connected to the apparatus via 5" flex duct and are maintained under slightly positive pressure. Air supplied to the process air inlet was humidified to 60% relative humidity using a Holmes Cool Mist Humidifier (model HM3655) to simulate a humid environment.

Air sampling within the environmental chambers was conducted using Andersen Microbial Impactors and Allergenco MK-3 Microbial Air samplers. Samples were analyzed by microscopic analysis for total count or visual count readings for culturable samples. All culturable samples were cultured with potato dextrose agar. Data were analyzed using a two-factor analysis of variance with equal replication. The treatments were temperatures ranging from 60°-360° F (16° -182° C) at intervals of 100° F (38° C). Temperature settings were determined based on the following information:

1. 60° F (16° C) –Setting the desiccant unit at this temperature allows the unit to operate without heat and isolates the desiccant wheel from the effect of temperature. At this temperature, the desiccant wheel can be saturated so that no desiccation is occurring.

2. 160° F (71° C) –Units utilizing low cost surplus heat generated from indirect sources such as steam boilers, engine-cooling jackets, refrigeration condensers, exhaust air, steam condensate, water heaters, etc., may operate at 160° F or less. The regeneration air must be either hotter or dryer than the process air (Harriman III 1999 p.28), allowing the desiccant unit to operate at lower temperatures. Using indirect

energy sources can accomplish this task without the energy penalty of heating the regeneration air. These units are not suitable for low dew-point applications.

3.   260° F (127° C) –This temperature was chosen within the range necessary for low dew point applications, which may require reactivation temperatures as high as 250° F (121° C) to 275° F (135° C) (Harriman III, Humidity Control Design Guide, p. 211).

4.   360° F (182° C) –Setting the desiccant unit at this temperature allowed the unit to maximize the heat applied to the regeneration air stream and isolated effects due to temperature. At this temperature, the desiccant wheel could not be saturated by humid air and desiccation was maximized. Commercial desiccant units typically operate between 180° F (82° C) and 225° F (107° C) (Harriman III 1999 p. 29) and do not typically operate at this temperature.

To conduct an ANOVA for data generated by the two instruments (Allergenco™ and Andersen), a total of [(10 replications x 4 temperature settings x 3 tests (P, T1, T2) x 2 instruments) + (2 negative control samples x 2 chambers x 4 temperature settings x 2 instruments)] = a minimum of 272 samples must be taken per organism to complete each experiment. To test the three microorganisms *Aspergillus niger, Cladosporium cladosporioides*, and *Penicillium chrysogenum*, 816 samples must be taken, analyzed, and interpreted. Many simulation studies suggest that, generally, the central limit theorem holds for n>30 (Ott 1993), allowing the use of normal distributions in the analysis of data. If these data were shown to be non-normal and did not meet the assumption of normality via transformation, a non-parametric two-factor ANOVA was to be used. Non-normal data obtained from a two-factor experimental design should be analyzed non-parametrically by an extension of the Kruskal-Wallis test for single-factor analysis (Zar 1974), allowing the research hypotheses to remain unchanged.

Hypotheses

The general hypothesis stated that the active desiccant wheel removes statistically significant concentrations of airborne IAQ-related bioaerosols from the air supplied to a building. The sampling mechanism dictated whether results are expressed as viable or total number of spores per cubic meter of air. Viable particulate samplers are used to collect and assay airborne concentrations of aerobic species of culturable bacteria and fungi. All inertial impactors that use solid media produce data in colony forming units (CFU) (Ness 1991). A CFU is a viable fungal spore or bacterium capable of producing a mass of organisms, or colony that originates from a single cell or spore.

Data generated from the Andersen Impactor are expressed in CFU/m^3. Table 1 summarizes the testable hypothesis that this research will investigate from the data generated using the Andersen Impactor. Intertial slit impactors like the Allergenco MK-3 Microbial Air Sampler collect total concentrations of both viable and non-viable bioaerosols from the air. No distinction can be made between viable and non-viable airborne concentrations using slit impactors. The data generated by the Allergenco MK-3 Microbial Air Sampler are expressed as a total concentration of $spores/m^3$. Table 1 summarizes the testable hypothesis that this research will investigate from the data generated using the Allergenco MK-3 Microbial Air Sampler.

RESULTS AND ANALYSES

A total of 816 samples were taken to complete the experiment. Tests for normality showed that several of the data sets did not meet the assumption of normality and several data sets contained outliers. When concerned over the normality effect and outliers, a separate ANOVA should be performed on both the original data and the ranks; should the two procedures differ, use of the rank transformation ANOVA is preferred because it is less likely to be distorted by non-normality and unusual observations (Montgomery 2001).

Table 1. Summary of Testable Hypotheses for Data Generated by the Andersen Impactor.

Testable Hypothesis for Data Generated by the Andersen Impactor		
Hypothesis	**Test Statistic**	Variables (airborne concentrations in colony forming units/m^3)
Hypothesis 1: No interaction between temperature and viable airborne fungal concentrations.	Interaction	Temperature is the temperature settings for each test. μ_p is the mean of the airborne concentrations of the positive control. μ_{T1} is the mean of the airborne concentrations of the dry-chamber test.
Hypothesis 2: No difference on the airborne concentration of viable spores due to temperature.	H1: $\mu T1 = \mu P$ Ha: $\mu T1 \neq \mu P$	μ_p is the mean of the airborne concentrations of the positive control. μ_{T1} is the mean of the airborne concentrations of the dry-chamber test.

| Hypothesis 4: No difference on the airborne concentration of viable spores due to exposure to the desiccant wheel. | H1: $\mu T1 = \mu P$
Ha: $\mu T1 \neq \mu P$ | μ_p is the mean of the airborne concentrations of the positive control.
μ_{T1} is the mean of the airborne concentrations of the dry-chamber test. |

Table 2. Summary of Testable Hypotheses for Data Generated by the Allergenco MK-3 Microbial Air Sampler.

Testable Hypothesis for Data Generated by the Allergenco MK-3 Microbial Air Sampler		
Hypothesis	**Test Statistic**	**Variables** (airborne concentrations in spores/m^3)
Hypothesis 1: No interaction between temperature and total airborne fungal concentrations.	Interaction	Temperature is the temperature settings for each test. μ_p is the mean of the airborne concentrations of the positive control. μ_{T1} is the mean of the airborne concentrations of the dry-chamber test.
Hypothesis 3: No difference on the airborne concentration of total spores due to temperature.	H1: $\mu T1 = \mu P$ Ha: $\mu T1 \neq \mu P$	μ_p is the mean of the airborne concentrations of the positive control. μ_{T1} is the mean of the airborne concentrations of the dry-chamber test.
Hypothesis 5: No difference on the airborne concentration of total spores due to exposure to the desiccant wheel.	H1: $\mu T1 = \mu P$ Ha: $\mu T1 \neq \mu P$	μ_p is the mean of the airborne concentrations of the positive control. μ_{T1} is the mean of the airborne concentrations of the dry-chamber test.

This experimental analysis was completed using a two-factor ANOVA for both the original and the ranked data. The results of the two analysis differed in several instances, specifically in differences in temperature and groups (control versus test).

Andersen Impactor

Tests for normality were conducted to help determine protocols necessary for data analysis. The Kolmogorov-Smirnov (K-S) goodness of fit test was employed because it is particularly useful when sample sizes are small and when no parameters have been estimated. A large p-value indicates a good fit, while a small p-value indicates a poor fit (Banks 2001). Nine of the 12 *Aspergillus niger* data sets failed to meet the assumption of normality. Eight of

12 *Cladosporium cladosporioides*-Andersen Impactor and 4 of 12 *Penicillium chrysogenum*-Andersen Impactor data sets did not meet the assumption of normality. One two-factor ANOVA per organism tested was conducted for the data generated using the Andersen Impactor. The ANOVAs for each organism are described below.

Aspergillus Niger-Andersen Impactor Data Analysis

Post hoc analysis using the Student-Newman-Keuls (SNK) test for rank value showed that measured airborne concentrations were not significantly different at any of the four temperature settings. The SNK grouping for the positive control (μ_p) vs. dry chamber test (μ_{T1}) vs. humid chamber test (μ_{T2}) showed that the controls were significantly different from the dry chamber and humid chamber results at each temperature setting. The power of the *Aspergillus niger*-Andersen Impactor analysis for control vs. dry and humid chamber results is greater than 0.99.

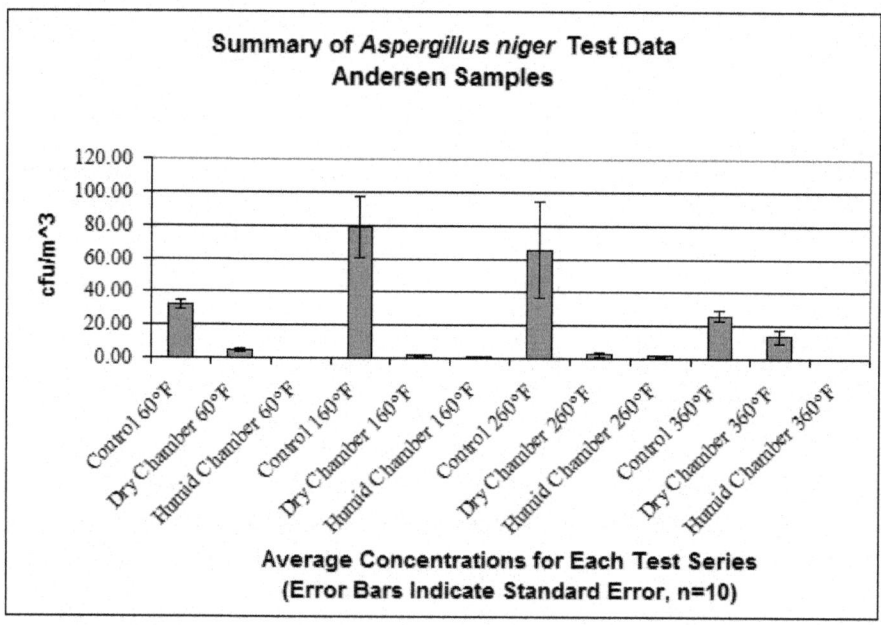

Figure 5. Average airborne concentrations for *Aspergillus niger*-Andersen Impactor tests.

The results showed that there was interaction between that humid chamber and dry chamber data sets, there was no difference in the ranks due to temperature, and there was a significant difference between the mean ranks of the positive control (μ_p) and the mean ranks of the dry chamber tests (μ_{T1}) and humid chamber tests (μ_{T2}). Interaction occurred at the 160° F (71° C) and 260° F (127° C) between the dry chamber data ranks and the humid chamber data ranks. See Fig. 6. The mean removal efficiencies are shown in Fig. 7 and can be described by the equation '$y = -0.0561x^3 + 0.2612x^2 - 0.2687x + 0.9153$' with a correlation coefficient (R^2) of one.

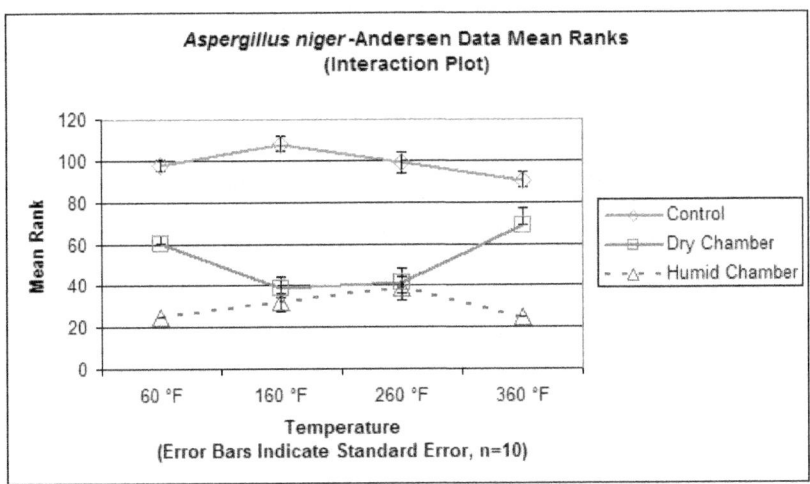

Figure 6. Interaction Plot for the *Aspergillus niger*-Andersen Data Mean Ranks.

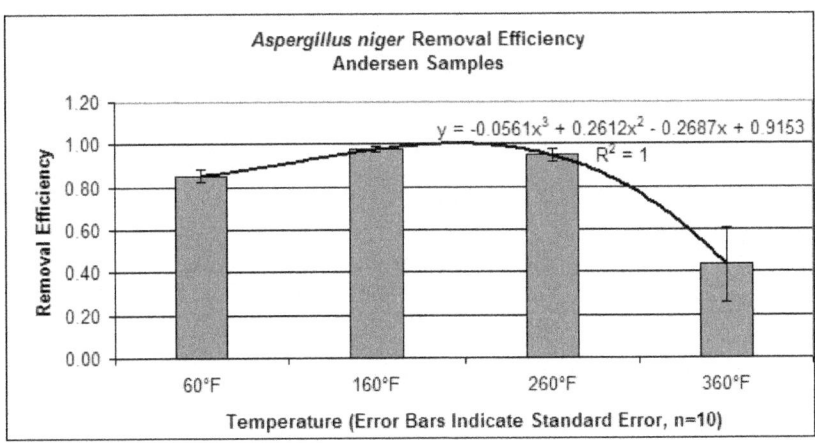

Figure 7. *Aspergillus niger* Removal Efficiencies for Andersen Samples.

Cladosporium Cladosporioides-Andersen Impactor Data Analysis

Post hoc analysis using the Student-Newman-Keuls (SNK) test for rank value showed that measured airborne concentrations were not significantly different at either of the four temperature settings. The SNK grouping for the positive control (μ_p) vs. dry chamber test (μ_{T_1}) vs. humid chamber test (μ_{T_2}) showed that the controls were significantly different from the dry chamber and humid chamber results at each temperature setting. The power of the *Cladosporium*-Andersen Impactor analysis for control vs. dry and humid chamber results is greater than 0.99. Fig. 8 summarizes the *Cladosporium cladosporioides*-Andersen Impactor data sets. These data show the mean airborne concentrations of the positive control versus the dry and humid chamber tests at each temperature setting. The results showed that there was no interaction, there was no difference in the mean ranks due to temperature, and there was a significant difference between the mean ranks of the positive control (μ_p) and the mean ranks of the dry chamber test (μ_{T_1}) and humid chamber results (μ_{T_2}). The mean removal efficiencies are shown in Fig. 9 and can be described by the equation '$y = -0.0053x^3 + 0.0475x^2 - 0.3278x + 1.18$' with a correlation coefficient (R^2) of one.

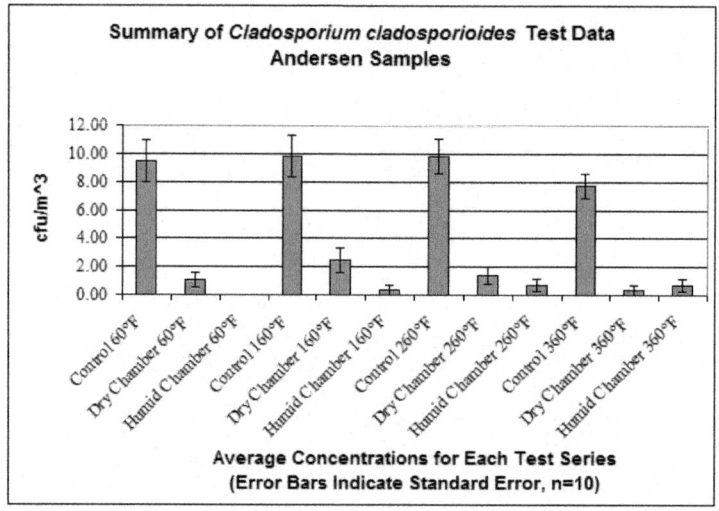

Figure 8. Average airborne concentrations for *Cladosporium cladosporioides*-Andersen Impactor tests.

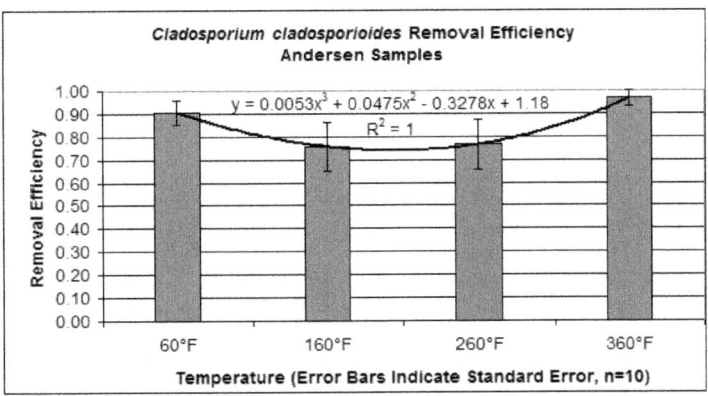

Figure 9. *Cladosporium* removal efficiencies for Andersen samples

Penicillium Chrysogenum-Andersen Impactor Data Analysis

Post hoc analysis using the Student-Newman-Keuls (SNK) test for rank value showed that measured airborne concentrations were not significantly different at any of the four temperature settings. The SNK grouping for the positive control (μ_p) vs. dry chamber test (μ_{T1}) vs. humid chamber test (μ_{T2}) showed that the controls were significantly different from the dry chamber and humid chamber results at each temperature setting. The power of the *Penicillium chrysogenum*-Andersen Impactor analysis for control vs. dry and humid chamber results is greater than 0.99. Fig. 10 summarizes the *Penicillium chrysogenum*-Andersen Impactor data sets. These data show the mean airborne concentrations of the positive control versus the dry and humid chamber tests at each temperature setting. The results showed that there was no interaction, there was no difference in the ranks due to temperature, and there was a significant difference between the mean ranks of the positive control (μ_p) and the mean ranks of the dry chamber test (μ_{T1}) and humid chamber tests (μ_{T2}). The mean removal efficiencies are shown in Fig. 11 and can be described by the equation 'y = -0.0728x³ + 0.5584x² – 1.3065x + 1.5574' with a correlation coefficient (R^2) of one.

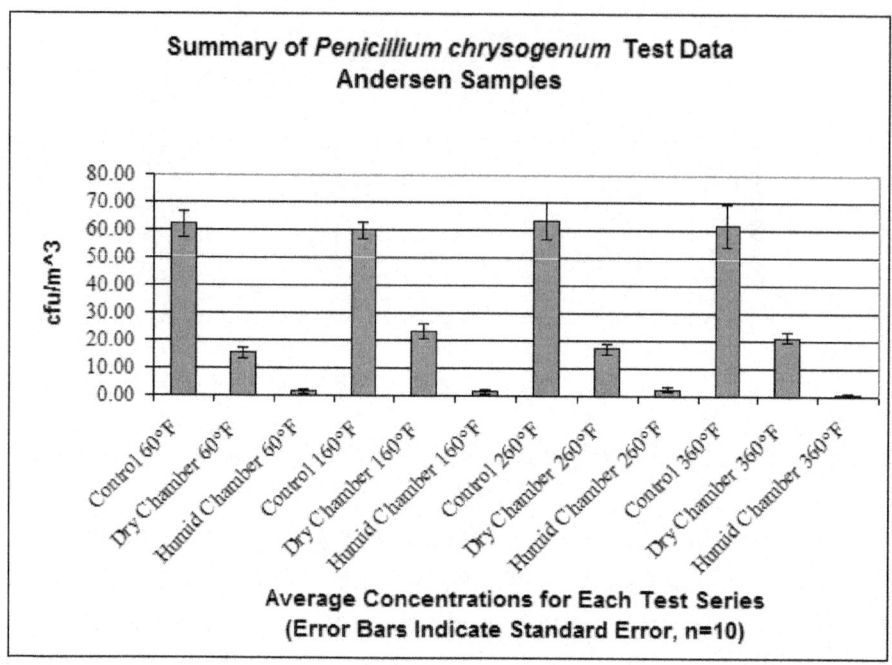

Figure 10. Average airborne concentrations for *Penicillium chrysogenum* Andersen Impactor tests.

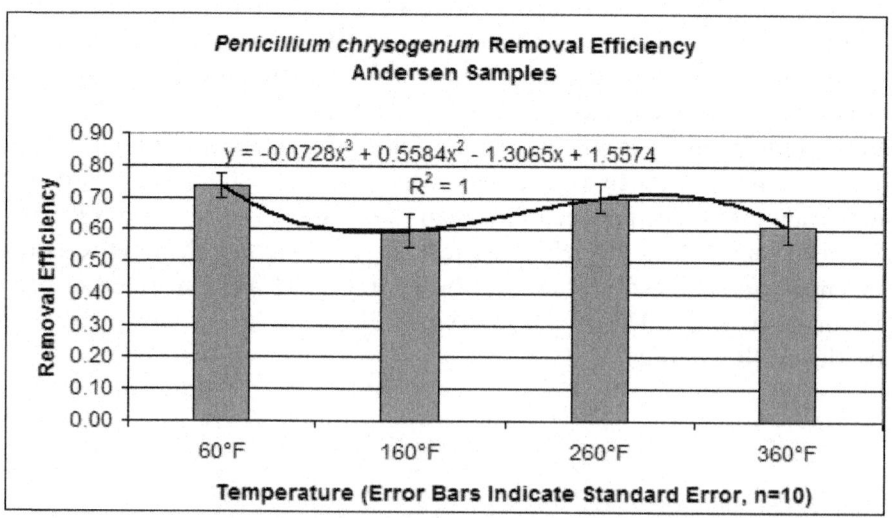

Figure 11. *Penicillium chrysogenum* Removal Efficiencies for Andersen Samples.

Combined Removal Efficiencies for Andersen Impactor Data

The combined removal efficiencies of the *Aspergillus niger, Cladosporium cladosporioides, and Penicillium chrysogenum* data sets are summarized in Fig. 12.

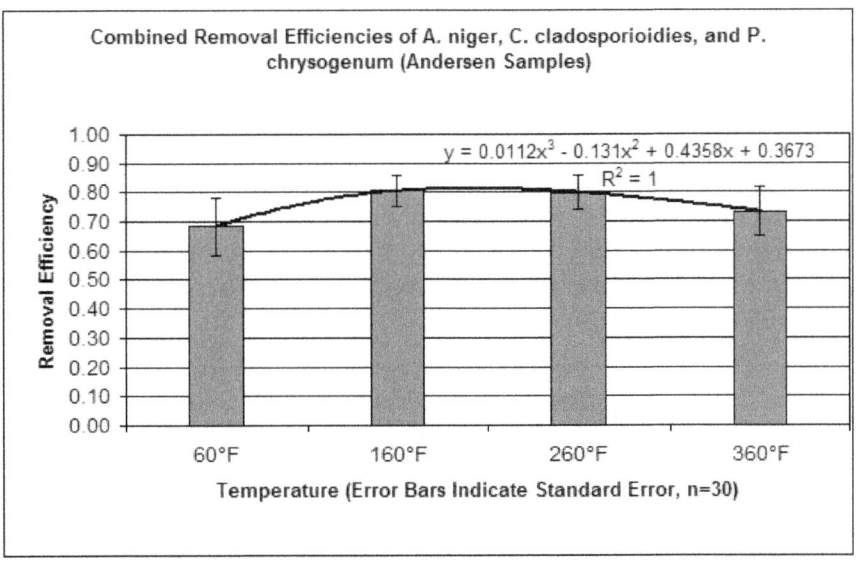

Figure 12. Combined removal efficiencies of the *Aspergillus niger, Cladosporium cladosporioides, and Penicillium chrysogenum* –Andersen Impactor data sets.

Allergenco MK-III Microbial Air Sampler

Tests for normality were conducted to help determine protocols necessary for data analysis. The K-S goodness of fit test was employed because it is particularly useful when sample sizes are small and when no parameters have been estimated. A large p-value indicates a good fit, while a small p-value indicates a poor fit (Banks 2001). Seven of the 12 *Aspergillus niger* data sets failed to meet the assumption of normality. Six of 12 *Cladosporium cladosporiodies*-Allergenco Microbial Air Sampler and 6 of 12 *Penicillium chrysogenum*- Allergenco Microbial Air Sampler data sets did not meet the assumption of normality. One two-factor ANOVA per organism tested was conducted for the data generated using the Allergenco Microbial Air Sampler. The ANOVAs for each organism are described below.

Aspergillus Niger-Allergenco Microbial Air Sampler Data Analysis

Post hoc analysis using the SNK test for rank value showed that measured airborne concentrations were not significantly different at 160° F (71° C) and 260° F (127° C) or at 60° F (16° C) and 360° F (182° C). The two temperature groupings (160° F (71° C) and 260° F (127° C), 60 F (16° C) and 360° F (182° C)) were also statistically different. The SNK grouping for the positive control (μ_p) vs. dry chamber test (μ_{T1}) vs. humid chamber test (μ_{T2}) showed that the controls were significantly different from the dry chamber and humid chamber results at each temperature setting. The power of the *Aspergillus niger*-Andersen Impactor analysis for a temperature effect is 0.825, while the power of the analysis for control vs. dry and humid chamber results is greater than 0.99. Fig. 13 summarizes the *Aspergillus niger*-Allergenco Microbial Air Sampler data sets.

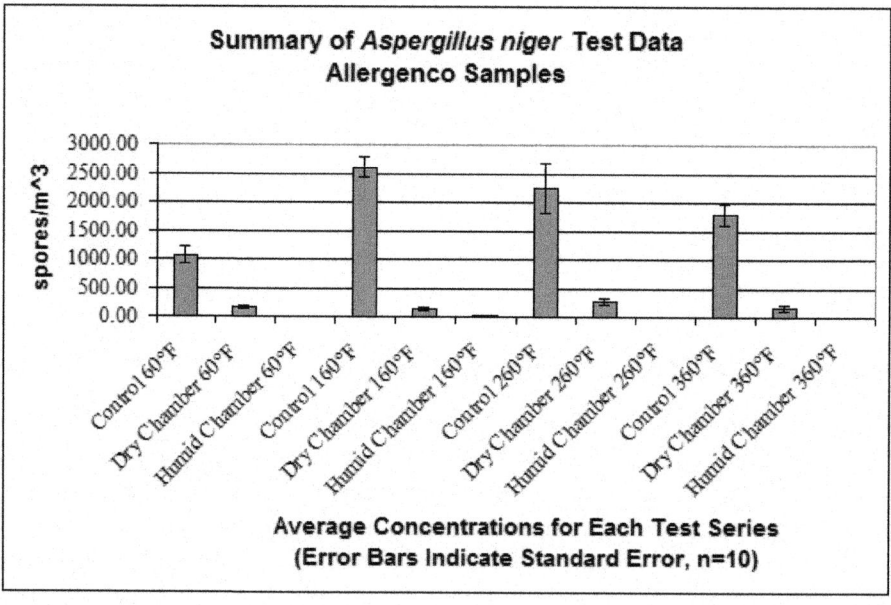

Figure 13. Average airborne concentrations for *Aspergillus niger*-Allergenco Microbial Air Sampler tests.

These data show the mean airborne concentrations of the positive control versus the dry and humid chamber tests at each temperature setting. The results showed that there was interaction between that humid chamber and dry chamber data sets, there were differences in the ranks due to temperature, and there was a significant difference between the mean ranks of the positive control (μ_p) and the mean ranks of the dry chamber test (μ_{T1}). The interaction occurs at 160° F (71° C) between the dry and humid chamber mean ranks. The mean removal efficiencies can be described by the equation 'y = 0.0662x³ - 0.5048 x² + 1.1476x + 0.1393' with a correlation coefficient (R²) of one. See Fig.s 4.10 AND 4.11.

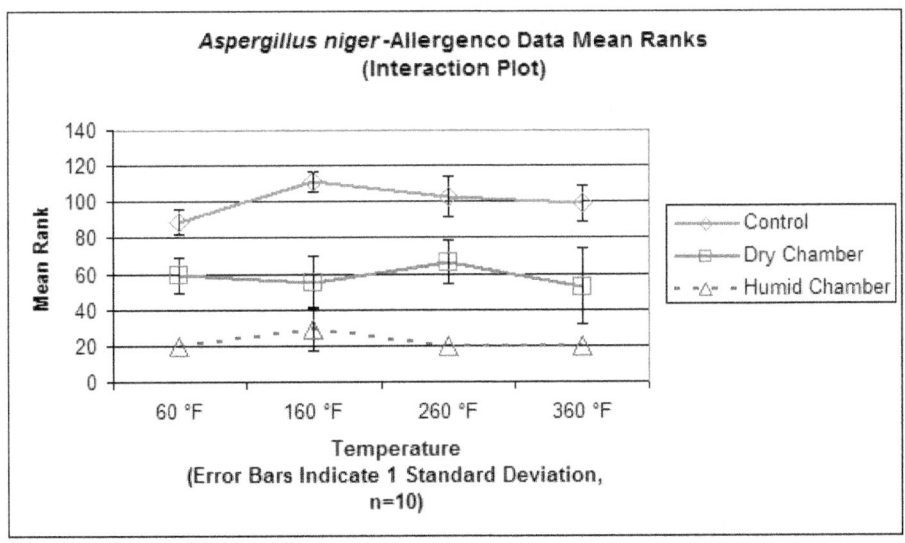

Figure 14. Interaction Plot for the *Aspergillus niger*-Allergenco Microbial Air Sampler Data Mean Ranks.

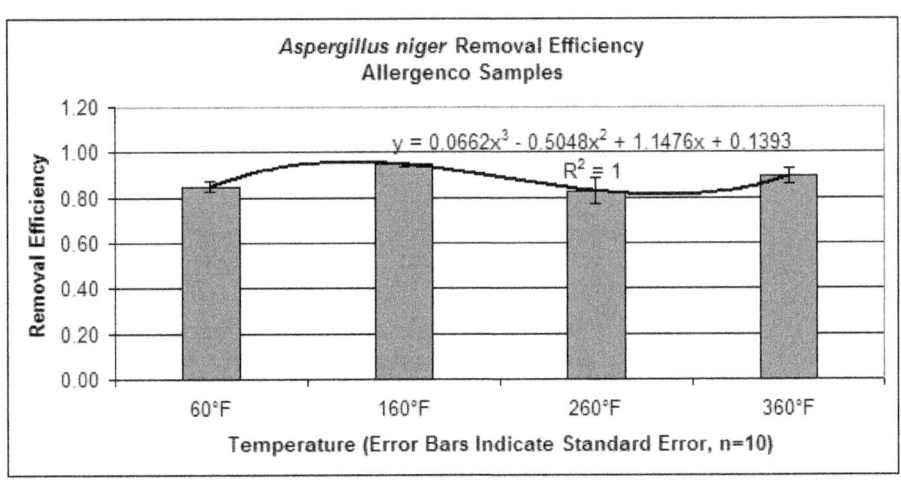

Figure 15. *Aspergillus niger* Removal Efficiencies for Allergenco samples.

Cladosporium Cladosporioides-Allergenco

Post hoc analysis using the SNK test for rank value showed that measured airborne concentrations were not significantly different. The SNK grouping for the positive control (μ_p) vs. dry chamber test (μ_{T1}) vs. humid chamber test (μ_{T2}) showed that the controls were significantly different from the dry chamber and humid chamber results at each temperature setting. The power of the *Cladosporium cladosporioides*-Allergenco Microbial Air Sampler analysis for control vs. dry and humid chamber results is greater than 0.99. Fig. 16 summarizes the *Cladosporium cladosporioides*-Allergenco Microbial Air Sampler data sets. Post hoc analysis using the SNK test for rank value showed that measured airborne concentrations were not significantly different at the four temperatures. The results showed that there was interaction between the dry chamber and humid chamber data sets, there was no difference in the ranks due to temperature, and there was a significant difference between the mean ranks of the positive control (μ_p) and the mean ranks of the dry (μ_{T1}) and humid chamber tests (μ_{T1}). The mean removal efficiencies are shown in Fig. 17 and is described by the equation '$y = -0.033x^3 + 0.22x^2 - 0.4149x + 0.9267$' with a correlation coefficient (R^2) of one.

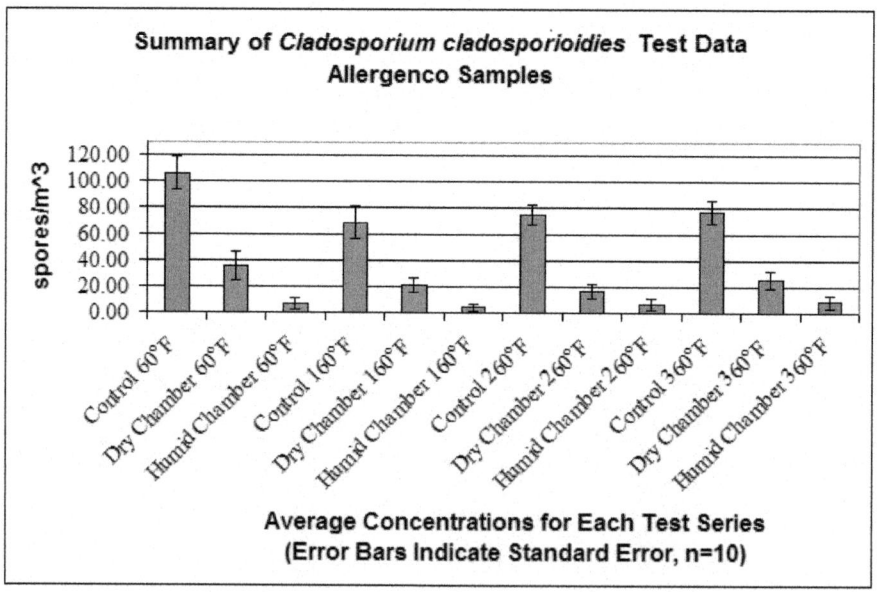

Figure 16. Average airborne concentrations for *Cladosporium cladosporioides*–Allergenco Microbial Air Sampler tests.

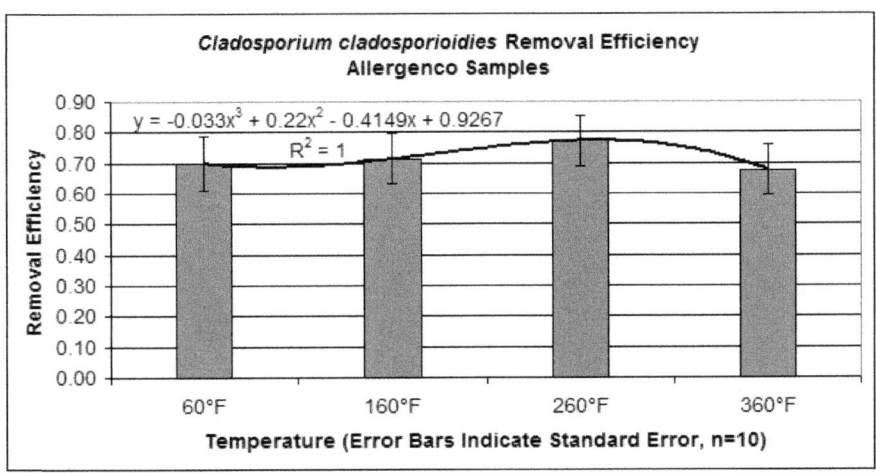

Figure 17. *Cladosporium cladosporioides* Removal Efficiencies for Allergenco Samples.

Penicillium Chrysogenum-Allergenco Microbial Air Sampler Data Analysis

Post hoc analysis using the SNK test for rank value showed that measured airborne concentrations were significantly different at 360° F (182° C). The three remaining temperatures (60° F (16° C), 160° F (71° C), and 260° F (126° C)) were not statistically different. The SNK grouping for the positive control (μ_p) vs. dry chamber test (μ_{T1}) vs. humid chamber test (μ_{T2}) showed that the controls were significantly different from the dry chamber and humid chamber results at each temperature setting. The power of the *Penicillium chrysogenum*-Andersen Impactor analysis for a temperature effect was 0.55, while the power of the analysis for control vs. dry and humid chamber results was greater than 0.99. Fig. 18 summarizes the *Penicillium chrysogenum*-Allergenco Microbial Air Sampler data sets. The results showed that there was no interaction between that humid chamber and dry chamber data sets, there was a difference in the ranks due to temperature, and there was a significant difference between the mean ranks of the positive control (μ_p) and the mean ranks of the dry chamber test (μ_{T1}). The mean removal efficiencies are shown in Fig. 19 and can be described by the equation '$y = 0.0844x^3 - 0.7017x^2 + 1.9039x - 0.9933$' with a correlation coefficient (R^2) of one.

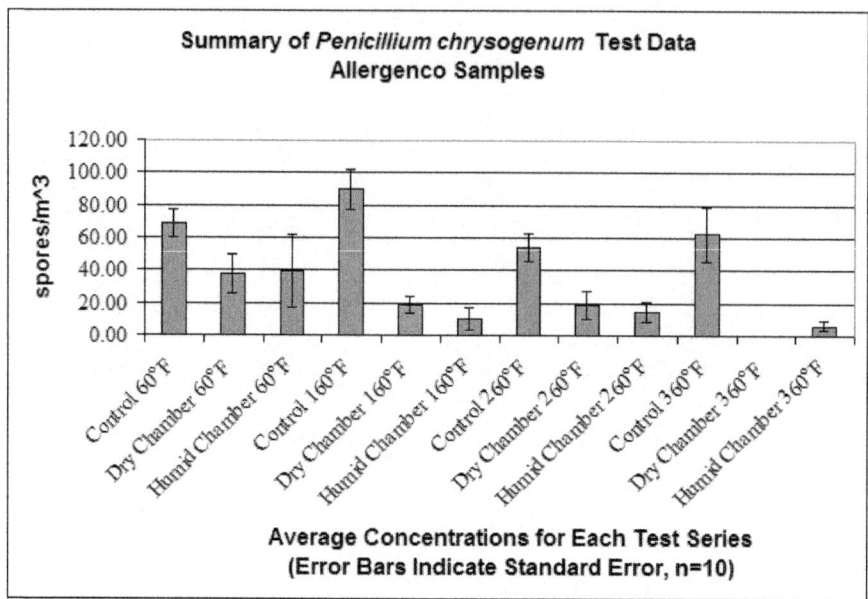

Figure 18. Average airborne concentrations for *Penicillium chrysogenum*–Allergenco Microbial Air Sampler tests.

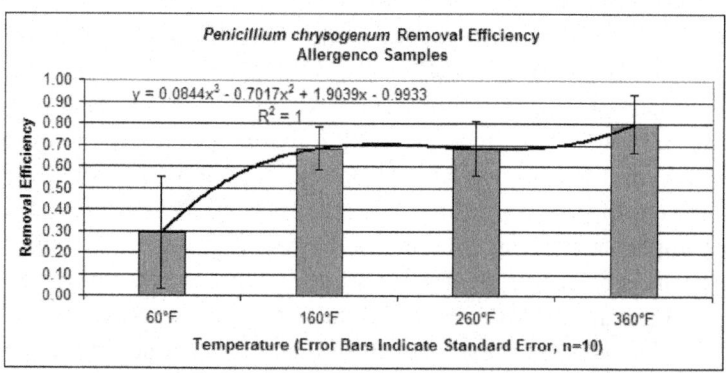

Figure 19. *Penicillium chrysogenum* Removal Efficiencies for Allergenco Samples.

Combined Removal Efficiencies for Andersen Impactor Data

The combined removal efficiencies of *Aspergillus niger, Cladosporium cladosporioides,* and *Penicillium chrysogenum* are summarized in Fig. 20 and can be explained by the polynomial regression equation '$y = -0.0132x^3 + 0.0912x^3 - 0.1913x + 0.8744$' with a correlation coefficient of one.

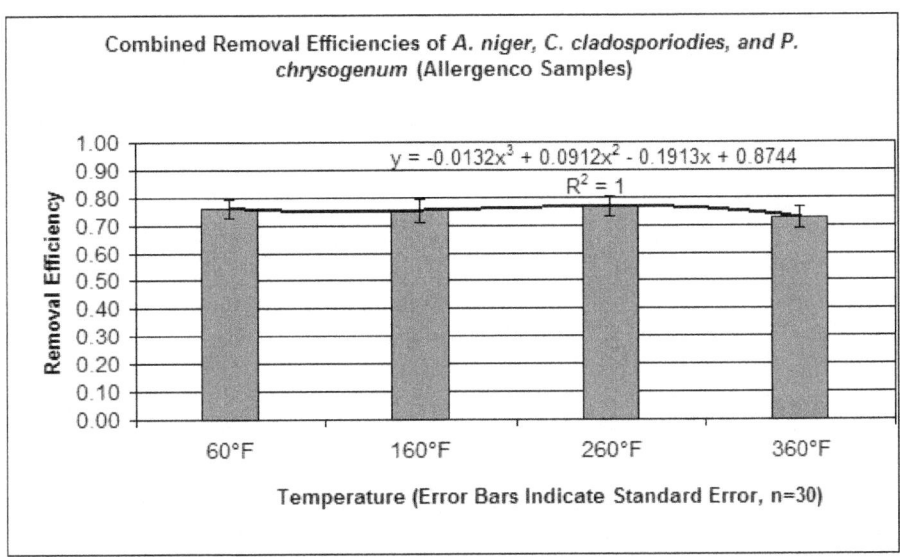

Figure 20. Combined removal efficiencies of the *Aspergillus niger, Cladosporium cladosporioides, and Penicillium chrysogenum*–Allergenco Microbial Air Sampler data sets.

SUMMARY OF FINDINGS

The overall results of the hypothesis testing are summarized in Table 3. In all cases, the positive control (μ_p) was significantly greater than the dry chamber (μ_{T1}) results. Differences in temperature were found with the *Aspergillus niger* and *Penicillium chrysogenum* Allergenco data sets. Interaction was found within the *Aspergillus niger* data sets for both the Andersen Impactor and Allergenco Microbial Air Sampler.

Interaction of the *Aspergillus niger* data sets occurred in both cases between the mean ranks of the dry and humid chambers (μ_{T1} and μ_{T2}). Significant interaction did not occur between the mean ranks of the positive control and the mean ranks of the dry chamber test ($\mu_{P>}\mu_{T1}$), which were the parameters of interest when determining removal efficiencies and airborne concentrations delivered indoors. Differences due to temperature were identified within the *Aspergillus niger*-Allergenco Microbial Air Sampler data sets. See Table 3 above for the SNK grouping. A difference in temperature was also identified by the *Penicillium chrysogenum*-Allergenco Microbial Air Sampler data analysis with a statistical power (P) is 0.55, which is less than the commonly accepted value of 0.8. This means that there is 55% confidence that the temperature effect exists and a 45% confidence that the temperature effect does not exist.

See the SNK grouping in Table 3 above. This indicates that the desiccant wheel may be more efficient at removing *Penicillium chrysogenum* spores from the airstream at 360° F (182° C). In both cases, the temperature difference occurred at one or both of the extremes of the temperature settings; the *Aspergillus niger*-Allergenco Microbial Air Sampler analysis showed that the total spore concentrations at 60° F (16° C) and 360° F (182° C) were different from those concentrations at 160° F (71° C) and 260° F (127° C), and the *Penicillium chrysogenum*-Allergenco Microbial Air Sampler analysis showed that the total spore concentrations at 360° F (182° C) were different from those concentrations found at 60 F (16° C), 160° F (71° C), and 260° F (127° C).

DISCUSSION

The purpose of this research was to determine if the desiccant wheel was effective at removing statistically significant concentrations of IAQ-related microorganisms from the air it supplies to a building. In addition, the capabilities of the desiccant wheel at removing airborne concentrations of IAQ-related microorganisms at four separate temperature settings were explored. The purpose of exploring the removal capabilities at different temperature settings was to establish the mechanism of spore removal and generate prediction models of airborne removal efficiencies for the two sampling methods and three organisms tested. In two of the six two-factor analyses of variance, a temperature effects were significant.

In one instance the *Aspergillus niger*-Allergenco Microbial Air Sampler analyses showed that the mean of the ranks at 60° F (16° C) and 360° F (182° C) (60° F (16° C) and 360° F (182° C) were not statistically different) were statistically different from the mean of the ranks at 160° F (71° C) and 260° F (127° C) (160° F (71° C) and 260° F (127° C) were not statistically different). Both the 60° F (16° C) and 360° F (182° C) temperature settings were designed as controls to help determine the mechanism of spore removal by the desiccant wheel. This lack of a temperature effect between the two control temperatures indicates that the mechanism of removal was likely a mechanical filtration effect unaffected by the increasing magnitude of adsorption created by an increase in the reactivation temperature. This premise was supported by five of the six analyses. In the second instance of a significant temperature difference, the *Penicillium chrysogenum*-Allergenco Microbial Air Sampler analyses showed that the mean rank of the airborne concentrations at 360° F (182° C) was significantly different from the mean ranks at 60° F (16° C), 160° F (71° C), and 260° F (127° °C), which were not statistically different ($P=0.55$). With a statistical power less than 0.8, however, the difference does not maintain the confidence necessary to validate the temperature effect.

Therefore, the mechanism of removal appeared to be a mechanical filtration of spores resulting in a decrease in the airborne concentrations of viable and total concentrations of *Aspergillus niger*, *Cladosporium cladosporioides*, and *Penicillium chrysogenum* through the process air stream of the desiccant unit. The desiccant unit removed significant airborne concentrations of both viable and total spores for the three organisms in each of the six analyses. The results of this study showed that the desiccant unit significantly reduces the airborne concentrations of these organisms introduced into the indoor environment.

CONCLUSION

This study aimed to quantify the removal capabilities of a rotary wheel (honeycomb) solid-desiccant dehumidifier at removing selected IAQ-related fungal organisms from the airstream. For each organism, the reductions in airborne concentrations delivered to the dry chamber were significant. These results support the findings of several authors who have stated that the use of active desiccant technology enhances the quality of the indoor air by helping to maintain comfort criteria (temperature, humidity and ventilation) (Meckler 1994; Kovak 1997; Fischer and Bayer 2003), removing particulates and bioaerosols from the air (Hines 1992a; Kovak 1997), and removing chemical pollutants from the air (Hines 1992c; Popescu and Ghosh 1999). This study demonstrates the ability of the desiccant unit to remove IAQ-related microorganisms from the air. In addition, the study shows the removal capabilities are significant at the four temperatures tested. The ability of active desiccants to remove particulates, bioaerosols, chemical pollutants, and water vapor from the airstream delivered to a building provides a unique opportunity to view active desiccant technology as a viable control strategy for enhancing and maintaining a favorable IAQ in cooling climates.

Mold and other factors related to damp conditions indoors are linked to increased asthma symptoms in asthmatics and coughing, wheezing, and upper respiratory tract symptoms in otherwise healthy people; and damp indoor conditions may be associated with the onset of asthma, as well as shortness of breath and lower respiratory illness in otherwise healthy children. The Institute of Medicine calls for studies that compare various ways to limit moisture or eliminate mold growth indoors and to evaluate whether interventions improve the health of occupants. This study provides a foundation for exploring the feasibility of integrating desiccation technologies into existing HVAC system design for cooling climates for improving the IAQ within the built environment.

REFERENCES

1.   D. M. Addington, 2000 The History and Future of Ventilation. Indoor Air

Quality Handbook. J. D. Spengler, Samet, J.M., McCarthy, J. New York, McGraw-Hill-Professional.

2. M. A. Andersson, M. Nikulin, U. Koljalg, M. C. Andersson, F. Rainey, F. Reijula, E. Hintikka, L. , M. Salkinoja-Salonen, 1997 "Bacteria, molds, and toxins in water-damaged building materials." Applied and Environmental Microbiology 63 387393 .

3. S. Andrae, O. Axelson, B. Bjorksten, M. Fredriksson, N. M. Kjellman, 1988 "Symptoms of bronchial hyperrreactivity and asthma in relation to environmental factors." Archives of Disease in Childhood 63 5 473478 .

4. M. O. Bachmann, J. E. Myers, 1995 "Influences on sick building syndrome symptoms in three buildings." Society, Science and Medicine. 40 245251 .

5. E. Bardana, Jr , A. Montanaro, et al. 1988 "Building related illness." Clinical Reviews in Allergy 6 1 6189 .

6. R. Bascom, 1997 Plenary paper: health and indoor air quality in schools- a spur to action or false alarm? Proc Healthy Buildings/IAQ' 97 American Society of Heating, Refrigerating and Air-conditioning Engineers, Bethesda, Md.

7. S. Batterman, H. Burge, 1995 "HVAC systems as emission sources affecting indoor air quality: A critical review." United States Environmental Protection Agency Research and Development(February): 139 .

8. C. W. Bayer, 2000 "Humidity control and ventilation in schools." ASHRAE Journal(Summer).

9. C. W. Bayer, S. A. Crow, 1992a Odorous volatile emissions from fungal contamination. Atlanta, Proc IAQ'92, American Society of Heating, Refrigerating and Air Conditioning Engineers: 99104 .

10. R. S. Bernstein, W. G. Sorenson, D. Garabant, C. Reaux, R. D. Treitman, 1983 "Exposure to respirable, airborne Penicillium from a contaminated ventilation system: Clinical environmental and epidemiological aspects." American Industrial Hygiene Association Journal 44 3 161169 .

11. H. S. Bernton, 1930 "Asthma due to a Mold- Aspergillus fumigatus." J.A.M.A 95 3 189192 .

12. B. Brunekreef, 1992 "Damp housing and adult respiratory symptoms." Allergy: European Journal of BrunekreefB. (1992). "Damp housing and adult respiratory symptoms." Allergy: European Journal of Allergy and Clinical Immunology. . and Clinical Immunology. 47 5 498502 .

13. B. Brunekreef, D. W. Dockery, F. E. Speizer, J. H. Ware, J. D. Spengler, B. G. Ferris, 1989 "Home dampness and respiratory morbidity in children."

American Review of Respiratory Disease 140 13631367 .

14. Bureau of National Affairs 1992 Indoor air- OSHA request for informatin for future regulation. Washington, D.C., Occupational Safety and Health Administration Report: 10971098 .

15. H. A. Burge, 1989 "Indoor air and infectious disease." Occupational Medicine 4 4 713721 .

16. N. F. Conant, H. C. Wagner, F. M. Rackemann, 1936 "Fungi Found in Pillows, Mattresses, and Furniture." The Journal of Allergy: 234237 .

17. W. Cookson, 1999 "The alliance of genes and environment in asthma and allergy." Nature 402: B5 -B11.

18. J. D. Cooley, W. C. Wong, C. A. Jumper, D. C. Straus, 1998 "Correlation between the prevalence of certain fungi and sick building syndrome." Occupational and Environmental Medicine 55 579584 .

19. B. A. Credille, 1933 "Report of a Case of Bronchial Asthma due to Molds." The Jouurnal of the Michigan State Medical Society 32(3): 167.

20. W. A. Croft, B. B. Jarvis, C. S. Yatawara, 1986 "Airborne outbreak of trichothecene toxicosis." Atmospheric Environment 20 3 549552 .

21. W. A. Croft, B. M. Jastromski, et al. 2002 "Clinical confirmation of trichothecene mycotoxicosis in patient urine." Journal of Environmental Biology 23 3 301320 .

22. R. Dales, H. Zwanenburg, R. Burnett, C. Franklin, 1991 "Respiratory health effects of home dampness and molds among Canadian children." American Journal of Epidemiology 134 2 196203 .

23. R. E. Dales, R. Burnett, H. Zwanenburg, 1991b "Adverse health effects among adults exposed to home dampness and molds." American Rev Respiratory Disease: 505509 .

24. B. S. Davanagere, D. B. Shirey, K. Rengarajan, F. Colacino, 1997 "Mitigating the impacts of ASHRAE Standard 621989 on Florida Schools." ASHRAE Trans: 241-258.

25. I. Dill, B. Niggeman, 1996 "Domestic fungal viable propagules and sensitization in children with IgE mediated allergic diseases." Pedriatric Allergy Immunology 7 151155 .

26. W. H. Dolan, 1989 "Desiccant Cooling Systems-A New HVAC Opportunity." Energy Engineering 86 4 69 .

27. Environmental Protection Agency 2010 "Managing asthma in the school environment."

28. EPA 2003 Managing asthma in the school environment. Washington, D.C., U.S. Environmental Protection Agency.

29. M. S. Finningan, C. A. C. Pickering, et al. 1984 "The sick building syndrome: prevalence studies." British Medical Journal 289 15731575 .

30. J. C. Fischer, 1996 Optimizing IAQ, humidity control, and energy efficiency in school environments through the application of desiccant-based total energy recovery systems. Proc IAQ'96, American Society of Heating, Refrigerating and Air Conditioning Engineers, Washington, D.C.

31. J. C. Fischer, C. W. Bayer, 2003 "Failing grade for most schools; report card on humidity control." ASHRAE Journal(May): 3039 .

32. W. J. Fisk, 2000 Estimates of potential nationwide productivity and health benefits from better indoor environments: an update. Indoor Air Quality Handbook. J. D. Spengler, Samet, J.M., McCarthy, J., Mcgraw-Hill.

33. B. Flannigan, E. M. Mc Gabe, F. Mc Garry, 1991 "Allergenic and toxigenic microrganisms in houses." Journal of Applied Bacteriology. 70: 61S-73S.

34. B. Flannigan, J. D. Miller, 1994 Health implications of fungi in indoor environments-an overview. Health Implications of Fungi in Indoor Environments, Elsevier. Air Quality Monographs 2 3-28.

35. C. A. Flood, 1931 "Observations on Sensitivity to Dust Fungi in Patients with Asthma." J.A.M.A 96 25 20942095 .

36. M. H. Garrett, P. R. Rayment, et al. 1998 "Indoor airborne fungal spores, house dampness and associations with environmental factors and respiratory health in children." Clinical and Environmental Allergy 28 459467 .

37. W. A. Gordon, E. Johanning, et al. 1999 Cognitive impairment associated with exposure to toxigenic fungi. Bioaerosols, Fungi, and Mycotoxins: Health Effects, Assessment, Prevention and Control. E. Johanning. Albany, NY, Eastern New York Occupational Health Program: 9498 .

38. M. Hagmann, 2000 "A mold's toxic legacy revisited." Science 288 243244 .

39. G. Handal, M. A. Leiner, et al. 2004 "Children symptoms before and after knowing about an indoor fungal contamination." Indoor Air 14 8791 .

40. I. I. I. L. G. Harriman, Ed, 2002c The Dehumidification Handbook Second Edition. Amesbury, MA, Munters Corporation.

41. I. I. I. L. G. Harriman, M. J. Witte, et al. 1999 "Evaluating active desiccant systems for ventilating commercial buildings." ASHRAE Journal(October 1999): 28 EOF -32.

42. L. Harriman, R. Kittler, 2001 Chapter 13 Dehumidifiers. Humidity

Control Design Guide for Commercial and Institutional Buildings. L. G. Harriman III, Brundrett, G.W., Kittler, R. Atlanta, American Society of Heating, Refrigerating and Air Conditioning Engineers: 195214 .

43. R. M. Heller, T. W. Heller, et al. 2003 ""tsara'at," Leviticus, and the history of confusion." Perspectives in Biology and Medicine 46 4 588591

44. A. L. Hines, T. K. Ghosh, 1992b Air Dehumidification and Removal of Indoor Air Pollutants by Liquid Desiccants: Investigation of Co-Sorption of Gases and Vapors as a Means to Enhance Indoor Air Quality- Phase II, Gas Research Institute GRI-92/0157.3.

45. A. L. Hines, T. K. Ghosh, S. K. Loyalka, R. C. Warder, Jr , 1992c A Summary of Pollutant Removal Capabilities of Solid and Liquid Desiccants From Indoor Air: Investigation of Co-Sorption of Gases and Vapors as a Means to Enhance Indoor Air Quality, Gas Research Institute GRI-92/0157.1.

46. A. L. Hines, T. K. Ghosh, S. K. Loyalka, 1992a Removal of Particulates and Airborne Microorganisms by Solid Adsorbents and Liquid Desiccants: Investigation of Co-Sorption of Gases and Vapors as a Means to Enhance Indoor Air Quality- Phase II, Gas Research Institute GRI-92/0157.5.

47. M. Hodgson, 1992 "Field studies on the sick building syndrome." Ann NY Acad Sci 641 2136 .

48. M. Hodgson, P. Morey, et al. 1985 "Pulmonary disease associated with cafeteria flooding." Archives of Environmental Health 40 2 96101 .

49. M. Hodgson, P. Morey, et al. 1998 "Building-Associated Pulmonary Disease From Exposure to Stachybotrys chartarum and Aspergillus versicolor." Journal of Occupational and Environmental Medicine 40 3 241249 .

50. T. Husman, T. Meklin, et al. 2002 Respiratory infection among children in moisture damaged schools. Indoor Air.

51. A. Hyvärinen, J. Pekkanen, et al. 2002 Moisture damage at home and childhood asthma- a case-control study. Indoor Air.

52. J. J. K. Jaakkola, N. Jaakkola, R. Ruotsalainen, 1993 "Home dampness and molds as determinants of respiratory symptoms and asthma in pre-school children." Journal of Experimental Analysis and Environmental Epidemiology. 3(S1): 129-142.

53. M. S. Jaakkola, H. Nordman, et al. 2002 "Indoor Dampness and Molds and Development of Adult-Onset Asthma: A population-Based Incident Case-Control Study." Environmental Health Perspectives 110 5 543547 .

54. J. Q. Jarvis, P. R. Morey, 2001 "Allergic Respiratory Disease and Fungal

Remediation in a Building in a Subtropical Climate." Occupational and Environmental Hygiene 16 3 380388 .

55. E. Johanning, R. Biagini, D. L. Hull, P. Morey, B. Jarvis, P. Landsbergis, 1996 "Health and immunology study following exposure to toxigenic fungi (Stachybotrys chartarum)." International Archives of Occupational Environmental Health 68 207218 .

56. A. Joki, V. Saano, T. Reponen, A. Nevalainen, 1993 Effect of indoor microbial metabolites on ciliary function in respiratory airways. Proc Indoor '93, Proceedings of the 6th International Conference on Air Quality and Climate, Helsinki, Findland.

57. E. Karunasena, 2005 "The mechanisms of neurotoxicity induced by a Stachybotrys chartarum tricothecene mycotoxin in an invitro model." A Dissertation in Microbiology and Immunology, Texas Tech University Health Science Center.

58. E. Karunasena, et. Larrañaga, al, 2010 "Building-associated neurological damaged modeled in human cells: a mechanism of neurotoxic effects by exposure to mycotoxins in the indoor environment." Mycopathalogia Dec(6): 377-390.

59. N. King, P. Auger, 2002 "Indoor air quality, fungi and health. How do we stand?" Canadian Family Physician 48 298302 .

60. M. R. Kinshella, M. V. Van Dyke, et al. 2001 "Perceptions of indoor air quality associated with ventilation system types in elementary schools." Applied Occupational and Environmental Hygiene 16 10 952960 .

61. O. Koskinen, 1995 "Reduced exposure to molds brings fewer respiratory symptoms." Indoor Air 5 39 .

62. B. Kovak, P. R. Heimann, J. Hammel, 1997 "The Sanitizing Effects of Desiccant-Based Cooling." ASHRAE Journal: 6064 .

63. S. Kumar, W. J. Fisk, 2002 "IEQ and the impact on Employee Sick Leave." ASHRAE Journal: 9798 .

64. G. Landrus, T. Axcel, 1990 "Survey of asthma, allergy, and environmental sensitivity in an urban Canadian school system". Proceedings of the 5th International Conference on Air Quality and Climate, Toronto, Canada.

65. M. Larrañaga, M. Beruvides, et al. 2008 "DOAS & Humidity Control." ASHRAE Journal May 2008 3440 .

66. S. B. Lehrer, L. Aukrust, J. E. Salvaggio, 1983 "Respiratory allergy induced by fungi." Clinical and Chest Medcine 4 2341 .

67. H. Levin, 1995 "Physical factors in th indoor environment." Occupational Medicine 10 5995 .

68. C. S. Li, L. Y. Hsu, 1997 "Airborne fungus allergen in association with residential characteristics in atopic and control children in a subtropical region." Archives of Environmental Health 52 7279 .

69. K. Licorish, H. S. Novey, P. Kozak, R. D. Fairshter, A. F. Wilson, 1985 "Role of Alternaria and Penicillium spores in the pathogenesis of asthma." Journal Allergy Clinic and Immunology 76 6 819825 .

70. C. J. Martin, S. D. Platt, S. M. Hunt, 1987 "Housing conditions and ill health." British Medical Journal 294 11251127 .

71. M. Meckler, 1994 "Desiccant-Assisted Air Conditioner Improves IAQ and Comfort." Heating/Piping/Air Conditioning(October 1994): 75-84.

72. W. J. Meggs, 1993 "Neurogenic inflammation and sensitivity to environmental chemicals." Environmental Health Perspectives 101 234238 .

73. T. Meklin, T. Husman, et al. 2002 "Indoor air microbes and respiratory symptoms of children in moisture damaged and reference schools." Indoor Air 12 3 175183 .

74. M. J. Mendell, 1993 "Non-specific symptoms in office workers: A review and summary of the epidemiologic literature." Indoor Air Quality 3 227236 .

75. J. D. Miller, 1992 "Fungi as contaminants in indoor air." Atmospheric Environment 26A(12): 2163-2172.

76. S. K. Mishra, L. Ajello, D. G. Ahearn, et al. 1992 ""Environmental mycology and its importance to public health." Journal of Medicine Veterinary Mycology 30(Supplement 1): 287-305.

77. S. A. Mumma, 2001 "Designing dedicated outdoor air systems." ASHRAE Journal(May): 2831 .

78. National Academy of Sciences 1981 Indoor Pollutants. Washington, D.C., National Academy Press.

79. National Academy of Sciences 1993 Indoor Allergens. Washington, D.C., National Academy Press.

80. S. A. Ness, 1991 Air Monitoring for Toxic Exposures-An Integrated Approach. New York, Van Nostrand Reinhold.

81. K. F. Nielsen, M. O. Hansen, et al. 1998 "Production of trichothecene mycotoxins on water damaged gypsum boards in Danish buildings." International Biodeterioration & Biodegradation: 17 .

82. R. L. Ott, 1993 An Introduction to Statistical Methods and Data Analysis. Belmont, Ca, Duxbury Press.

83. T. J. Passon, J. W. Brown, S. Mante, 1996 "Sick-building syndrome and

building- related illness." New and emerging pathogens(July): 8495 .

84. A. A. Pesaran, 1994 A Review of Desiccant Dehumidification Technology. Golden, Colorado, National Renewable Energy Laboratory NREL/TP-472-7010: 1-8.

85. B. Peyronel, 1919 "I germi atmosferici dei funghi con misselio." Diss. Padua 1914. f. Bakt. (2 Abt.) 49: 465.

86. S. D. Platt, C. J. Martin, S. M. Hunt, C. W. Lewis, 1989 "Damp housing, mould growth, and symptomatic health state." British Medical Journal 298 16731678 .

87. M. Popescu, T. K. Ghosh, 1999 "Dehumidification and Simultaneous Removal of Selected Pollutants from Indoor Air by a Desiccant Wheel Using a 1M Type Desiccant." Journal of Solar Energy Engineering 121(February 1999): 1-13.

88. S. C. Redd, 2002 "State of the Science on Molds and Human Health." Center for Disease Control and Prevention, U.S. Department of Health and Human Services.: 111 .

89. K. Reijula, 1996 "Buildings with moisture problems- a new challenge to occupational health care." Scandinavian Journal of Work, Environment and Health 22 13 .

90. J. L. Richard, R. D. Platnerr, et al. 1999 "The occurence of ochratoxin A in dust collected from a problem household." Mycopathologia 146 99103 .

91. W. Richards, 1986 "Allergy, asthma, ans school problems." Journal of School Health 56 4 151152 .

92. O. Rostrup, 1908 "Nogle Undersogelser over Luftens af Svampekim." Bot. Tidsskr 29: 32.

93. K. Saito, 1904 "Untersuchungen über die atmospharischen Pilzkeime. I. Mittheilung." J. Coll. Sci. Tokyo. 18(Art. 5): 58.

94. A. Sakula, 1984 "Sir John Floyer's A Treatise of the Asthma (1698)." Thorax 39 248254 .

95. J. E. Salvaggio, 1994a "Inhaled particles and respiratory disease." Journal Allergy and Clinical Immunology 94 304309 .

96. J. E. Salvaggio, 1994b "Psychological aspects of environmental illness, multiple chemical sensitivity and building related illness." Journal Allergy and Clinical Immunology 94 366370 .

97. J. E. Salvaggio, L. Aukrust, 1981 "Mold induced asthma." Journal Allergy and Clinical Immunology 68 327346 .

98. J. Samet, 1990 "Environmental controls and lung disease." American

Review of Respiratory Diseases 142 915938 .

99. T. A. Seitz, 1990 NIOSH indoor air quality investigations: 1971 through 1988. The practitioner's approach to indoor air quality investigations. D. M. Weekes, Gammage, R.B. Cincinnati, OH, American Industrial Hygiene Association: 163171 .

100. R. C. Smith, 1990 "Controlling sick building syndrome." Journal of Environmental Health 53 3 2223 .

101. W. R. Solomon, 1975 "Assessing fungus prevalence in domestic interiors." The journal of allergy and clinical immunology 56 3 235242 .

102. J. Spengler, L. Neas, S. Nakai, 1994 "Respiratory symptoms and housing characteristics." Indoor Air 4 7282 .

103. B. Stenberg, N. Eriksson, J. Hoog, J. Sundell, S. Wall, 1994 "The sick building syndrome (SBS) in office workers: a case-referent study of personal, psychosocial and building-related risk indicators." International Journal of Epidemiology. 23 11901197 .

104. T. D. Sterling, A. Arundel, 1984 "Possible carcinogenic components of indoor air quality, combustions by-products, formaldehyde, mineral fibers, radiation, and tobacco smoke." Journal of Environmental Science and Health(62): 185-230.

105. D. L. Sudakin, 1998 "Toxigenic fungi in a water-damaged building: An intervention study." American Journal of Industrial Medicine 34 183190 .

106. R. C. Summerbell, F. Staib, R. Dales, N. Nolard, J. Kane, H. Zwanenburg, R. Burnett, S. Krajden, D. Fung, D. Leong, 1992 "Ecology of fungi in human dwellings." Journal of Medical Veterinary Mycology 30 1 279285 .

107. S. M. Tarlo, A. Fradkin, R. S. Tobin, 1988 "Skin testing with extracts of fungal species derived from the homes of allergy clinic patients in Toronto, Canada." Clinical Allergy 18 1 4552 .

108. K. Y. Teichman, 1995 "Indoor Air Quality: Research Needs." Occupational Medicine 10 1 217227 .

109. R. S. Tobin, E. Baranowski, A. P. Gilman, T. Kuiper-Goodman, J. D. Miller, M. Giddings, 1987 "Significance of Fungi in Indoor Air: Report of a Working Group." Canadian Journal of Public Health. 78(S1): S1 -S14.

110. A. P. Verhoeff, R. T. Van Strien, J. H. Van Wijnen, B. Brunekreef, 1995 "Damp housing and childhood respiratory symptoms: the role of sensitization to dust mites and molds." American Journal of Epidemiology 141 103110 .

111. A. P. Verhoeff, J. H. Van Wijnen, B. Brunekreef, P. Fischer, E. S. Van Reenen-Hoekstra, R. A. Samson, 1992 "Presence of viable mold propagules in indoor air-relation to house dampness and outdoor air." Allergy: European Journal of Allergy and Clinical Immunology 47(2 Pt 1): 83-91.

112. G. Vogel, 1997 "New clues to asthma therapies." Science 276 13 16431646 .

113. K. B. Weiss, P. J. Gergen, T. A. Hodgson, 1992 "An economic evaluation of asthma in the United States." New England Journal of Medicine 326: 862.

114. K. B. Weiss, S. D. Sullivan, et al. 2000 "Trends in the cost of illness for asthma in the United States, 1985-1994." J Allergy Clin Immunol 106 3 493499 .

115. J. H. White, 1995 Moldy Houses: Why they are and why we care. Ottawa, Ontario, Morrison Hershfield Limited: 165 .

116. World Health Organization 1983 Indoor air pollutants: exposure and health effects. Copenhagen, World Health organization.

117. J. H. Zar, 1974 Biostatistical Analysis. Englewood Cliffs, NJ, Prentice-Hall, Inc.

Chapter 5

MUNICIPAL WASTE PLASTIC CONVERSION INTO DIFFERENT CATEGORY OF LIQUID HYDROCARBON FUEL

Moinuddin Sarker

Natural States Research, Inc., USA

INTRODUCTION

Plastics were first invented in 1860, but have only been widely used in the last 30 years. Plastics are light, durable, modifiable and hygienic. Plastics are made up of long chain of molecules called polymers. Polymers are made when naturally occurring substances such as crude oil or petroleum are transformed into other substances with completely different properties. These polymers can then be made into granules, powders and liquids, becoming raw materials for plastic products. Worldwide plastics production increases 80 million tons every year. Global production and consumption of plastics have increased, from less than 5 million tons in the year 1950 to 260 million tons in the year 2007. Of those over one third is being used for packaging, while rest is used for other sectors. Plastic production has increased by more than 500% over the past 30 years. Per capita consumption of plastics will increase by more than 50% during the next decades. In the Western Europe total annual household waste generation is approximately 500 kg per capita and 750 kg per capita in the United States; 12% of this total waste is plastics. The global total waste plastic generation is estimated to be over 210 million tons per year. US alone generate 48 million tons per year (Stat data from EPA). The growth in plastics use is due to their beneficial characteristics; 21st century Economic growth making them even more suitable for a wide variety of applications, such as: food and product packaging, car manufacturing, agricultural use, housing products and etc. Because of good safety and hygiene properties for food packaging, excellent thermal and electrical insulation properties, plastics are more desirable among consumers. Low production cost, lower energy consumption and CO_2 emissions during production of plastics are relatively lower than making alternative materials, such as glass, metals and etc. Yet for all their advantages, plastics have a considerable downside in terms of their environmental impact. Plastic production requires large amounts of resources, primarily fossil fuels and 8% of the world's annual oil production is used in the production of plastics. Potentially harmful chemicals are added as stabilizers or colorants. Many of these have not undergone environmental risk assessment and their impact on human health and environment is currently uncertain. Worldwide municipal sites like shops or malls had the largest proportion of plastic rubbish items. Ocean soup swirling the debris of plastics trash in the Pacific Ocean has now grown to a size that is twice as large as the continental US. In 2006, 11.5 million of tons of plastics were wasted in the landfill. These types of disposal of the waste plastics release toxic gas; which has negative impact on environment. Most plastics are non-biodegradable and they take long time to break down in landfill, estimated to be more than a century. Plastic waste also has a detrimental impact on wild life; plastic waste in the oceans is estimated to cause the death of more than a million seabirds and more than 100,000 marine mammals every year (UN Environmental Program Estimate). Along with this hundreds of thousands of sea turtles, whales and other marine mammals die every year eating discarded waste plastic bags mistaken for food. Setting up intermediate treatment plants for waste plastic, such as: plastic incineration, recycle, or obtaining the landfill for reclamation is difficult. The types of the waste plastics are LDPE, HDPE, PP, PS, PVC, PETE, PLA and etc. The problems of waste plastics can't be solved by landfilling or incineration, because the safety deposits are expensive and incineration stimulates the growing emission of harmful greenhouse gases, e.g COx, NOx, SOx and etc. By using NSR's new technology we can convert all types of waste plastics into liquid hydrocarbon fuel by setting temperature profile 370° C to 420° C, we can resolve all waste plastic problems including land, ocean, river and green house effects. Many of researcher and experts have done a lot of research and work on waste plastics; some of the thesis's are on thermal degradation process [1-10], pyrolysis process [11-20] and catalytic conversion process [21-30]. Producing fuels can be alternative of heating oil, gasoline, naphtha, aviation, diesel and fuel oil. We also produce light gaseous (natural gas) hydrocarbon compound (C_1-C_4), such as: methane, ethane, propane and butane. This process is profitable because it requires less production cost per gallon. We can produce individual plastic to fuel, mixed waste plastic to fuel and that produced fuel can make different category fuels by using further fractional distillation process. This NSR technology will not only reduce the production cost of fuel, but it will also reduce 9% of foreign oil dependency, create more electricity and new jobs all over the world. To mitigate the present world market demand, we can substitute this method as a potential source of new renewable energy.

EXPERIMENTAL SECTION

Waste Plastics Properties

A plastic has physical and chemical properties. Different types of plastics displayed distinguishable characteristics and properties. Many kinds of plastics are appeared like LDPE, HDPE, PP, PS, PVC &PETE etc. Several individual plastics properties are elaborated in shortly, that's given below inTable-1, Table-2, Table-3and Table-4.

Table 1. HDPE-2 Plastic Properties

Quantity	Value	Units
Thermal expansion	110 - 130	e-6/K
Thermal conductivity	0.46 - 0.52	W/m.K
Specific heat	1800 - 2700	J/kg.K
Melting temperature	108 - 134	°C
Glass temperature	-110 - -110	°C
Service temperature	-30 - 85	°C
Density	940 - 965	kg/^{m}3
Resistivity	5e+17 - 1e+21	Ohm.mm²/m
Shrinkage	2 - 4	%
Water absorption	0.01 - 0.01	%

Table 2. LDPE-4 Plastic Properties

Quantity	Value	Units
Thermal expansion	150 - 200	e-6/K
Thermal conductivity	0.3 - 0.335	W/m.K
Specific heat	1800 - 3400	J/kg.K
Melting temperature	125 - 136	°C
Glass temperature	-110 - -110	°C
Service temperature	-30 - 70	°C
Density	910 - 928	kg/^{m}3
Resistivity	5e+17 - 1e+21	Ohm.mm²/m
Breakdown potential	17.7 - 39.4	kV/mm
Shrinkage	1.5 - 3	%

| Water absorption | 0.005 - 0.015 | % |

Table 3. PP-5 Plastic Properties

Quantity	Value	Units
Thermal expansion	180 - 180	e-6/K
Thermal conductivity	0.22 - 0.22	W/m.K
Melting temperature	160 - 165	°C
Glass temperature	-10 - -10	°C
Service temperature	-10 - 110	°C
Density	902 - 907	kg/^{m}3
Resistivity	5e+21 - 1e+22	Ohm.mm^2/m
Breakdown potential	55 - 90	kV/mm
Shrinkage	0.8 - 2	%

Table 4. PS-6 Plastic Properties

Quantity	Value	Units
Thermal expansion	60 - 80	e-6/K
Thermal conductivity	0.14 - 0.16	W/m.K
Specific heat	1300 - 1300	J/kg.K
Glass temperature	80 - 98	°C
Service temperature	-10 - 90	°C
Density	1040 - 1050	kg/^{m}3
Resistivity	1e+22 - 1e+22	Ohm.mm^2/m
Breakdown potential	100 - 160	kV/mm
Shrinkage	0.3 - 0.7	%

Pre Analysis of Gas Chromatography & Mass Spectrometer (GC/ MS) Analysis

Before starting the fuel production experiment, we have analyzed each of the individual raw waste plastics. Types of analyzed raw waste plastics are following, HDPE-2 (High Density Polyethylene), LDPE-4 (Low Density Polyethylene), PP-5 (Polypropylene) and PS-6 (Polystyrene)

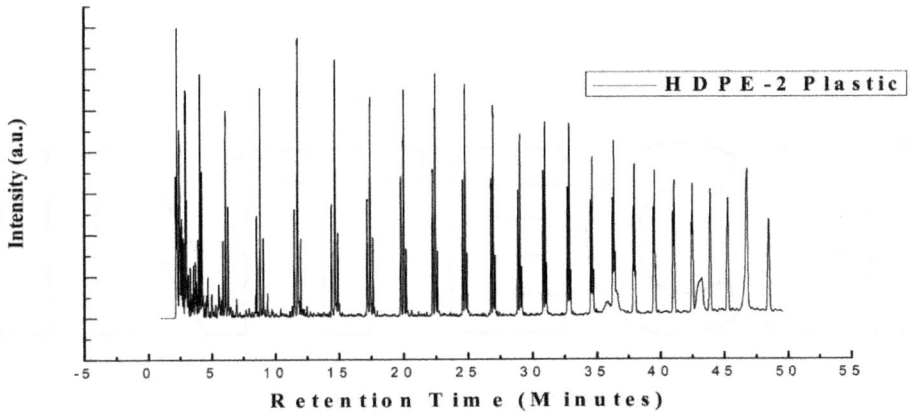

Figure 1. GC/MS Chromatogram of HDPE-2 Raw Waste Plastic

Table 4. GC/MS Compound List of HDPE-2 Waste Plastic

Retention Time	Compound Name	Formula	Retention Time	Compound Name	Formula
2.14	Propane	C3H8	22.62	Tetradecane	C14H30
2.23	3-Butyn-1-ol	C4H6O	24.57	1,13-Tetradeca-diene	C14H26
17.61	Dodecane	C12H26	40.94	1,19-Eicosa-diene	C20H38
19.78	1,13-Tetradeca-diene	C14H26	41.02	1-Docosene	C22H44
20.00	1-Tridecene	C13H26	42.48	1-Docosene	C22H44
20.19	Tridecane	C13H28	43.89	1-Tetracosanol	C24H50O
22.24	1,13-Tetradeca-diene	C14H26	45.28	9-Tricosene, (Z)-	C23H46
22.45	Cyclotetradec-ane	C14H28	46.76	17-Pentatriac-ontene	C35H70

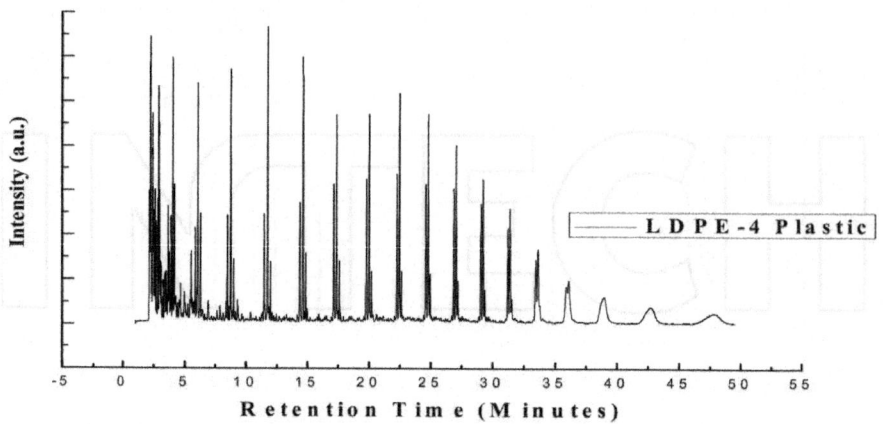

Figure 2. GC/MS Chromatogram of LDPE-4 Raw Waste Plastic

Table 5. GC/MS Chromatogram Compound list of LDPE-4 Raw Waste Plastic

Retention Time (Minutes)	Compound Name	Formula	Retention Time (Minutes)	Compound Name	Formula
2.11	Propane	C3H8	17.13	1,11-Dodecadiene	C12H22
2.19	Cyclopropyl carbinol	C4H8O	17.37	Cyclododecane	C12H24
11.44	1,9-Decadiene	C10H18	33.62	1-Nonadecene	C19H38
11.73	Cyclodecane	C10H20			
11.95	Decane	C10H22	35.87	1,19-Eicosadiene	C20H38
14.35	1,10-Undeca-diene	C11H20	36.08	1-Heneicosyl formate	C22H44O2
14.61	1-Undecene	C11H22	42.76	1-Docosanol	C22H46O
14.84	Undecane	C11H24	47.91	9-Tricosene, (Z)-	C23H46

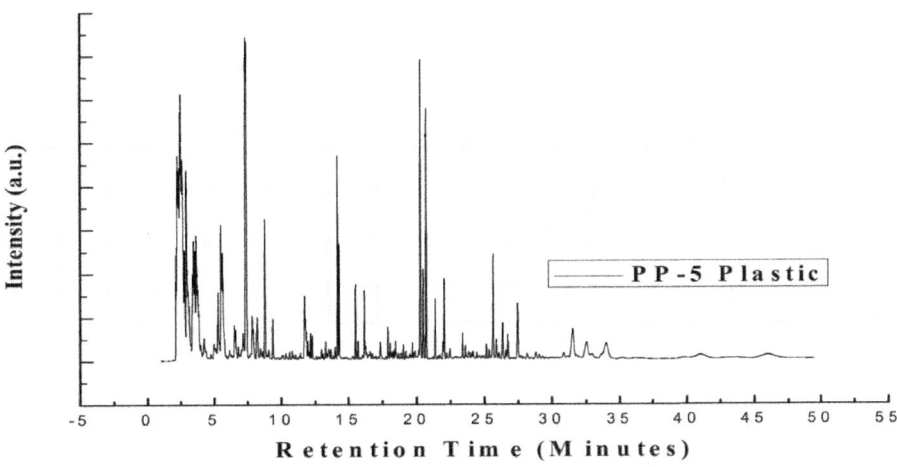

Figure 3. GC/MS Chromatogram of PP-5 Raw Waste Plastic

Table 6. GC/MS Chromatogram Compound List of PP-5 Raw Waste Plastic

Reten-tion Time (Minutes)	Compound Name	Formula	Reten-tion Time (Min-utes)	Compound Name	Formula
2.13	Cyclopropane	C3H6	12.29	Decane, 4-methyl-	C11H24
2.26	1-Butyne	C4H6	14.18	2-Dodecene, (E)-	C12H24
9.36	1,6-Octadiene, 2,5-dimethyl-, (E)-	C10H18	26.35	1-Hexadecanol, 3,7,11,15-tetra-methyl-	C20H42O
11.71	Nonane, 2-meth-yl-3-methylene-	C11H22	31.52	1-Heneicosyl formate	C22H44O2
11.78	1-Ethyl-2,2,6-trimethylcyclo-hexane	C11H22	32.51	1-Nonadecanol	C19H40O
12.17	Nonane, 2,6-di-methyl-	C11H24	33.98	1,22-Docosane-diol	C22H46O2

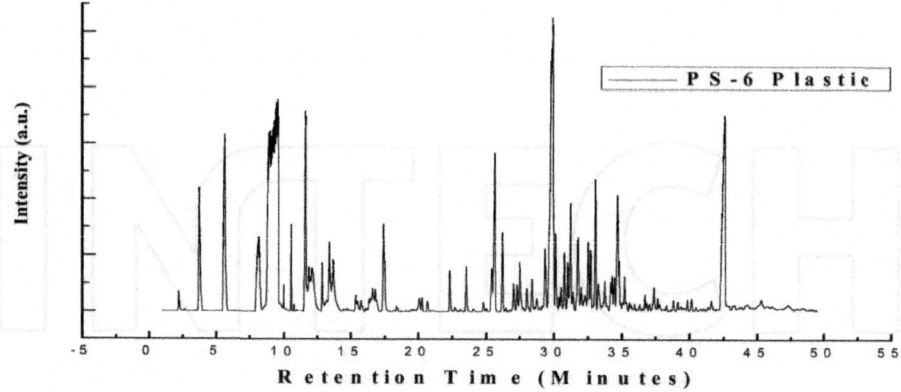

Figure 4. GC/MS Chromatogram of PS-6 Raw Waste Plastic

Table 7. GC/MS Chromatogram of PS-6 Raw Waste Plastic Compound List

Reten-tion Time (Minutes)	Compound Name	Formula	Reten-tion Time (Minutes)	Compound Name	Formula
2.17	Cyclopropane	C3H6	24.78	1,1'-Biphenyl, 3-methyl-	C13H12
2.24	Methylenecy-clopro-pane	C4H6	25.64	1,2-Diphenylethylene	C14H12
5.52	Toluene	C7H8	27.30	1,2-Diphenylcyclo-propane	C15H14
20.09	1,4-Methanon-aphthalene, 1,4-dihydro-	C11H10	37.35	Naphthalene, 1-(phenylmethyl)-	C17H14
20.28	Benzocyclo-hepta-triene	C11H10	37.63	p-Terphenyl	C18H14
20.67	Naphthalene, 1-methyl-	C11H10	38.79	Fluoranthene, 2-methyl-	C17H12
22.32	Biphenyl	C12H10	39.83	Benzene, 1,1'-[1-(eth-ylthio)propylidene] bis-	C17H20S
23.52	Diphenylmeth-ane	C13H12	40.13	Benzene, 1,1',1'',1'''-(1,2,3,4- butanetet-rayl) tetrakis-	C28H26

Individual raw waste plastics of GCMS pre-analysis in accordance with their numerous retention times many compound are found, some of them are mentioned shortly. In HDPE-2 raw waste plastics on retention time 2.14, compound is Propane (C_3H_8), on retention time 22.45, compound is

Cyclotetradecane and finally on retention time 46.76 obtained compound is Pentatriacotene ($C_{35}H_{70}$) [Shown above Fig.1 and Table-4]. In LDPE-4 raw waste plastics on retention time 2.11, compound is Propane (C_3H_8), on retention time 14.84, compound is Undecane ($C_{11}H_{24}$) and finally on retention time 47.91 obtained compound is 9-Tricosene (Z)-($C_{23}H_{46}$) [Shown above Fig.2 and Table-5]. In PP-5 initially on retention time 2.13 compound is Cyclopropane (C_3H_6) and finally on retention time 33.98 obtained compound is 1, 22-Docosanediol ($C_{22}H_{46}O_2$) [Shown above Fig.3 and Table-6]. Accordingly in PS-6 on retention time 2.17 found compound is Cyclopropane and eventually on retention time 40.13 obtained compound is Benzene, 1,1',1'',1'''-(1,2,3,4-butanetetryl)tetrakis[Shown above Fig.4 and Table-7].

Sample Preparation

We take municipal mixed waste plastics or any other source of mixed waste plastics; we initially sort out the foreign particles, clean the waste plastics and clean wash them with detergent. After clean up all waste plastics spread in the open air for air dry. When dried out we shred them by scissors, now shredded plastics are grinded by grinding machine. Grinded samples structure are granular form small particles and that easy to put into the reactor. In our laboratory facility we can utilize 400g to 3kg of grinding sample for any experimental purposes.

PROCESS DESCRIPTION

Individual Plastic to Fuel Production Process

The process has been conducted in small scales with individual plastics in laboratory, on various waste plastics types; High-density polyethylene (HDPE, code 2), low-density polyethylene (LDPE, code 4), polypropylene (PP, code 5) and polystyrene (PS, code 6). These plastic types were investigated singly. For small-scale laboratory process the weight of input waste plastics ranges from 400 grams to 3kg. These waste plastics are collected, optionally sorted, cleaned of contaminants, and shredded into small pieces prior to the thermal liquefaction process. The process of converting the waste plastic to alternative energy begins with heating the solid plastic with or without the presence of cracking catalyst to form liquid slurry (thermal liquefaction in the range of 370-420 °C), condensing the vapor with standard condensing column to form liquid hydrocarbon fuel termed "NSR fuel". Preliminary tests on the produced NSR fuel have shown that it is a mixture of various hydrocarbons range. The produced fuel density varies based on individual plastic types. In equivalent to obtaining the liquid hydrocarbon fuel we also receive light gaseous hydrocarbon compounds (C_1-C_4) which resembles natural gas. Further fractional distillation based on different temperature is producing different category fuels; such as heating oil, gasoline, Naphtha (chemical), Aviation, Diesel and Fuel Oil. Experiment diagram given below in Fig.5.

Mixed Waste Plastic to Fuel Production Process

Mixed waste plastics to fuel production process performed in the laboratory on various waste plastics types; High-density polyethylene (HDPE, code 2), low-density polyethylene (LDPE, code 4), polypropylene (PP, code 5) and polystyrene (PS, code 6).

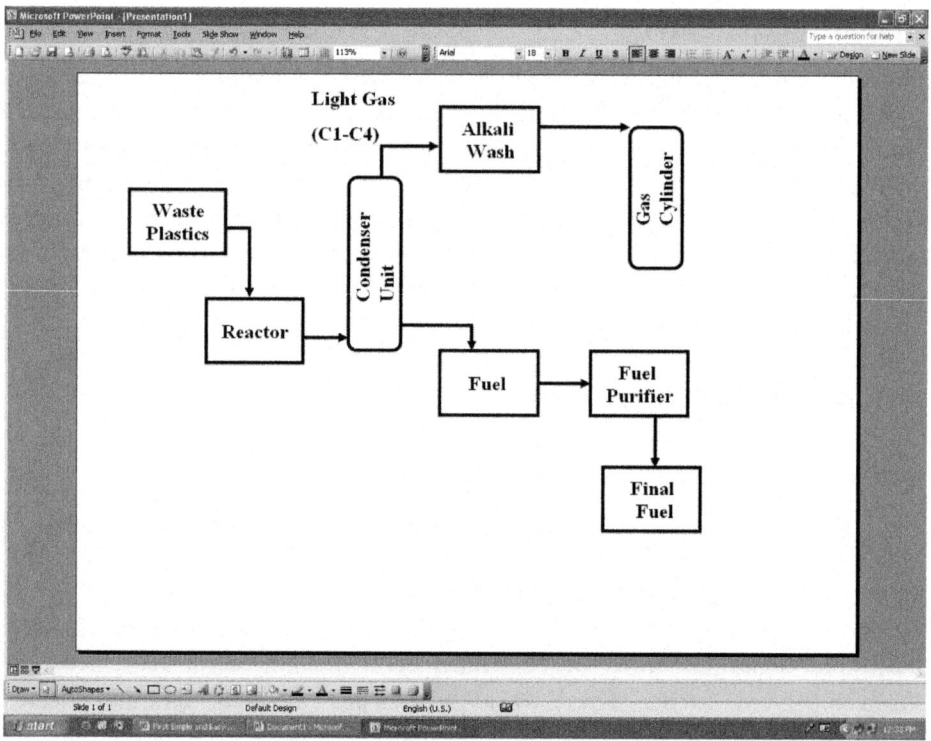

Figure 5. Individual & Waste Plastic to Fuel Production Process Diagram

These processes were investigated with mixture of several plastics such as HDPE-2, LDPE-4, and PP-5 &PS-6. These waste plastics are collected, optionally sorted, cleaned of contaminants, and shredded into small pieces prior to the thermal degradation process. The experiment could be randomly mixture of waste plastics or proportional ratio mixture of waste plastics. For small-scale laboratory process the weight of input waste plastics ranges from 300 grams to 3kg. In the laboratory processes our present reactor chamber capacity is 2-3 kg. We put 2 kg of grinding sample into the reactor chamber to expedite the experiment process. At the starting point of experiment reactor temperature set up at 350 °C for quick melting, after melted temperature maintained manually from "reactor temperature profile menu option" by increasing and decreasing depending to the rate of reaction. The optimum temperature (steady & more fuel production state) is 305 °C. From 2kg of waste plastics obtained fuel amount is 2 liter 600 ml (2600 ml), fuel density is 0.76 g/ml. We defined the fuel as heating oil named "NSR fuel". The experiment additionally produced light gases Methane, Ethane, Propane and Butane as well as few amount of carbon ashes as a remaining residue. These light gases would be the alternative source of natural gases. Mixed waste plastic to produced fuel preliminary test indicated that the hydrocarbon compound rage from C_3 to C_{27}.

Fractional Distillation Process

Fractional distillation process has been conducted according to the laboratory scale. We measured 700 ml of NSR fuel called heating fuel and took the weight of 1000 ml boiling flask (Glass Reactor). Subsequently fuel poured into the boiling flask, after that we put filled boiling flask in 1000 ml heat mantle as well as connected variac meter with heat mantle. Attached distillation adapter, clump joint, condenser and collection flask with high temperature apiezon grease and insulated by aluminum foil paper. Initially we ran the experiment at 40 °C to collect gasoline grade, after gasoline collection subsequently we raised the temperature to 110 °C for naphtha (Chemical), 180 °C for aviation fuel, 260 ° C for diesel fuel and eventually at 340 °C we found fuel oil. At the end of the experiment remaining residual fuel was less, approximately amount 10-15 ml. Out of 700 ml NSR fuel we collected 125 ml of gasoline; density is 0.72 g/ml, 150 ml of naphtha; density is 0.73, 200 ml of aviation fuel; density is 0.74, 150 ml of diesel fuel; density is 0.80 g/ml and 50-60 ml of fuel oil; density is 0.84.

FUEL PRODUCTION YIELD PERCENTAGE

After all experiment done on behalf of each experiment we calculated the yield percentages of fuel production, light gases and residue. In addition described the physical properties of each fuel such as fuel density, specific gravity, and fuel color and fuel appearance respectively. Similarly, individual fuel production yield percentages & properties are given below in Table 8 (a) & 9 (a) and Mixed Waste Plastics to fuel Yield percentages & properties are also given below in Table 8(b) & 9 (b).

Table 8A. Individual Fuel Production Yield Percentage

Waste Plastic Name	Fuel Yield %	Light Gas %	Residue %
HDPE-2	89.354	5.345	5.299
LDPE-4	87.972	5.806	6.221
PP-5	91.981	2.073	5.944
PS-6	85.331	4.995	9.674

Table 8B. Mixed Waste Plastic to Fuel Yield Percentage

Sample Name	Fuel Yield %	Light Gas %	Residue %
HDPE,LDPE,PP&PS	90	5	5

Table 9A. Individual Plastic to Fuel Properties

Name of Waste Plastic Fuel	Fuel Density gm/ml	Specific Gravity	Fuel Color	Fuel Appearance
LDPE-4	0.771	0.7702	Yellow, light transparent	Little bit wax and ash content
HDPE-2	0.782	0.7812	Yellow, no transparent	Wax, cloudy and little bit ash content
PP-5	0.759	0.7582	Light brown, light transparent	Little bit wax and ash content
PS-6	0.916	0.9150	Light yellow, not transparent	Wax, cloudy and little bit ash content

Table 9B. Mixed Waste Plastic to Fuel Properties

Name of Fuel	Density g/ml	Specific Gravity	Fuel Color	Fuel
Mixed Plastic to Fuel	0.775	0.7742	Yellow light transparent	Ash contain present

Fuel Analysis and Result Discussion

Gas Chromatography and Mass Spectrometer (GC/MS) Analysis

Analysis of Individual waste plastics (HDPE-2, LDPE-4, PP-5, and PS-6) to individual fuel:

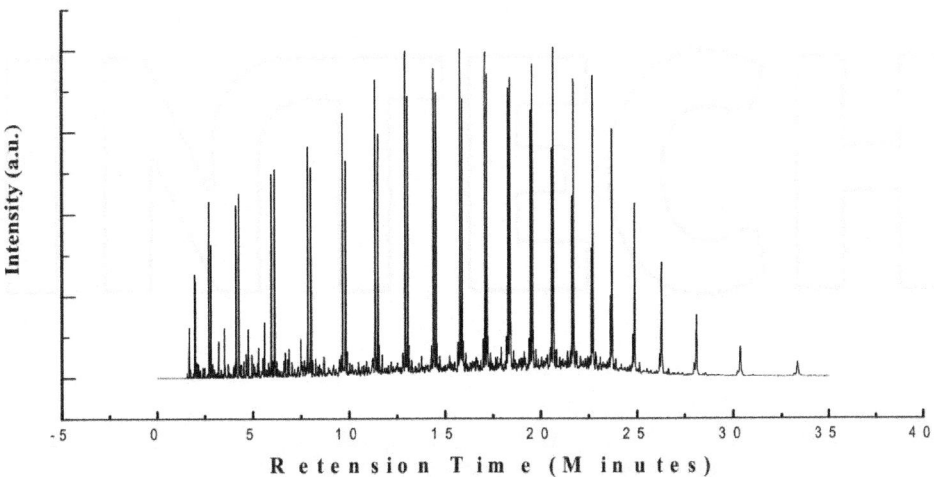

Figure 6. GC/MS Chromatogram of HDPE-2 Waste Plastic to Fuel

Table 10. GC/MS Chromatogram Compound List of HDPE-2 Waste Plastic to Fuel

Retention Time (Minutes)	Compound Name	Formula	Retention Time (Minutes)	Compound Name	Formula
1.56	Propane	C3H8	12.18	Cyclopentane, hexyl-	C11H22
1.66	2-Butene, (E)-	C4H8	12.92	1-Dodecene	C12H24
1.68	Butane	C4H10	13.05	Dodecane	C12H26
1.96	Cyclopropane, 1,2-dimethyl-, cis-	C5H10	13.76	Cyclododecane	C12H24
9.65	1-Decene	C10H20	27.98	1-Docosene	C22H44
9.80	Decane	C10H22	28.09	Tetracosane	C24H50
11.35	1-Undecene	C11H22	30.24	1-Docosene	C22H44
11.49	Undecane	C11H24	30.38	Octacosane	C28H58

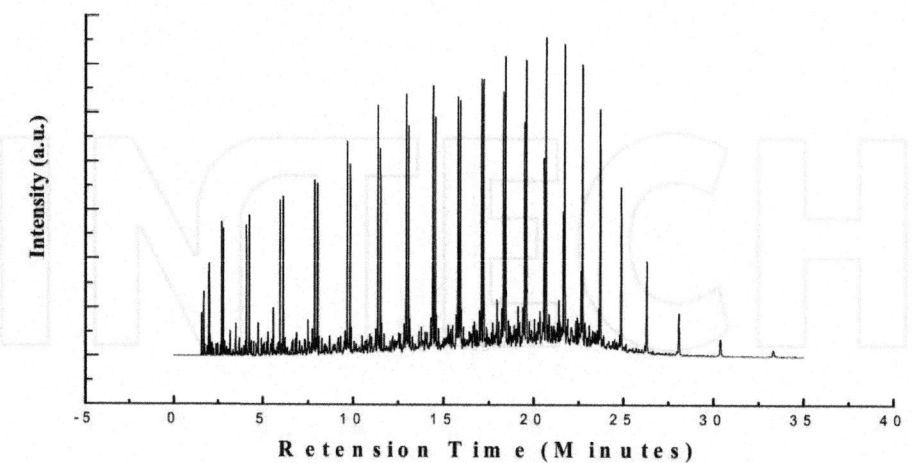

Figure 7. GC/MS Chromatogram of LDPE-4 Waste Plastic to Fuel

Table 11. GC/MS Chromatogram Compound List of LDPE-4 Waste Plastic to Fuel

Retention Time (Minutes)	Compound Name	Compound Formula	Retention Time (Minutes)	Compound Name	Compound Formula
1.55	Cyclopropane	C3H6	12.92	1-Dodecene	C12H24
1.68	Butane	C4H10	13.06	Dodecane	C12H26
1.96	2-Pentene, (E)-	C5H10	13.76	Cyclododecane	C12H24
1.99	Pentane	C5H12	14.40	1-Tridecene	C13H26
10.48	Cyclodecane	C10H20	24.88	Heneicosane	C21H44
10.89	Cyclohexene, 3-(2-methylpropyl)-	C10H18	26.31	Heneicosane	C21H44
11.35	1-Undecene	C11H22	28.09	Tetracosane	C24H50
11.49	Undecane	C11H24	33.21	Octacosane	C28H58

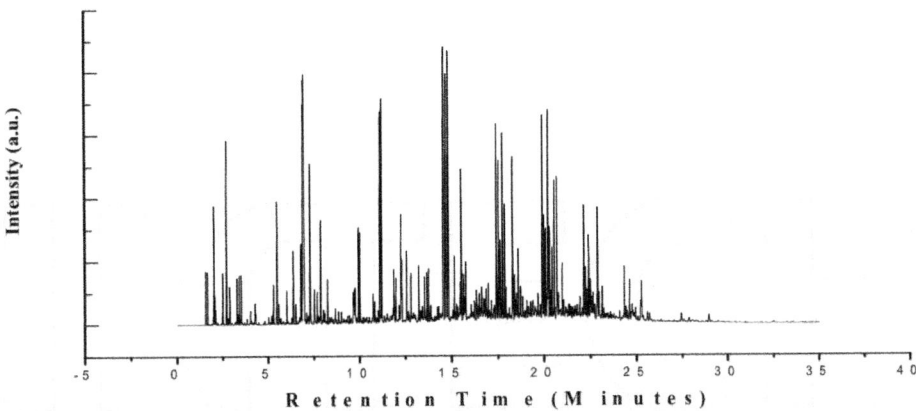

Figure 8. GC/MS Chromatogram of PP-5 Waste Plastic to Fuel

Table 12. GC/MS Chromatogram Compound List of PP-5 Waste Plastic to Fuel

Retention Time (Minute)	Compound Name	Formula	Retention Time (Minute)	Compound Name	Formula
1.55	Cyclopropane	$C_{3H}6$	11.13	Cyclooctane, 1,4-di-methyl-, cis-	$C_{10H2}0$
1.66	1-Propene, 2-methyl-	C4H8	11.20	1-Tetradecene	$C_{14H2}8$
1.99	Pentane	C5H12	11.86	1-Dodecanol, 3,7,11-trimethyl-	$C_{15}H_{32}O$
2.48	Pentane, 2-methyl-	C6H14	12.25	(2,4,6-Trimethylcy-clohexyl) methanol	$C_{10}H_{20}O$
9.64	Nonane, 2-methyl-3-methylene-	C11H22	23.13	Dodecane, 1-cyclo-pentyl-4-(3-cyclo-pentylpropyl)-	$C_{25H4}8$
9.74	3-Undecene, (Z)-	C11H22	25.72	Cyclotetradecane, 1,7,11-trimethyl-4-(1-methylethyl)-	$C_{20H4}0$
9.92	Octane, 3,3-dimethyl-	C10H22	28.95	Dodecane, 1-cyclo-pentyl-4-(3-cyclo-pentylpropyl)-	$C_{25H4}8$
10.73	3-Decene, 2,2-di-methyl-, (E)-	C12H24			

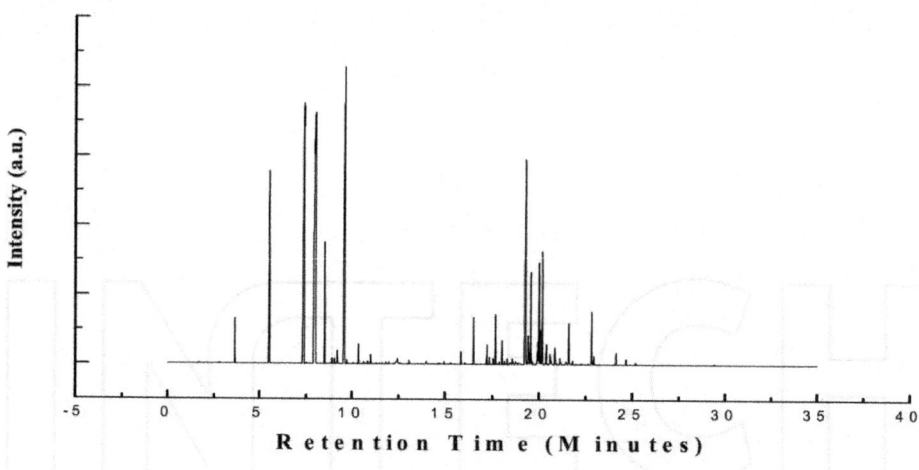

Figure 9. GC/MS Chromatogram of PS-6 Waste Plastic to Fuel

Table 13. GC/MS Chromatogram Compound List of PS-6 Waste Plastic to Fuel

Retention Time (Minute)	Compound Name	Formula	Retention Time (Minute)	Compound Name	Formula
3.65	1,5-Hexadiyne	$C_{6H}6$	17.68	Benzene, 1,1'-(1,2-ethanediyl)bis-	$C_{14H1}4$
5.54	Toluene	C7H8	18.03	Benzene, 1,1'-(1-methyl-1,2-ethanediyl)bis-	$C_{15H1}6$
7.94	Styrene	C8H8	19.30	Benzene, 1,1'-(1,3-propanediyl)bis-	$C_{15H1}6$
11.00	Acetophenone	C8H8O	21.61	Naphthalene,1-phenyl-	$C_{16H1}2$
13.07	Naphthalene	C10 H8	21.81	o-Terphenyl	$C_{18H1}4$
15.84	Biphenyl	C12H10	22.83	2-Phenylnaphthalene	$C_{16H1}2$
16.51	Diphenylmethane	C13H12	24.14	9-Phenyl-5H-benzocycloheptene	$C_{17H1}4$
17.22	Benzene,1,1'-ethylidenebis-	C14H14	24.67	p-Terphenyl	$C_{18H1}4$

From GCMS analysis of Individual HDPE-2, LDPE-4, PP-5, and PS-6 fuel, in accordance with their numerous retention times many compounds are found, some of them are mentioned shortly. In HDPE-2 fuel at retention time

1.56, compound is Propane (C_3H_8), and finally at retention time 30.38 obtained compound is Octacosane ($C_{28}H_{58}$), [Shown above, Fig.6 & Table-10]. In LDPE-4 fuel at retention time 1.55, compound is Cyclopropane (C_3H_6), and finally at retention time 33.21 obtained compound is Octacosane ($C_{28}H_{58}$) [Shown above, Fig.7 & Table-11]. In PP-5 initially at retention time 1.55 compound is Cyclopropane (C_3H_6) and finally at retention time 28.95 obtained compound is Dodecane,-1-Cyclopentyl-4-(3-Cyclopentylpropyl) ($C_{22}H_{46}O_2$) [Shown above, Fig.8 & Table-12]. Accordingly in PS-6 at retention time 3.65 found compound is 1, 5-Hexadiyne and eventually at retention time 24.67 obtained compound is p-Terphnyl ($C_{18}H_{14}$) [Shown above, Fig.9 & Table-13].

Analysis of Mixed Waste Plastics to Fuel (Heating Oil):

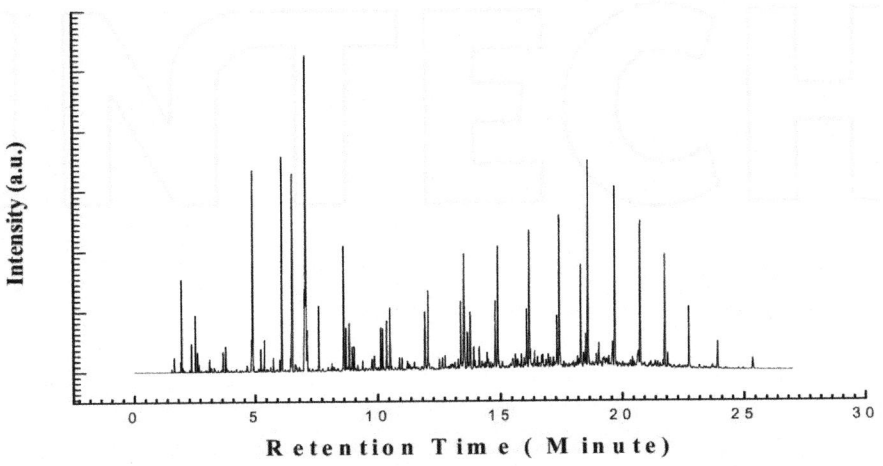

Figure 10. GC/MS Chromatogram of Mixed Waste Plastic to Fuel (Heating Oil)

Table 14. GC/MS Chromatogram Compound List of Mixed Waste Plastic to Fuel (Heating Oil)

Compound Name	Formula	Compound Name	Formula
Cyclopropane	(C3H6)	Dodecane	(C12H26)
2-Butene, (E)-	(C4H8)	Decane, 2,3,5,8-tetra-methyl-	(C14H30)
Pentane	(C5H12)	1-Tridecene	(C13H26)
Pentane, 2-methyl-	(C6H14)	Tridecane	(C13H28)
Cyclopropane, 1-heptyl-2-methyl-	(C11H22)	Heneicosane	(C21H44)
Undecane	(C11H24)	Nonadecane	(C19H40)

1-Dodecanol, 3,7,11-tri-methyl-	(C15H32 O)	Benzene, hexadecyl-	(C22H38)
1-Dodecene	(C12H24)	Heptacosane	(C27H56)

From GCMS analysis of NSR fuel (Called Heating Fuel) primarily we found long chain hydrocarbon of compound. In the GCMS data we have noticed that the obtained compounds are Cyclopropane (C_3H_6) to Heptacosane ($C_{27}H_{56}$) including long and short chain of hydrocarbon compound [Shown above, Fig.10 & Table-14].

GCMS Analysis of Mixed Waste Plastics to Fractional Distillation Fuel:

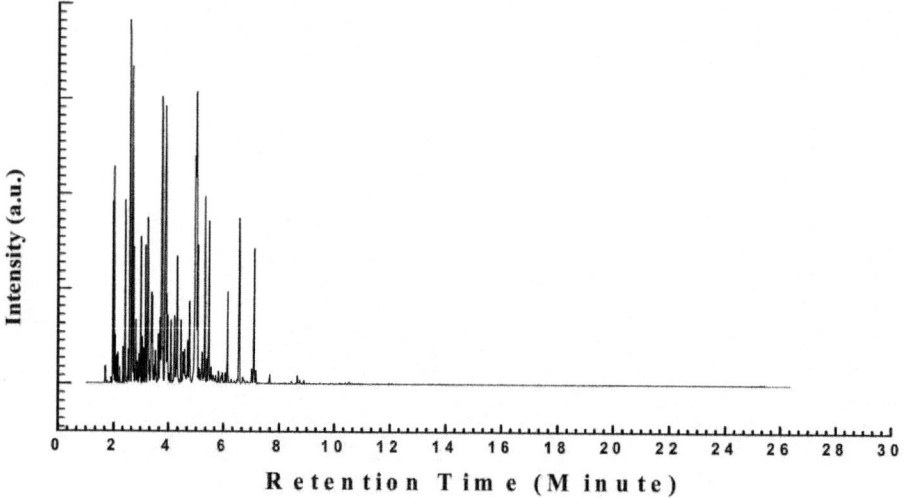

Figure 11. GC/MS Chromatogram of Mixed Waste Plastic Fuel to 1st Fractional Fuel (Gasoline)

Table 15. GC/MS Chromatogram compound list of Mixed Waste Plastic Fuel to 1st Fractional Fuel (Gasoline)

Compound Name	Formula	Compound Name	Formula
1-Propene,2-methyl-	(C4H8)	Heptane	(C7H16)
Butane	(C4H10)	1,4-hexadiene,4-methyl-	(C7H12)
2-Pentene	(C5H10)	1,4-Heptadiene	(C7H12)
2-Pentene,(E)	(C5H10)	Cyclohexane,methyl-	(C7H14)
Cyclohexane	(C6H12)	1-Nonane	(C9H18)
Hexane,3-methyl	(C7H16)	Styrene	(C8H8)
Cyclohexene	(C6H10)	Nonane	(C9H20)

1-Hexene,2-methyl-	(C7H14)	Benzene,(1-methylethyl)-	(C9H12)
1-Heptane	(C7H14)		

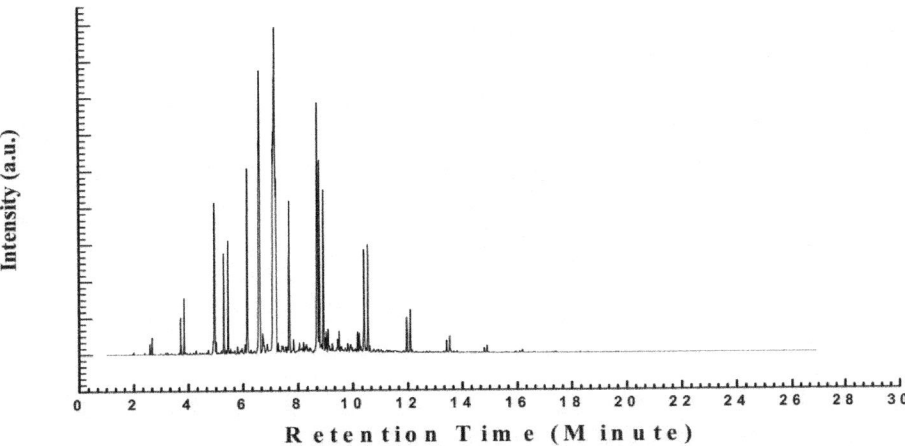

Figure 12. GC/MS Chromatogram of Mixed Waste Plastic Fuel to 2nd Fractional Fuel (Naphtha, Chemical)

Table 16. GC/MS Chromatogram Compound List of Mixed Waste Plastic Fuel to 2nd Fractional Fuel (Naphtha, Chemical)

Compound Name	Formula	Compound Name	Formula
1-Hexene	(C6H12)	Cyclopentane-butyl-	(C9H8)
Hexane	(C6H14)	Benzene,propyl	(C9H12)
1-Heptene	(C7H14)	a-methylsyrene	(C9H10)
Heptane	(C7H16)	1-Decene	(C10H20)
2,4-dimethyl-1-heptene	(C9H18)	Cyclopropane,1-heptyl-2-methyl-	(C11H22)
Ethylbenzene	(C8H10)	Undecane	(C11H24)
1-Nonene	(C9H18)	1-Dodecene	(C12H24)
Styrene	(C8H8)	Dodecane	(C12H26)
1,3,5,7-Cyclooctatetraene	(C8H8)	Tridecane	(C13H28)
Nonane	(C9H20)	Tetradecdane	(C14H30)

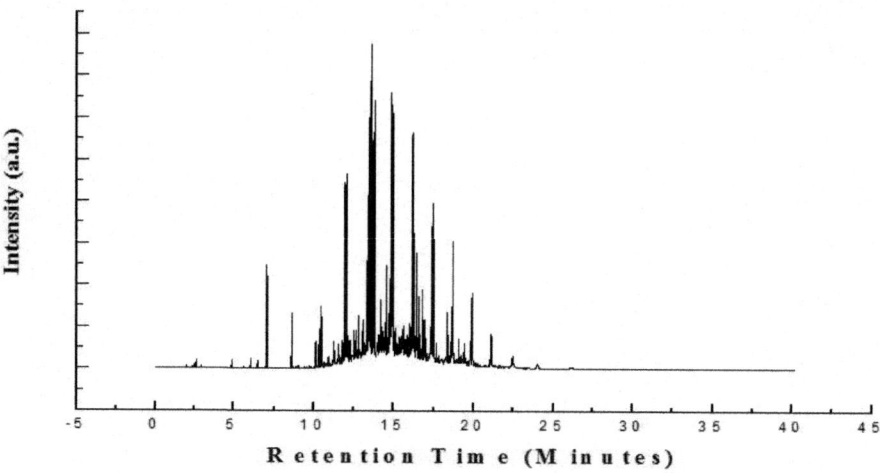

Figure 13. GC/MS Chromatogram of Mixed Waste Plastic Fuel to 3rd Fractional Fuel (Aviation)

Table 17. GC/MS Chromatogram Compound list of Mixed Waste Plastic Fuel to 3rd Fractional Fuel (Aviation)

Reten-tion Time (Min.)	Compound Name	Formula	Reten-tion Time (Min.)	Compound Name	Formula
7.04	Styrene	C8H8	14.93	Tetradecane	C14H30
8.60	a-Methylstyrene	C9H10	16.12	Cyclopen-tadecane	C15H30
10.18	Cyclooctane,1,4-dimethyl-,cis-	C10H20	16.23	Pentadecane	C15H32
10.38	1-Undecene	C11H22	17.37	1-Hexadecene	C16H32
12.07	Dodecane	C12H26	19.80	E-15-Heptade-canal	C17H32O
13.42	1-Tridecene	C13H26	19.89	Octadecane	C18H38
13.56	Tridecane	C13H28	21.13	Nonadecane	C19H40
14.81	Cyclotetradecane	C14H28	22.45	Eicosane	C20H42

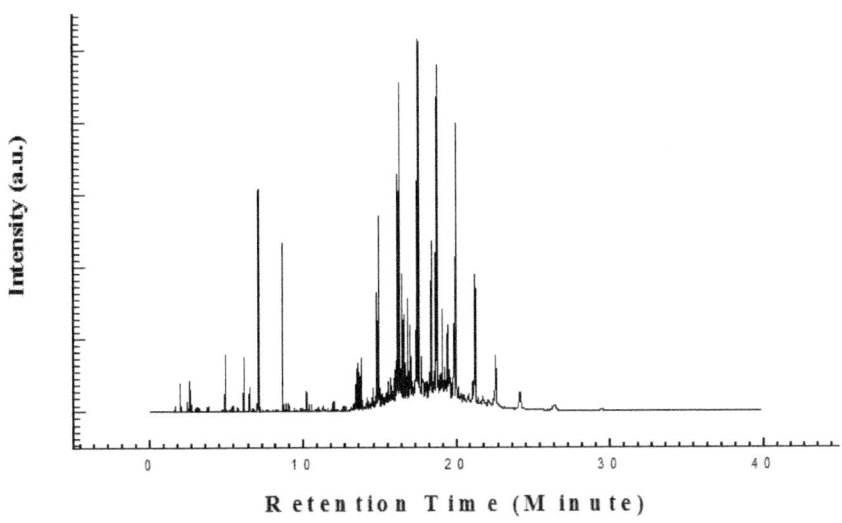

Figure 14. GC/MS Chromatogram of Mixed Waste Plastic Fuel to 4th Fractional Fuel (Diesel)

Table 18. GC/MS Chromatogram Compound List of Mixed Waste Plastic Fuel to 4th Fractional Fuel (Diesel)

Compound Name	Formula	Compound Name	Formula
Pentane	(C5H12)	1-Pentadecene	(C15H30)
1-Pentene, 2-methyl-	(C6H12)	Pentadecane	(C15H32)
Heptane, 4-methyl-	(C8H18)	1-Nonadecanol	(C19H40 O)
Toluene	(C7H8)	1-Hexadecene	(C16H32)
E-14-Hexadecenal	(C16H30 O)	Eicosane	(C20H42)
4-Tetradecene, (E)-	(C14H28)	Heneicosane	(C21H44)
Tetradecane	(C14H30)	Octacosane	(C28H58)

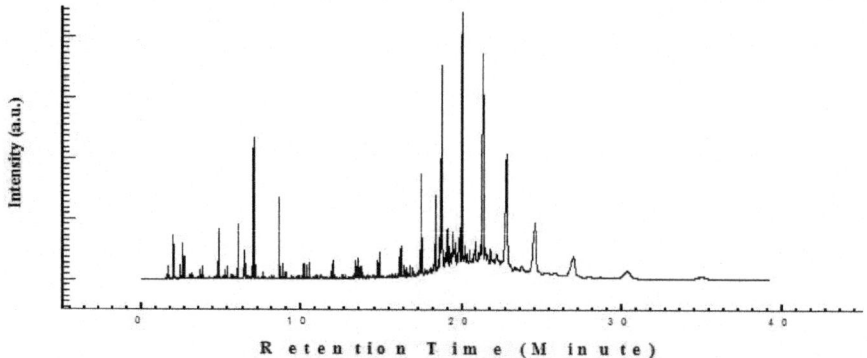

Figure 15. GC/MS Chromatogram of Mixed Waste Plastic Fuel to 5th Fractional Fuel (Fuel Oil)

Table 19. GC/MS Chromatogram Compound list of Mixed Waste Plastic Fuel to 5th Fractional Fuel (Fuel Oil)

Compound Name	Formula	Compound Name	Formula
1) 1-Propene, 2-methyl-	(C4H8)	16) Tridecane	(C13H28)
2) Pentane	(C5H12)	17) Tetradecane	(C14H30)
3)1-Pentene, 2-methyl-	(C6H12)	18) Pentadecane	(C15H32)
4) Hexane	(C6H14)	19) Hexadecane	(C16H34)
5) Heptane	(C7H16)	20) Benzene, 1,1'-(1,3-pro-panediyl)bis-	(C15H16)
6) à-Methylstyrene	(C9H10)	27) Heneicosane	(C21H44)
7) Decane	(C10H22)	28) Tetracosane	(C24H50)
8) Undecane	(C11H24)	29) Heptacosane	(C27H56)

　　　GC/MS analysis of fractional distillation fuel, a lot of compound is appeared in each individual fuel. Some of those compounds are mentioned, such as in Gasoline (1ST Fraction) we found Carbon range C_4 to C_9 and compound is 1-Propene-2-Methyl (C_3H_8) to Benzene, (1-methylethyl) - (C_9H_{12}) [Shown above, Fig.11 & Table-15]. In naphtha (2nd Fraction) Carbon range is C_6 to C_{14} and compound is 1- Hexene (C_6H_{12}) to Tetradecane ($C_{14}H_{30}$) [Shown above, Fig.12 & Table-16]. In Aviation fuel (3^{rd}Fraction) Carbon range is C_8 to C_{20} and compound is Styrene (C_8H_8) to Eicosane ($C_{20}H_{42}$) [Shown above, Fig.13 & Table-17]. In Diesel (4th Fraction) Carbon range is C_5 to C_{28} and compound is pentane (C_5H_{12}) to Octacosane ($C_{20}H_{58}$) [Shown above, Fig.14 & Table-18]. Eventually in Fuel oil (5th Fraction) Carbon range is C_4 to C_{27}, and compound is 1-Propene-2-methyl (C_4H_8) to Heptacosane ($C_{27}H_{56}$) [Shown above, Fig.15 & Table-19].

FTIR (Spectrum-100) Analysis

Analysis of Individual waste plastics (HDPE-2, LDPE-4, PP-5, and PS-6) to individual fuel:

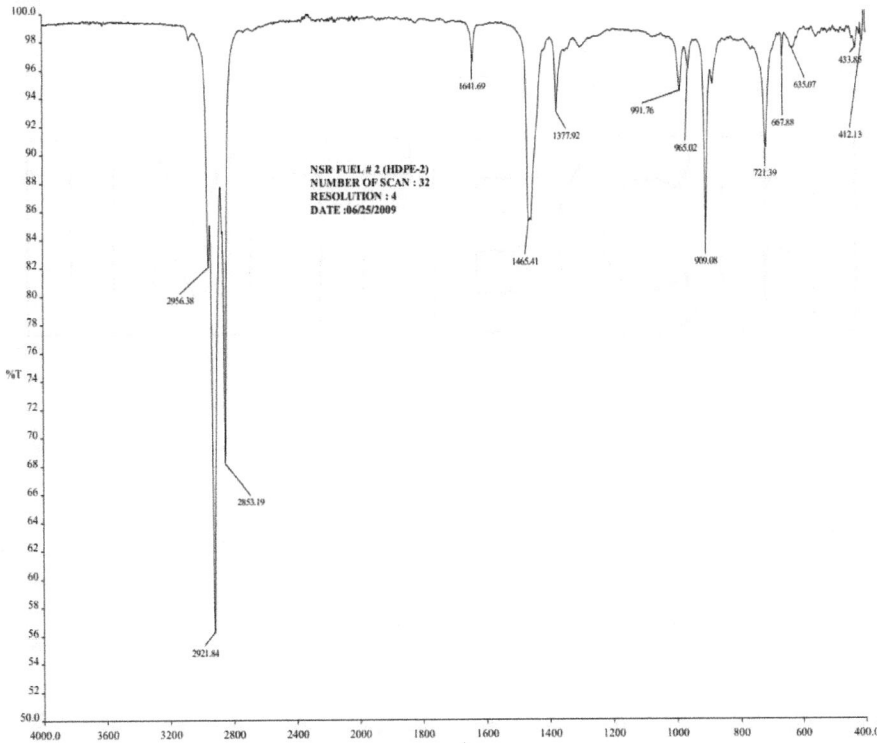

Figure 16. FTIR Spectra of HDPE-2 Plastic to Fuel

Table 20. FTIR Spectra of HDPE-2 Plastic to Fuel Functional Group Name

Band Peak Number	Wave Number (cm⁻¹)	Compound Group Name
1	2956.38	C-C$_H$3
2	2921.84	C-C$_H$3
3	2853.19	C$_H$2
4	1641.69	Non-Conjugated
5	1465.41	C$_H$3

6	1377.92	C_H3
7	991.76	-CH= C_H2
8	965.02	-CH=CH-(Trans)
9	909.08	-CH= C_H2
10	721.39	-CH=CH-(Cis)
11	667.88	-CH=CH-(Cis)

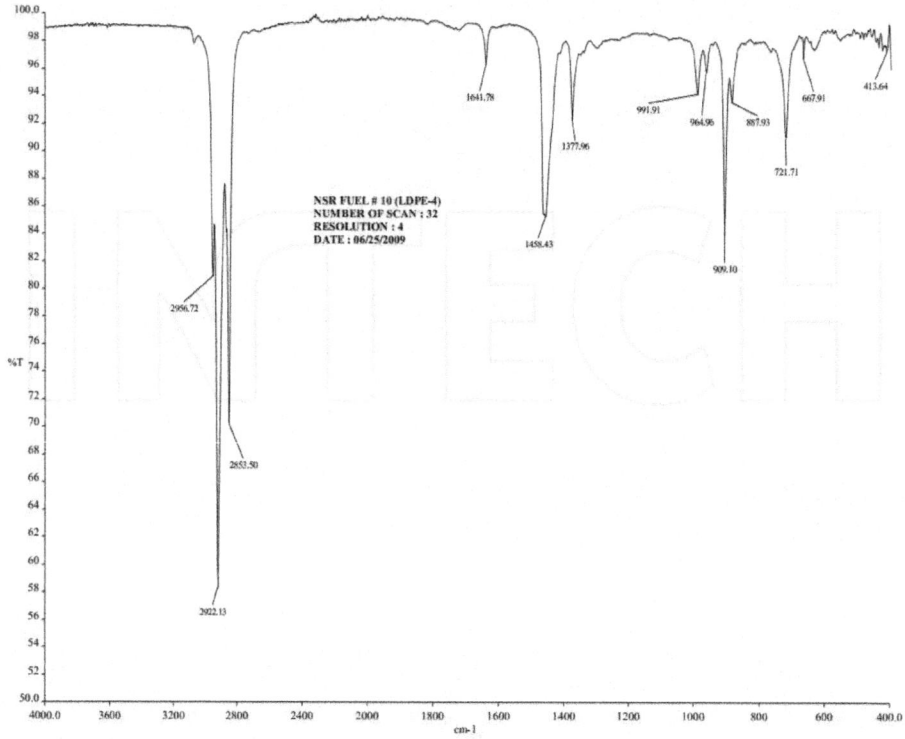

Figure 17. FTIR Spectra of LDPE-4 Plastic to Fuel

Table 21. FTIR Spectra of LDPE-4 Plastic to Fuel Functional Group Name

Band Peak Number	Wave Number (cm^{-1})	Functional Group Name
1	2956.72	C-C_H3
2	2922.13	C-C_H3
3	2853.50	C_H2
4	1641.78	Non-Conjugated

5	1458.43	C_H3
6	1377.96	C_H3
7	964.96	-CH= C_H2
8	909.10	-CH=CH-(Trans)
9	887.93	-CH= C_H2
10	721.71	-CH=CH-(Cis)
11	667.91	-CH=CH-(Cis)

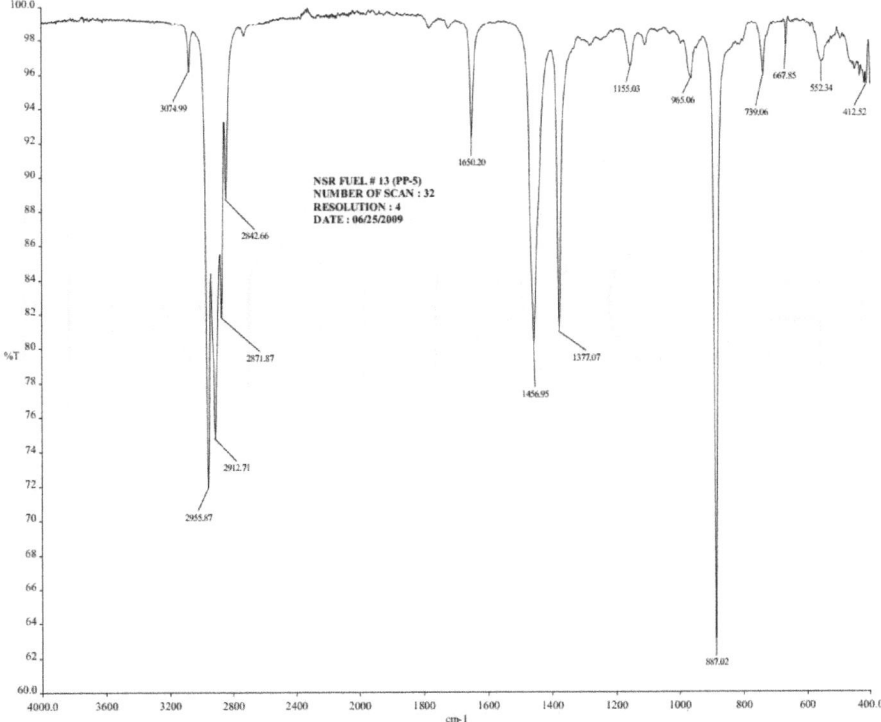

Figure 18. FTIR Spectra of PP-5 Plastic to Fue.

Table 22. FTIR Spectra of PP-5 Plastic to Fuel Functional Group Name

Band Peak Number	Wave Number (cm⁻¹)	Compound Group Name	Band Peak Number	Wave Number (cm⁻¹)	Compound Group Name
1	3074.99	H Bonded NH	8	1377.07	C_H3
2	2955.87	C-CH3	9	1155.03	

3	2912.71	C-CH3	10	965.06	-CH=CH- (Trans)
4	2871.87	C-CH3	11	887.02	$C=C_H 2$
5	2842.66	C-CH3	12	739.06	-CH=CH- (Cis)
6	1650.20	Amides	13	667.85	-CH=CH- (Cis)
7	1465.95	CH2			

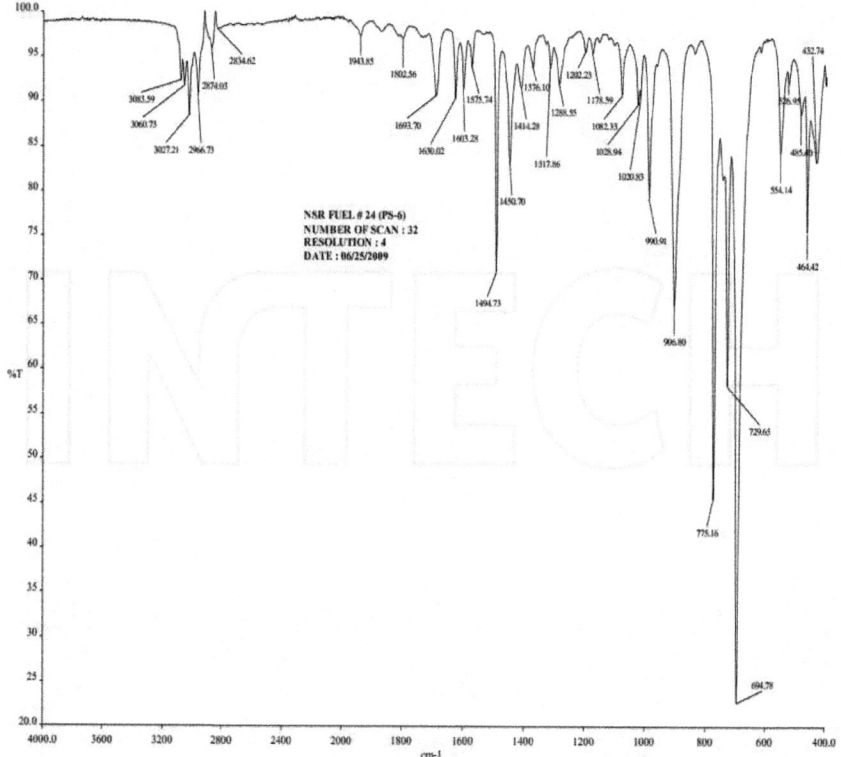

Figure 19. FTIR Spectra of PS-6 Plastic to Fuel

Table 23. FTIR Spectra of PS-6 Plastic to Fuel Functional Group Name

Band Peak Number	Wave Number (cm⁻¹)	Compound Group Name	Band Peak Number	Wave Number (cm⁻¹)	Compound Group Name
1	3083.59	=C-H	15	1414.28	$C_H 2$

2	3060.73	=C-H	16	1376.10	C_H3
3	3027.21	=C-H	17	1317.86	
4	2966.73	C-CH3	18	1288.55	
5	2874.03	C-CH3	19	1202.23	
6	2834.62	C-CH3	20	1178.59	
7	1943.85		21	1082.33	
8	1802.56	Non-Con-jugated	22	1028.94	Acetates
9	1693.70	Conjugated	23	1020.83	Acetates
10	1630.02	Conjugated	24	990.91	$-CH=C_H2$
11	1603.28	Conjugated	25	906.80	$-CH=C_H2$
12	1575.74		26	775.16	
13	1494.73		27	729.65	-CH=CH-(Cis)
14	1450.70	CH3	28	694.78	-CH=CH-(Cis)

In FTIR analysis of HDPE-2 fuel obtained functional groups are C-CH$_3$, CH$_2$, Non-Conjugated, CH$_3$,-CH=CH$_2$,-CH=CH- (Cis) and –CH=CH-(Trans) [Shown above, Fig.16& Table-20].In LDPE-4 analysis functional groups are C-CH$_3$, CH$_2$, Non-Conjugated, CH$_3$,-CH=CH$_2$,-CH=CH- (Cis) and –CH=CH-(Trans)[Shown above, Fig.17 &Table-21].In PP-5 analysis functional groups are CH$_3$,C-CH$_2$,-CH=CH- (Cis) and,-CH=CH- (Trans). [Shown above, Fig.18 & Table-22] Subsequently in PS-6 analysis obtained functional groups are CH2, CH3, Acetates,-CH=CH2 and –CH=CH-(Cis) etc. [Shown above, Fig.19 & Table-23].

FTIR Analysis of Mixed Waste Plastics to Fuel:

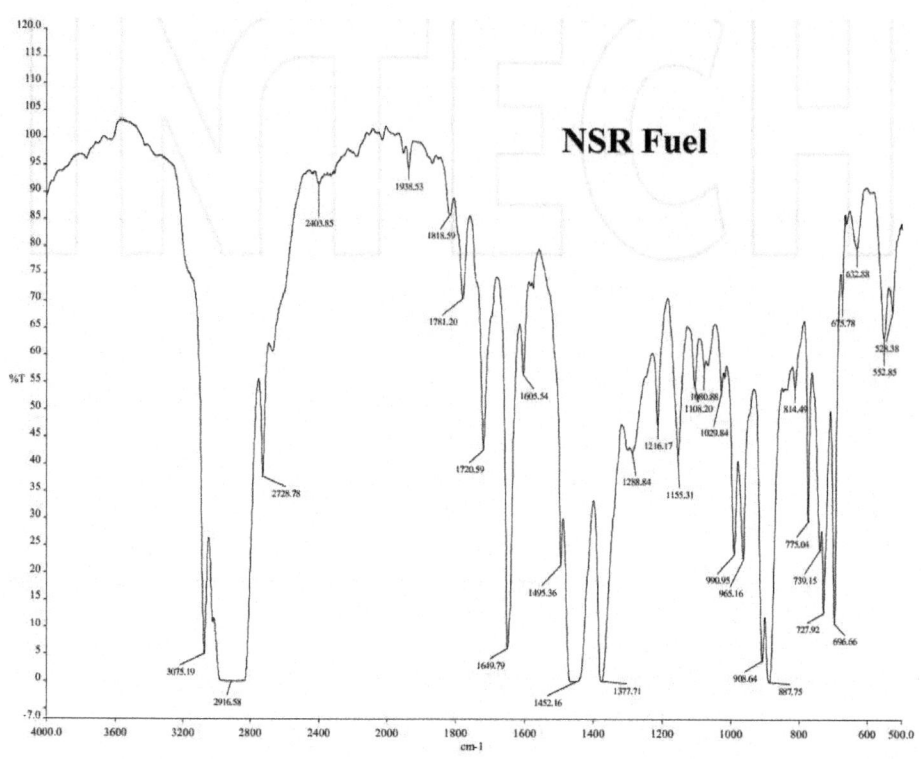

Figure 20. FTIR Spectra of Mixed Waste Plastic to Fuel

Table 24. FTIR Spectra of Mixed Waste Plastic to Fuel Functional Group Name

Band Peak Number	Wave Number (cm⁻¹)	Functional Group Name	Band Peak Number	Wave Number (cm⁻¹)	Functional Group Name
1	3075.19	H Bonded NH	13	1377.71	C_H3
2	2916.58	CH2	19	1029.84	Acetates
3	2728.78	C-CH3	20	990.95	Secondary Cyclic Alcohol
5	1938.53	Non-Conjugated	21	965.16	-CH=CH- (trans)
6	1818.59	Non-Conjugated	22	908.64	$-CH=C_H2$
7	1781.20	Non-Conjugated	23	887.75	$C=C_H2$
8	1720.59	Non-Conjugated	26	739.15	-CH=CH- (cis)
9	1649.79	Amides	27	727.92	-CH=CH- (cis)

| 10 | 1605.54 | Non-Conjugated | 28 | 696.66 | -CH=CH- (cis) |
| 12 | 1452.16 | CH2 | 29 | 675.78 | -CH=CH- (cis) |

In FTIR analysis of mixed waste plastics to NSR fuel obtained functional groups are: CH_3, Acetates, Secondary Cyclic Alcohol,-$CH=CH_2$, $C=CH_2$,-CH=CH-(Cis) and -CH=CH-(Trans) etc. [Shown above, Fig. 20 & Table-24].

FTIR Analysis of Mixed Waste Plastics to Fractional Distillation Fuel:

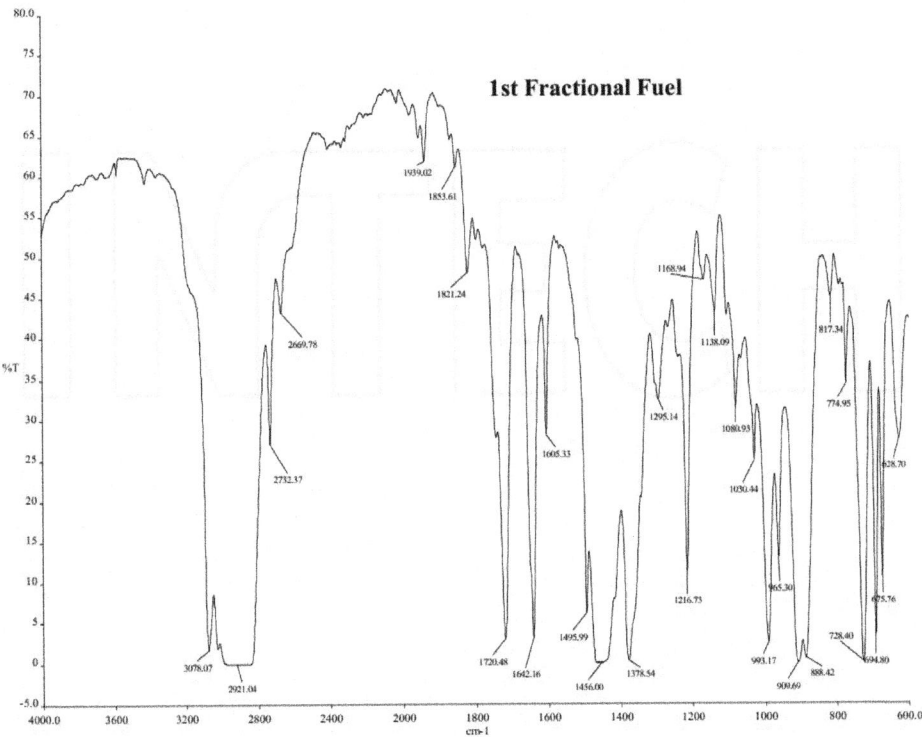

Figure 21. FTIR Spectra of Mixed Waste Plastic Fuel to 1st Fractional Fuel (Gasoline)

Table 25. Mixed Waste Plastic Fuel to 1st Fractional Fuel (Gasoline) FTIR Functional Group List

Band Peak Number	Wave Number (cm⁻¹)	Functional Group Name	Band Peak Number	Wave Number (cm⁻¹)	Functional Group Name
1	3078.07	H Bonded NH	13	1378.54	C_H3
2	2921.04	C-CH3	19	1030.44	Acetates

3	2732.37	C-CH3	20	993.17	Secondary Cyclic Alcohol
4	2669.78	C-CH3	21	965.30	-CH=CH- (trans)
6	1853.61	Non-Conjugated	22	909.69	-CH=C$_H$2
7	1821.24	Non-Conjugated	23	888.42	C=C$_H$2
8	1720.48	Non-Conjugated	26	728.40	-CH=CH- (cis)
9	1642.16	Conjugated	27	694.80	-CH=CH- (cis)
10	1605.33	Conjugated	28	675.76	-CH=CH- (cis)
12	1456.00	CH3	29	628.70	-CH=CH- (cis)

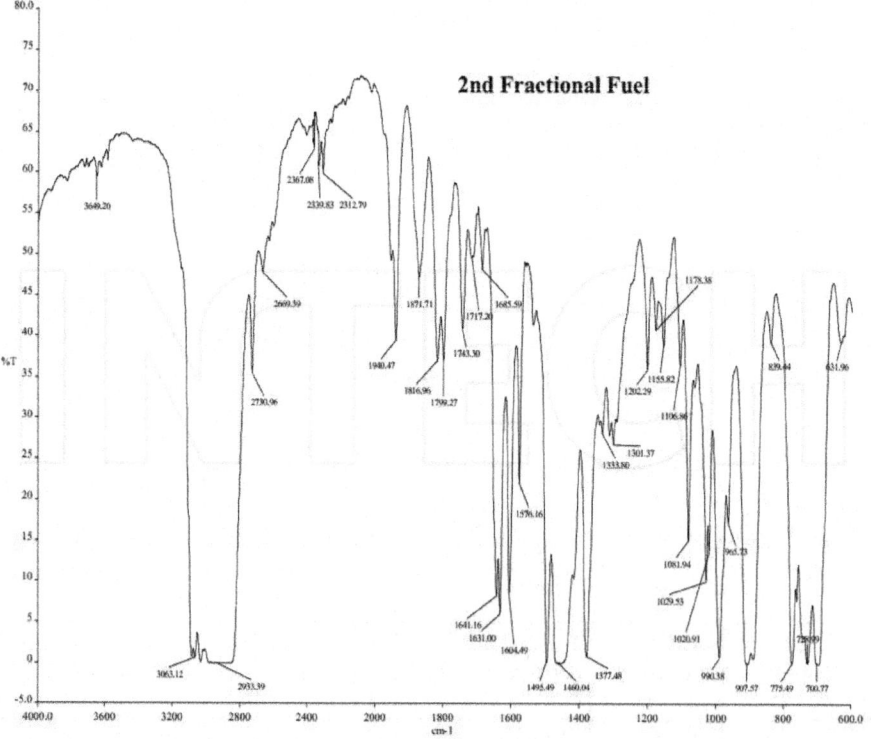

Figure 22. FTIR Spectra of Mixed Waste Plastic Fuel to 2nd Fractional Fuel (Naphtha,

Chemical)

Table 26. Mixed Waste Plastic Fuel to 2nd Fractional Fuel (Naphtha) FTIR Functional Group List

Band Peak Number	Wave Number (cm⁻¹)	Functional Group Name	Band Peak Number	Wave Number (cm⁻¹)	Functional Group Name
2	3063.12	=C-H	16	1641.16	Non-Conjugated
3	2933.39	$C-C_H3$	17	1631.00	Non-Conjugated
4	2730.96	$C-C_H3$	21	1460.04	C_H3
5	2669.39	C-CH3	22	1377.48	C_H3
9	1940.47	Non-Conjugated	30	1029.53	Acetates
10	1871.71	Non-Conjugated	31	1020.91	Acetates
11	1816.96	Non-Conjugated	32	990.38	$-CH=C_H2$
12	1799.27	Non-Conjugated	33	965.73	-CH=CH- (trans)
13	1743.30	Conjugated	34	907.57	$-CH=C_H2$
14	1717.20	Non-Conjugated	37	728.99	-CH=CH- (cis)
15	1685.59	Conjugated	38	700.77	-CH=CH- (cis)

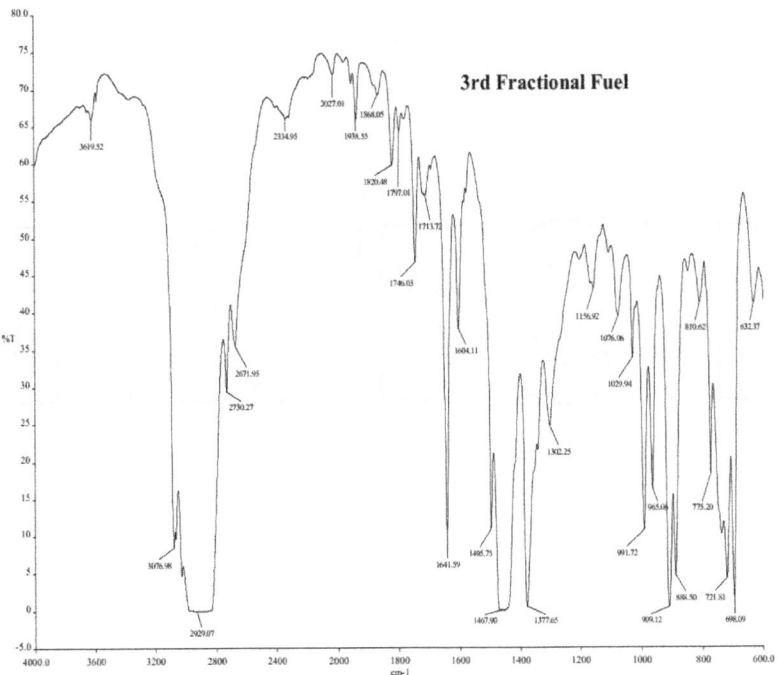

Figure 23. FTIR Spectra of Mixed Waste Plastic Fuel to 3rd Fractional Fuel (Aviation)

Table 27. Mixed Waste Plastic Fuel to 3rd^t Fractional Fuel (Aviation) FTIR Functional Group List

Band Peak Number	Wave Number (cm^{-1})	Functional Group Name	Band Peak Number	Wave Number (cm^{-1})	Functional Group Name
3	2929.07	C-C$_H$3	17	1467.90	C$_H$3
4	2730.27	C-CH3	18	1377.65	C$_H$3
5	2671.93	C-CH3	22	1029.94	Acetates
8	1938.55	Non-Conjugated	23	991.72	-CH=C$_H$2
9	1868.05	Non-Conjugated	24	965.06	-CH=CH- (trans)
10	1820.48	Non-Conjugated	25	909.12	CH=C$_H$2
11	1797.01	Non-Conjugated	26	888.50	C=C$_H$2
12	1746.03	Non-Conjugated	29	721.81	-CH=CH- (cis)
13	1713.72	Non-Conjugated	30	698.09	-CH=CH- (cis)
14	1641.59	Non-Conjugated			

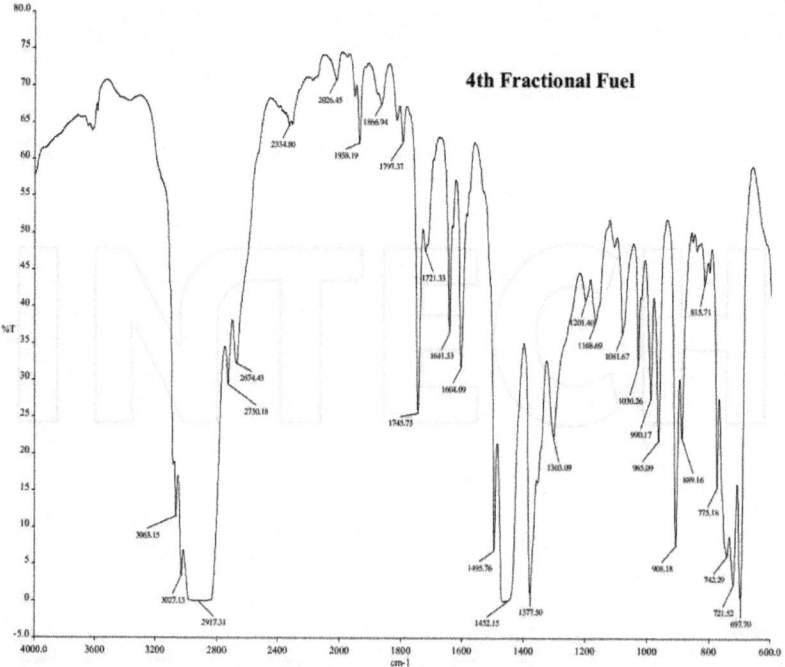

Figure 24. FTIR Spectra of Mixed Waste Plastic to Fuel (Diesel)

Table 28. Mixed Waste Plastic Fuel to 4th Fractional Fuel (Diesel) FTIR Functional Group List

Band Peak Number	Wave Number (cm⁻¹)	Functional Group Name	Band Peak Number	Wave Number (cm⁻¹)	Functional Group Name
1	3063.15	=C-H	16	1452.15	C_H2
2	3027.13	=C-H	17	1377.50	C_H3
3	2917.31	CH2	22	1030.26	Acetates
4	2730.18	C-CH3	23	990.17	-CH=C_H2
5	2674.43	C-CH3	24	965.09	-CH=CH- (trans)
8	1938.19	Non-Conjugated	25	908.18	-CH=C_H2
9	1866.94	Non-Conjugated	26	889.16	C=C_H2
10	1797.37	Non-Conjugated	29	742.29	-CH=CH- (cis)
11	1745.73	Non-Conjugated	30	721.52	-CH=CH- (cis)
12	1721.33	Non-Conjugated	31	697.70	-CH=CH- (cis)
13	1641.33	Non-Conjugated			

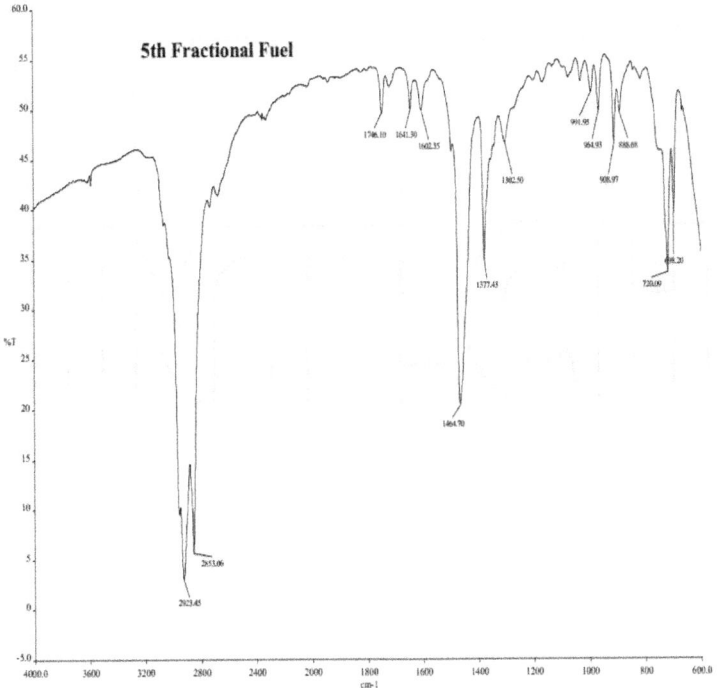

Figure 25. FTIR Spectra of Mixed Waste Plastic to Fuel (Fuel Oil)

Table 29. Mixed Waste Plastic Fuel to 5th Fractional Fuel (Fuel Oil) FTIR Functional Group List

Band Peak Number	Wave Number (cm^{-1})	Functional Group Name	Band Peak Number	Wave Number (cm^{-1})	Functional Group Name
1	2923.45	C$_H$2	9	991.95	Secondary Cyclic Alcohol
2	2853.06	C$_H$2	10	964.93	-CH=CH- (trans)
3	1746.10	Non-Conjugated	11	908.97	-CH=C$_H$2
4	1641.30	Non-Conjugated	12	888.68	C=C$_H$2
5	1602.35	Non-Conjugated	13	720.09	-CH=CH- (cis)
6	1464.70	CH2	14	698.20	-CH=CH- (cis)
7	1377.43	CH3			

In FTIR analysis of fractional distillation fuel such as in 1ST Fraction Fuel (Gasoline) obtained functional groups are CH$_3$, Acetates, Secondary Cyclic Alcohol, -CH=CH$_2$, C=CH$_2$,nad -CH=CH- (Cis). [Shown above, Fig.21 & Table-25]. In 2nd Fraction Fuel (Naphtha) analysis functional groups are CH$_3$, Non-Conjugated, Acetates,-CH=CH$_2$,-CH=CH- (Cis) and –CH=CH-(Trans). [Shown above,Fig.22& Table-26]. In 3rd Fraction Fuel (Aviation) analysis functional groups are CH$_3$, Acetates, C-CH$_2$,-CH=CH- (Cis) and -CH=CH- (Trans) [Shown above, Fig.23& Table-27]. In 4th Fraction Fuel (Diesel) analysis functional groups are CH$_2$, CH$_3$, Acetates,-CH=CH$_2$, C=CH$_2$ and,-CH=CH- (Cis) [Shown above, Fig.24 & Table-28]. Subsequently in 5th Fraction Fuel (Fuel Oil) analysis obtained functional groups are Secondary Cyclic Alcohol,-CH=CH$_2$, C=CH$_2$, –CH=CH (Trans) and –CH=CH-(Cis) etc. [Shown above, Fig.25& Table-29].

ELECTRICITY PRODUCTION FROM WASTE PLASTIC FUEL

Both NSR fractional fuels (NSR fractional 1st Fractional Fuel and NSR 4th Fractional Fuel) have been used to produce electricity by the help of conventional internal combustion generator. A flow diagram illustrating the process of energy production and consumption from NSR Fuel (Heating Oil) is shown below in Fig.26.

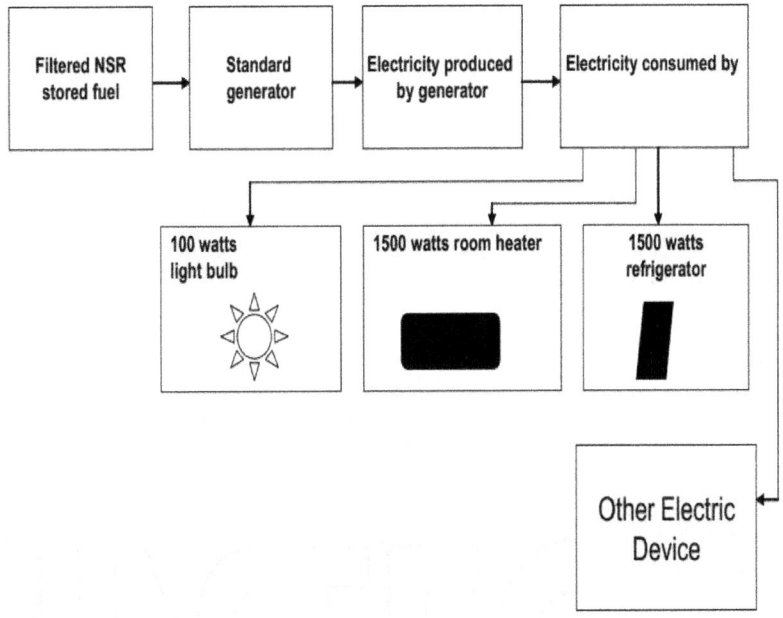

Figure 26. Flow diagram of electricity generation consumption

NSR fractional 1st collection fuel was used in a gasoline generator with max 4.0 kW and volt output of 120. ~1 litter of fractional fuel was injected in the generator and with ~2900 watt constant demand; the generator ran a total of 42 minutes. A similar test was performed with commercial gasoline (87). ~1 litter of commercial gasoline (87) was injected and with the same ~ 2900 watt, constant demand the generator ran a total of 38 minutes. The difference in time occurs because NSR fraction 1st collection fuel has longer Carbon content than that of the commercial gasoline (87).

NSR fractional 4th collection fuel was used in a diesel generator with a max 4.0 kW and an output of 120 volt. ~1 litter of NSR fractional 2nd collection fuel was injected in the generator and with a constant demand of 3200 watt; the generator ran a total of 42 minutes. The same test was conducted with commercial diesel, and with the same demand the generator ran for 34 minutes.

A diagram [Fig.27] is provided below showing the produced electricity consumption of commercial gasoline (87) and NSR fractional fuel 1st collection.

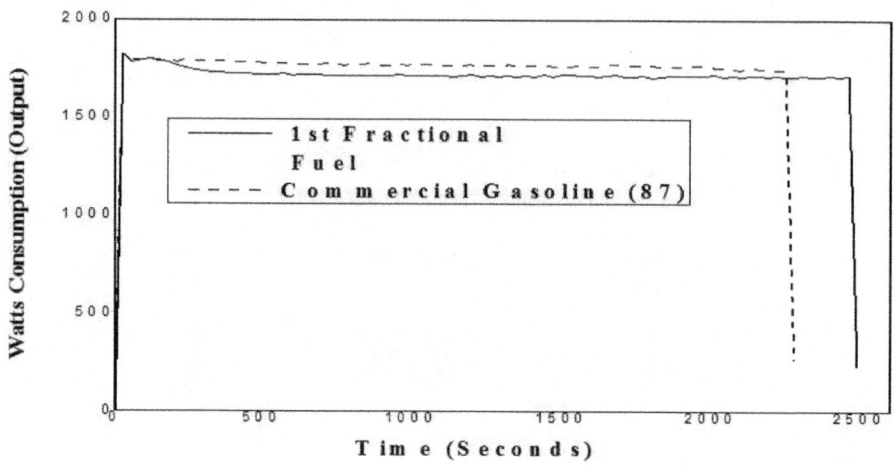

Figure 27. Electricity Consumption and run time monitored by EML 2020 logger system for 1st Fractional Fuel (Gasoline) and Commercial Gasoline87.

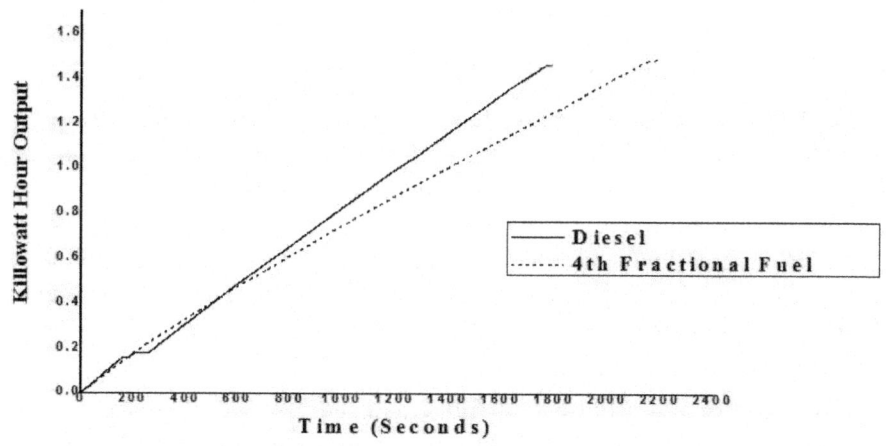

Figure 28. Electricity Output Comparison Graph of Waste Plastic Fuel to 4th Fractional Fuel and Commercial Diesel Fuel

Table 30. Comparison Table of 4th Fractional Fuel (Diesel) and Commercial Diesel

Fuel Name	Generator	Fuel Amount	Duration	kWh
4th Fractional Fuel (Diesel)	AMCO	1 Liter	37 min	2.028

Comparison of NSR 4th fraction fuel and commercial diesel was conduced using an AMCO Diesel Generator. Above, Fig. 28 and Table 30 demonstrate the comparative results between the two fuels. The results indicate that the NSR-2 fuel provided a longer run time of the generator than the diesel. This is due to the NSR fuel having longer carbon chains than the diesel fuel.

AUTOMOBILE TEST DRIVING

Both NSR fractional 5th collection fuel and commercial gasoline (87) was used for a comparison automobile test. A 1984 Oldsmobile vehicle (V-8 powered engine) was used for the test-drive and one gallon of fuel was used for both cases after complete drainage of the pre-existing fuel in the fuel tank. The test-drive was done on a rural highway with an average speed of 55 mph.

Based on the preliminary automobile test-drive, the NSR fuel has offered a competitive advantage in mileage over the commercial gasoline-87. NSR fuel showed better mileage performance of 21 miles per gallon (mpg) compared to 18 mpg with commercial gasoline (87).

It is expected that NSR double condensed fuel will show even higher performance with more fuel-efficient car such as V-4 engine and hybrid vehicles. Additional test-driving is going to be conducted in the near future to verify the results.

CONCLUSION

The conversion of municipal waste plastics to liquid hydrocarbon fuel was carried out in thermal degradation process with/without catalyst. Individually we ran our experiment on waste plastics such as: HDPE-2, LDPE-4, PP-5 & PS-6. Each of those experiment procedures are maintained identically, every ten (10) minutes of interval experiment was monitored and found during the condensation time changes of individual waste plastics external behavior different because of their different physical and chemical properties. Similarly, we ran another experiment with 2kg of mixture of waste plastics in stainless steel reactor. Initial temperature is 350 °C for quick melting and optimum temperature is 305 °C. For glass reactor every experiment temperature was maintained by variac meter, when experiment started variac percent was 90% (Tem-405 °C) for quick melting, after melted variac percent decreased to 70% (Tem- 315 °C) due to smoke formation. Average (optimum) used variac percent in this experiment 75% (337.5 °C).Gradually temperature range was maintained by variacmeter with proper monitoring. In fractional distillation process we separated different category of fuel such as gasoline, naphtha, jet fuel, diesel and fuel oil in accordance with their boiling point temperature profile.

ACKNOWLEDGEMENTS

The author acknowledges the support of Dr. Karin Kaufman, the founder and President of Natural State Research, Inc (NSR). The author also acknowledges the valuable contributions NSR laboratory team members during the preparation of this manuscript.

REFERENCES

1.   J. Aguado, D. P. Serrano, G. Vicente, N. Sa´nchez, EnhancedProduction. of-Olefinsr. by Thermal. Degradation of. High-Density Polyethylene. . H. D. P. E. Decalin in Solvent Effect of the Reaction Time Temperature Ind Eng. Chem. Res. 2007 46 34973504

2.   Marcilla. Antonio, A. Ä. ngela, N. Garcı´a, Remedio. Maria del, Thermal. Herna´ndez, of. L. D. P. E. Degradation-Vacuum, Oil. Gas, for. Mixtures, Wastes. Plastic, Energy. Valorization, Fuels, 2007 21 870880

3.   K. Achyut, A.B. Panda, R. K. Singh, 2010 D.K. Mishra b,2, Thermolysis of waste plastics to liquid fuel A suitable method for plastic waste management and manufacture of value added products-A world prospective, Renewable and Sustainable Energy Reviews 14 233248

4.   N. Miskolczia, L. Barthaa, G. Dea´ka, B. Jo´, Thermal. verb, of. degradation, plastic. municipal, for. waste, of. production, hydrocarbons. fuel-like, Degradation. Polymer, . Stability, 2004 357-366

5.   Miranda. Miguel, a,, Pinto. a. I. Filomena, a. I. Gulyurtlu, a. C. A. Cabrita, a. Nogueira, Matos. b. Arlindo, surface. Response, optimization. methodology, to. applied, tyre. rubber, wastes. plastic, conversion. thermal, . Fuel, 2010 2217-2229

6.   M. Stelmachowski, . , conversion. Thermal, waste. of, to. polyolefins, mixture. the, hydrocarbons. of, the. in, with. reactor, metal. molten, Conversion. Energy, . Management, 2010 2016 EOF2024 EOF -2024

7.   Karishma Gobin, George Manos*, Polymer degradation to fuels over microporous catalysts as a novel tertiary plastic recycling method, Polymer Degradation and Stability 83 2004 2004 267279

8.   Ding. Weibing, Liang. Jing, L. Larry, Anderson, Hydrocracking, of. Hydroisomerization, Polyethylene. High-Density, Plastic. Waste, Zeolite. over, Ni. Silica-Alumina-Supported, Sulfides. Ni-Mo, Energy, Fuels, 1997 1997 11 12191224

9.   Warren. Anthony, . Mahmoud-Halwagi El, economic. An, for. study, co-generation. the, liquid. of, fuel, from. hydrogen, coal, solid. municipal, Fuel. waste, Technology. . Processing, 1996 157 EOF166 EOF -166

10. Wei-Chiang Huang a,b,c, Mao-Suan Huang 1 Chiung-Fang Huang a,b,c, Chien-Chung Chen c,e,*, Keng-Liang Ou c,e,f,**, Thermochemical conversion of polymer wastes into hydrocarbon fuels over various fluidizing cracking catalysts, Fuel 89 2010 23052316

11. Valerio Cozzani, † Cristiano Nicolella,‡ Mauro Rovatti,‡ and Leonardo Tognotti*,†, Influence of Gas- Phase Reactions on the Product Yields Obtained in the Pyrolysis of Polyethylene, Ind. Eng. Chem. Res. 1997 1997 36 342348

12. A. Marcilla, M. I. Beltra´n, R. Navarro, with. Evolution, Temperature. the, the. of, Obtained. Compounds, the. in, Pyrolysis. Catalytic, Polyethylene. of, H. U. S. Y. over, Ind, Eng. Chem. Res. 2008 2008 47 68966903

13. Herna´ndez. A. Ä. del Remedio, N. ngela, Amparo. Garcı´a, Javier. Go´mez, Agullo´, Marcilla. Antonio, of. Effect, Time. Residence, Volatile. on, Obtained. Products, the. H. D. P. E. in, in. Pyrolysis, Presence. the, of. H. Z. S. Absence, 5 Ind. Eng. Chem. Res. 2006 45 87708778

14. A. Paula, Filomena. J. Costa,*,†, Ana. M. Pinto,†, Ibrahim. K. Ramos,‡, Isabel. A. Gulyurtlu,†, Cabrita,†, S. Maria, Kinetic. Bernardo‡, of. Evaluation, Pyrolysis. the, Polyethylene. of, Energy. Waste, Fuels, 2007 2007 21 24892498

15. Saha†. Biswanath, K. Aloke, Ghoshal*, Hybrid Genetic Algorithm and Model-Free Coupled Direct Search Methods for Pyrolysis Kinetics 5 ZSM-5 Catalyzed Decomposition of Waste Low-Density Polyethylene, Ind. Eng. Chem. Res. 2007 46 54855492

16. R. W. J. Westerhout, J. Waanders, J. A. M. Kuipers, W. P. M. van Swaaij, of. Recycling, Polyethene, in. a. Polypropene, Bench. Novel-Scale, Cone. Rotating, by. Reactor, Pyrolysis. High-Temperature, Ind, Eng. Chem. Res. 1998 1998 37 22932300

17. R. W. J. Westerhout, J. Waanders, J. A. M. Kuipers, W. P. M. van Swaaij, of. a. Development, Rotating. Continuous, Reactor. Cone, Plant. Pilot, the. for, of. Pyrolysis, Polyethene, Ind. Polypropene, Eng. Chem. Res. 1998 1998 37 23162322

18. R. W. J. Westerhout, R. H. P. Balk, R. Meijer, J. A. M. Kuipers, W. P. M. van Swaaij, Examination, of. Evaluation, Use. the, Screen. of, for. Heaters, Measurement. the, the. of, Temperature. High, Kinetics. Pyrolysis, Polyethene. of, Ind. Polypropene, Eng. Chem. Res. 1997 1997 36 33603368

19. Tang. Lan, H. Huang, Zhao. C. Z. Zengli, Wu, Y. Chen, of. Pyrolysis, in. a. Polypropylene, Plasma. Nitrogen, Ind. Reactor, Eng. Chem. Res. 2003

2003 42 11451150

20. George Manos,*,† Arthur Garforth,‡ and John Dwyer§, Catalytic Degradation of High-Density Polyethylene on an Ultrastable-Y Zeolite. Nature of Initial Polymer Reactions, Pattern of Formation of Gas and Liquid Products, and Temperature Effects, Ind. Eng. Chem. Res. 2000 39, 1203-1208,

21. George Manos,*,† Arthur Garforth,‡ and John Dwyer§, Catalytic Degradation of High-Density Polyethylene over Different Zeolitic Structures, Ind. Eng. Chem. Res. 2000 2000 39 11981202

22. Yoshio Uemichi,* Junko Nakamura, Toshihiro Itoh, and Masatoshi Sugioka, Conversion of Polyethylene into Gasoline-Range Fuels by Two-Stage Catalytic Degradation Using Silica-Alumin and 5 Zeolite, Ind. Eng. Chem. Res. 1999 38 385390

23. Manos,. George, †. Isman, Y. Yusof, ‡. Nikos, Papayannakos,§, H. Nicolas, Gangas§, Catalytic Cracking of Polyethylene over Clay Catalysts. Comparison with an Ultrastable Y Zeolite,Ind. Eng. Chem. Res. 2001 2001 40 22202225

24. M. Jose´, Arandes,*,† In˜ aki Abajo,† Danilo Lo´ pez-Valerio,§ Inmaculada Ferna´ ndez,† Miren J. Azkoiti,‡ Martı´n Olazar,† and Javier Bilbao†, Transformation of Several Plastic Wastes into Fuels by Catalytic Cracking, Ind. Eng. Chem. Res. 1997 1997 36 45234529

25. Selhan Karago¨z,†,§ Jale Yanik,*,‡ Suat Ucüar,† and Chunshan Song§, Catalytic Coprocessing of Low-Density Polyethylene with VGO Using Metal Supported on Activated Carbon, Energy & Fuels 2002 2002 16 13011308

26. Toshiyuki Kanno, Masahiro Kimura, Na-oki Ikenaga, and Toshimitsu Suzuki*, Coliquefaction of Coal with Polyethylene Using Fe (CO) 5 as Catalyst, Energy & Fuels 2000 14 612617

27. Nahid. Mohammad, a,. Siddiqui, Hamid. Halim, b. Redhwi, coprocessing. Catalytic, waste. of, plastics, residue. petroleum, liquid. into, oils. J. fuel, Anal. Appl. Pyrolysis 86 2009 2009 141147

28. Nakamura. . Ikusei, Fujimoto. Kaoru, of. Development, disposable. new, for. catalyst, plastics. waste, for. treatment, quality. high, fuel. transportation, Today. . Catalysis, 1996 175-179

29. A. G. Buekens, . , H. Huang, plastics. Catalytic, for. cracking, of. recovery, hydrocarbons. gasoline-range, municipal. from, wastes. plastic, Conservation. Resources, . Recycling, 1998 163 EOF -181

Chapter 6

LAB-SCALE EVALUATION OF TWO BIOTECHNOLOGIES TO TREAT VOC AIR EMISSIONS: COMPARISON WITH A BIOTRICKLING PILOT UNIT INSTALLED IN THE PLASTIC COATING SECTOR

F. Javier Álvarez-Hornos, Feliu Sempere, Marta Izquierdo and Carmen Gabaldón

GI2AM Research Group, Department of Chemical Engineering, University of Valencia, Burjassot, Spain

INTRODUCTION

Volatile organic compounds (VOCs) are one of the top five atmospheric pollutants, and, according to an EC directive, are defined as "all organic compounds arising from human activities, other than methane, which are capable of producing photochemical oxidants by reactions with nitrogen oxides in the presence of sunlight" (Council Directive 2001/81/EC). This definition highlights the fact that VOCs play a vital role in the formation of tropospheric ozone, which causes photochemical smog. Short-term exposure to photochemical smog affects respiratory function and has adverse effects on plants (World Health Organization, 2004). The distinction between biogenic and anthropogenic VOCs in the atmosphere is far from straightforward, because many VOC species are produced by both sources (Popescu&Ionel, 2010). Anthropogenic sources of VOCs include air emissions from wastewater treatments plants, motor vehicles, gasoline storage facilities and transportation, dry cleaning and other industrial sources (D.J. Kim & H. Kim, 2005). In this sense, the main sectors involved in non-methane VOC emissions in the EU-27 are solvent and product use (41%), road and non-road transportation (18%), and commercial, institutional, and household associated emissions (14%) (European Environment Agency, 2010). Regarding the industrial sources, Fig. 1 illustrates the contributions from various industrial sectors to EU-27 non-methane VOC industrial emissions in 2008 (European Pollutant Release and Transfer Register, 2008). The three most important industrial sources are: energy (41%); the chemical industry (22%); and coating and surface treatment activities (18%). In fact, over the past decade, emerging European Union environmental policy has focused on abatement of VOCs from industrial emissions, in an effort to protect environmental and public health. As a result of these initiatives, new European VOC emission limits have been established in the VOC Solvent Emissions Directive (Council Directive 1999/12/EC) for a wide range of industrial sectors. Currently, VOC concentration limits range from 50 to 150 mg C/Nm3, depending on the application and solvent consumption.

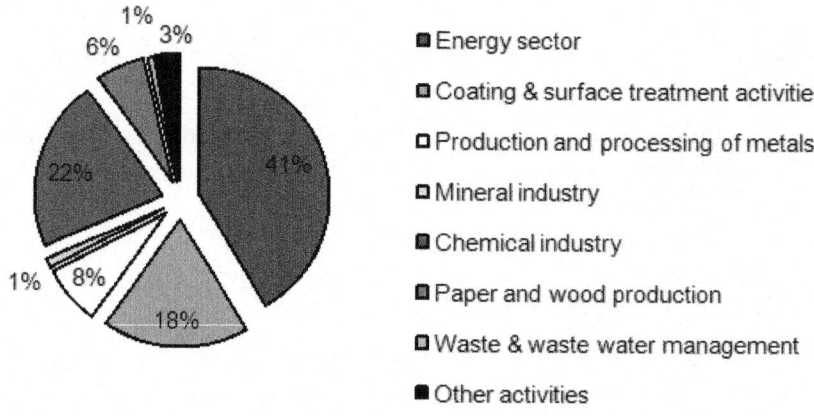

Figure 1. Distribution of non-methane VOC industrial emissions in the EU-27 in 2008 by industrial sector (adapted from European Pollutant Release and Transfer Register, 2008).

Although process changes andthe substitution of solvent-based products for water-based ones have the potential to minimise VOC emissions, stringent VOC emission limits require additional treatment technologies, better known as 'end-of-pipe' techniques. For years, these techniques have been primarily based on non-biological methods, such as condensation, adsorption, absorption/scrubbing and thermal destruction. However, over the last decade, vapour-phase biotechnologies, including biofilters, biotrickling filters and bioscrubbers, have proven to be both efficient and environmentally-friendly for the treatment of VOC emissions, and have been classified as best available technologies (BATs) for the reduction of VOC emissions in the chemical sector by the European IPPC Bureau (European Commission, 2003). Thus, biotechnologies are a potential alternative to conventional physicochemical processes for the removal of VOCs from high flow rate emissions streams with relatively low VOC concentrations: conditions which are common in painting, coating and printing processes. For example, replacing a conventional thermal oxidiser with a biotreatment system for the control of VOC emissions from a panel board press reduced greenhouse gas emissions by 60 to 80 percent, and operating costs by 90 percent (Boswell, 2009).

Vapor-phase biotechnologies are based on the capability of microorganisms of utilizing their metabolism to transform the organic pollutants to less toxic compounds. However, becauseVOC pollutants are in the air, they must first be transferred from the gas phase to an aqueous phase, where biodegradation can occur. Through biodegradation, contaminantsare used as energy and carbon sources for microbial growth, and are converted to carbon dioxide and water. The main vapour-phase biotechnologies available for the treatment of VOC emissions include biofilters (BF), biotrickling filters (BTF) and bioscrubbers (BS). Although the basic VOC removal mechanisms are similar in these systems, there are notable differences with respect to the aqueous phase and microorganism growth, which are summarised in Table 1.

Table 1. Classification of vapour-phase biotechnology systems.

Biotechnology system	Microorganism growth	Aqueous phase
Biofilter	Attached growth	Stationary
Biotrickling filter	Attached growth	Flowing
Bioscrubber	Suspended growth	Flowing

Biofilters work by passing polluted pre-humidified air through a porous packed bed of natural organic material,in which a culture of pollutant-degrading microorganisms is developed (Fig. 2 a). The packing material is a key factor for the successful application of biofilters: it is necessary to choose media with adequate physical and chemical properties, such as high surface area, long-term stability, low pressure drop, low-cost, good moisture retention, pH buffering capacity, appropriate adsorbent capacity and nutrients (Shareefdeen& Singh, 2005). In addition, because moisture content control is also critical, biofiltersystems usually incorporate some kind of water addition. Use of an occasional nutrient supply is also advisable. For biotrickling filters (Fig. 2 b), the polluted air is passed through inert packing material while a liquid stream is re-circulated over the bed. In this case the biofilm is developed on the packing surface, with the liquid phase providing nutrients to the biofilm, allowing for greater pH control and yielding a more stable operation in comparison with biofilters. These characteristics, along with a larger air/liquid specific surface area, lead to higher VOC removal rates than those obtained with conventional biofilters (Koutinas et al., 2005), suggesting that smaller biotrickling filters can be installed with lower capital investment for industrial applications.

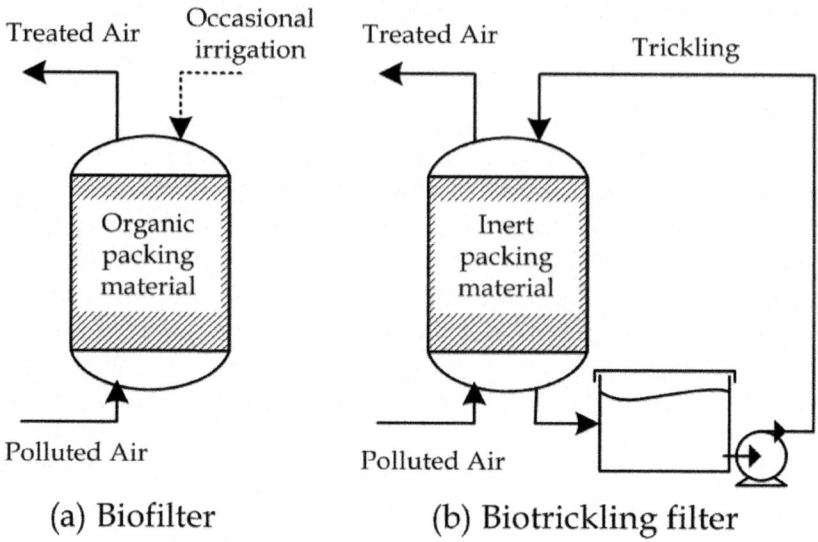

Figure 2. Schematic of vapour-phase biotechnology systems.

The number of laboratory studies on the control of continuous VOC emissions using biofilters (BF) and biotrickling filters (BTF) has significantly increased over the last two decades. These studies have mainly focused on evaluating process performance as a function ofthe operational conditions (e.g. gas empty bed residence time, inlet load or nutrient formulation and concentration), and/or packing material characteristics, with the aim of obtaining valuable information from these biological technologies. BFs and BTFs have been applied in the treatment of streams contaminated with a wide variety of pollutants, including oxygenated compounds (Cox &Deshusses, 2001; Steele et al., 2005), aromatics (Hwang et al., 2008; Kennes et al., 1996), or its mixtures (Paca et al., 2006; Sempere et al., 2008).

However, studies dealing with the removal of complex VOC mixtures from painting and coating processes using BFsor BTFs are limited, especially for pilot/full-scale units. Emissions from these processes typically result in streams with high flow rates and relatively low VOC concentrations, which usually contain a complex mixture of hydrophobic (e.g. toluene and xylenes) and hydrophilic (e.g. n-butyl acetate, ethyl acetate and methyl propyl ether) compounds. Concerning laboratory scale experiments, Mathur & Majumder (2008) investigated methyl ethyl ketone (MEK), toluene, n-butyl acetate and o-xylene elimination using a coal-based BTF, and reported a maximum elimination capacity (EC) of 185 g m^{-3} h^{-1} for an inlet load (IL) of 278 g m^{-3} h^{-1}, working at an empty bed residence time (EBRT) of 42.4 s. Similarly, Higuchi et al. (2010) observed a maximum EC of 87.5 g m^{-3} h^{-1} for removal of 2-butanone, butyl acetate, butoxyl, toluene, ethylbenzene and xylene, using a BTF packed with poly-vinyl formal (PVF) material operated at an inlet concentration of 0.43 g m^{-3}and an EBRT of 12 s. These data indicate that, although biofiltration is suitable in terms of overall VOC removal and has shown significant potential, it still requires additional optimization. It is thus essential to assess BFs and the BTFs in real situations using pilot scale units, with the aim of obtaining valuable and useful information necessary for scale-up. Webster et al. (1999) installed a 0.47 m^3-volume BTF pilot unit to treat off-gases from two spray paint booths, achieving removal efficiency (RE) higher than 70% working at an EBRT between 11 and 39 s for the target pollutants: MEK, methyl isobutyl ketone, o-xylene, m-xylene, p-xylene and n-butyl acetate. Martinez-Soria et al. (2009)evaluated introduction of an activated carbon prefilter in the treatment of VOC emissions from spray paint booths in the wood furniture industry, using a 0.75 m^3-volume BTF pilot unit: the prefilter buffered VOC fluctuations, ensuring that legal limits were met while working at an EBRT of 24 s.

This chapter presents studies conducted to assess environmentally friendly biotechnologies,such asbiofilters and biotrickling filters, for VOC abatement in air at two scales. First, a laboratory-scale study was designed to investigate the use of a biofilter (BF) and a biotrickling filter (BTF)under continuous feeding conditions,for VOC removal from air contaminated with three compounds commonly found in air emissions from paint and coating processes: n-butyl acetate, toluene and m-xylene (a 2:1:1 weight mixture was usedtosimulate exhaust gases). These compounds have been previously identified as representativeVOCs in paints (Boswell et al., 2001). Second, a biotricklingfilter pilot unit was usedto assess treatment of exhaust gases from a robotic spray paint booth at a plastic coating facility (located in Soria, Spain),which is a supplier of car mirrors to the automotive sector.The performance of this pilot-scale BTF was compared with results obtained from the lab-scale systems.

MATERIALS AND METHODS

Two experimental phases were carried out during this research: (1) laboratory scale experiments using both biofilter and biotrickling filter, and (2) pilot-scale operation of a biotrickling unit connected to a robotic spraypaint booth at a plastic coating facility.

Laboratory-Scale Systems

Treatment of air polluted with a 2:1:1 (wt) n-butyl acetate:toluene:m-xylene mixture was studied both in a biofilter and in a biotrickling filter at the laboratory scale.

A schematic of the laboratory-scale biofilter set-up is shown in Fig. 3. The lab-scale BF was made of methacrilate, with a total length of 97 cm and an internal diameter of 13.6 cm. The BF was equipped with 5 equidistant sampling ports to measure VOC concentrations, and four additional ports for temperature measurements and filter bed sampling. Fibrous peat (ProEcoAmbiente, Spain) was used as the filter material. Because the peat was acidic, the pH was adjusted to neutral using a diluted sodium hydroxide solution. Compressed, filtered and dried-air was passed through two serial-humidifiers, to assure a relative humidity of $\geq 90\%$. The EBRT was adjusted using a mass flow controller (Bronkhorst Hi-Tec, The Netherlands). Pollutant was introduced to the air stream using a syringe pump (New Era, infusion/withdraw NE 1000 model, USA) and then, air polluted with the VOC mixture was flowed downward into the bed.

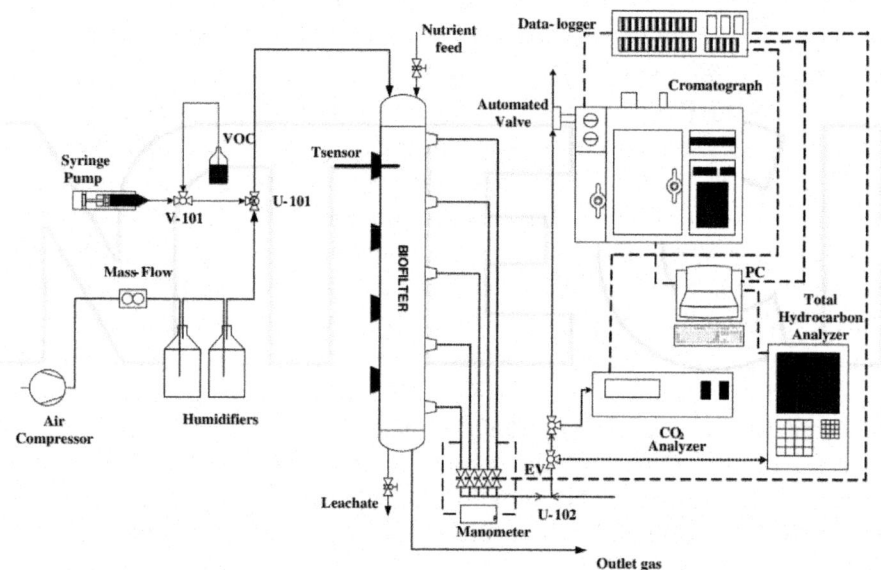

Figure 3. Schematic of the laboratory-scalebiofilter (BF).

A schematic of the laboratory-scale biotrickling filter is shown in Fig. 4. The lab-scale BTF was composed of 3 cylindrical modules of Plexiglas, with a total bed length of 120 cm and an internal diameter of 14.4 cm, and was equipped with a recirculation tank (effective volume, 10 L). The BTF was randomly filled with 1-inch of nominal diameter Flexiring™polypropylene rings (Koch-Glistch B.V.B.A., Belgium). Compressed, filtered and dried-air air was polluted with VOCs as above, using a syringe pump (New Era, infusion/

withdraw NE 1000 model, USA). Then, VOC contaminated air was introduced through the bottom of the column, andthe flow rate was adjusted using mass flow controllers (Bronkhorst Hi-Tec, The Netherlands). The recirculation stream was introduced counter to the air flow, and was partially renewed (50 – 100% of total volume) every week.

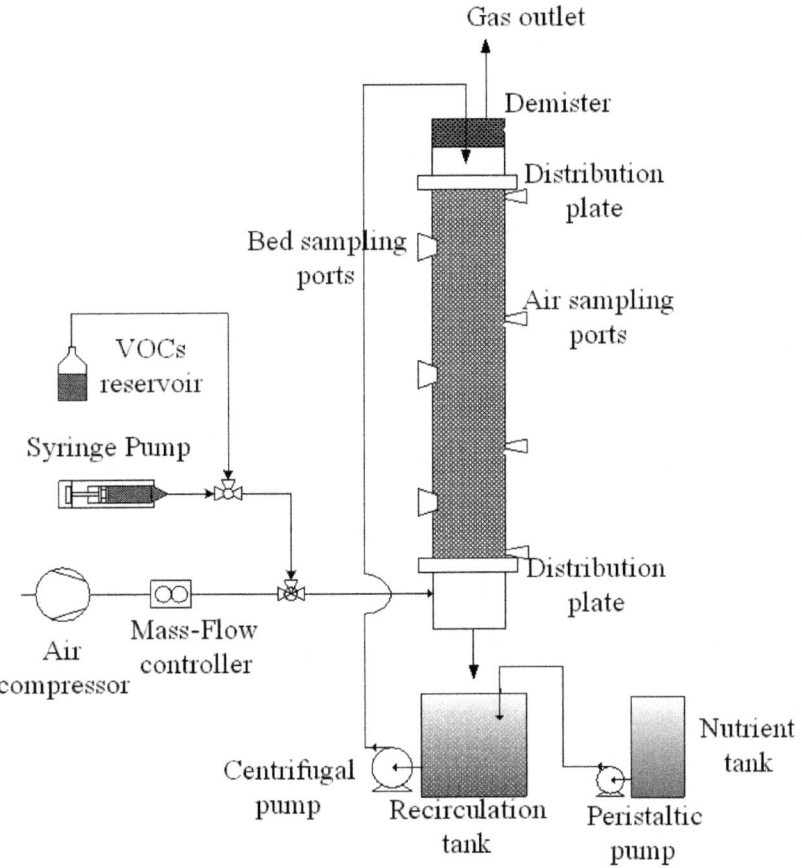

Figure 4. Schematic of the laboratory-scalebiotrickling filter (BTF).

The main properties of the peat and Flexiring™ ring packing materials used in the BF and the BTF set-ups, respectively, are shown in Table 2.

Table 2. Physical and chemical properties of the packing materials.

Fibrous peat	
Organic content, % wt	95

BET specific surface area, m2 g-1	13.4
pH	4.8
Bulk density, kg m-3	133
Water-holding capacity, % wt	88
Shape	Fibres
Fibre length, cm	2 – 8
FlexiringTM	
Specific surface area, m2 m-3	207
Void fraction, %	92
Bulk density, kg m-3	71
Shape	Rings
Ring diameter, cm	24.5

The necessary macro and micronutrients were incorporated using a pH buffered nutrient solution containing N (3 g L^{-1}), P (0.6 g L^{-1}), Ca, Fe, Zn, Co, Mn, Mo, Ni and B at trace doses. The nutrient solution was incorporated in the biofilter by directly pouring 100 mL per day on the top the column and was supplied to the biotrickling filter using a peristaltic pump, whose flow rate was adjusted to keep nitrogen concentrations>10 mg L^{-1} in the recirculation solution.

The two bioreactors were inoculated witha mixed microbial culture (obtained from activated sludgewhich was adapted to the compounds to be treated). For acclimation of the microbes, the VOC mixture was continuously fed to an activated sludge from the secondary clarifier of Carraixet Wastewater Treatment Plant (located in Alboraya, Spain), for a period of at least two months in order to obtain an adapted inoculum. The Carraixettreatment plant receives urban sewage from Alboraya town and pollutants from the Alboraya industrial site. Inoculation of the BF was performed by mixing the peat with a 1 L of the inoculum. For the BTF, 1.5 L of the inoculum was added into the recirculation tank.

The BF and the BTF were operated under continuous feeding conditions for a total period of 4 months and 3 months, respectively. For the BF, an EBRT of 60 s (gas flow rate of 0.85 m^3h^{-1}) and an IL of 19 g C m^{-3} h^{-1}, corresponding to an inlet total VOC concentration of 300 mg C Nm^{-3}, were applied. For the BTF, 6 experiments, each with duration of 2 weeks, were performed at an EBRT between 60 and 15 s (gas flow rate between 1.2 and 4.7 m^3 h^{-1}) for IL varying from 11 to 72 g C m^{-3} h^{-1}, corresponding to VOC concentrations between 160 mg C Nm^{-3} and 350 mg C Nm^{-3}.

Analytical Techniques

The total VOC concentration was measured by using a total hydrocarbon analyser equipped with an FID detector (Nira Mercury 901 model, Spirax-Sarco, Spain). The composition of the gas streams were monitored using a gas chromatograph (7890 model, Agilent Technologies, USA) equipped with a 1.0 mL automated gas valve injection system, a flame ionisation detector and an HP-5 capillary column (30 m × 0.32 mm × 0.25 µm, Agilent Technologies, USA). The gas carrier was helium,and a flow-rate of 9.4 ml min^{-1} was used. The injector, oven and detector temperatures were 180, 50 and 250 °C, respectively.

CO$_2$ concentrationsin the influent and effluent gas streams were periodically determined using a CARBOCAP® carbon dioxide analyser (GM70 model, Vaisala, Finland). The pressure drop was monitored daily with a digital manometer (KIMO, MP101 model, Spain). Temperature and pH were also measured daily for the biofilter leachate. The moisture content of the biofilter media was determined once a week at two locations (upper and lower), using the dry weight method. Similarly, the conductivity and pH of the BTF recirculation solution were analyseddaily. Soluble chemical oxygen demand (COD), suspended solids, and nitrate concentrations in the recirculation solution were periodically measured.

Pilot-Scale Unit

The pilot-scale plant was supplied by Pure Air Solutions B.V. (The Netherlands). Pure Air Solutions has developed a biotrickling pilot-scale unit, using its innovative abatement technology (VOCUS™Biotrickling Filter System). A schematic of this unit is presented in Fig. 5. The biological reactor, a column with a volume of 0.75 m^3, was randomly filled with two inches of nominal diameter Flexiring™ propylene rings (Koch-Glistch B.V.B.A., Belgium), with a 93% void fraction. The bioreactor was operated in counter-current mode. VOC polluted air from the factory was introduced below the column at a flow rate between 34 and 90 m^3 h^{-1}. Recirculated water was poured on the top of the filter media at a flow rate of 1.2 m^3 h^{-1} and the spraying frequency was fixed at 20 minutes per hour.

The trickled water was collected in a 0.4 m^{3}recirculation tank. The liquid level of this tank was controlled, so that fresh water was added when the level of water decreased as a consequence of evaporation. Recycled water was fully drained and replaced with fresh water once a month. A nutrient dosing system composed of a 100 L nutrient vessel and a dosing pump was set-up in the pilot unit. The nutrient solution (pH = 7, 3 g N L^{-1}, 0.6 g P L^{-1}, trace elements) was added to the recirculation tank at a maximum rate of 0.15 L h^{-1} in order to achieve a $C_{degraded}/N_{supplied}$ mass ratio above 30 – 40. The system was equipped with a programmable logic controller, and a set of sensors and devices enabling control and monitoring of the plant via modem communication.

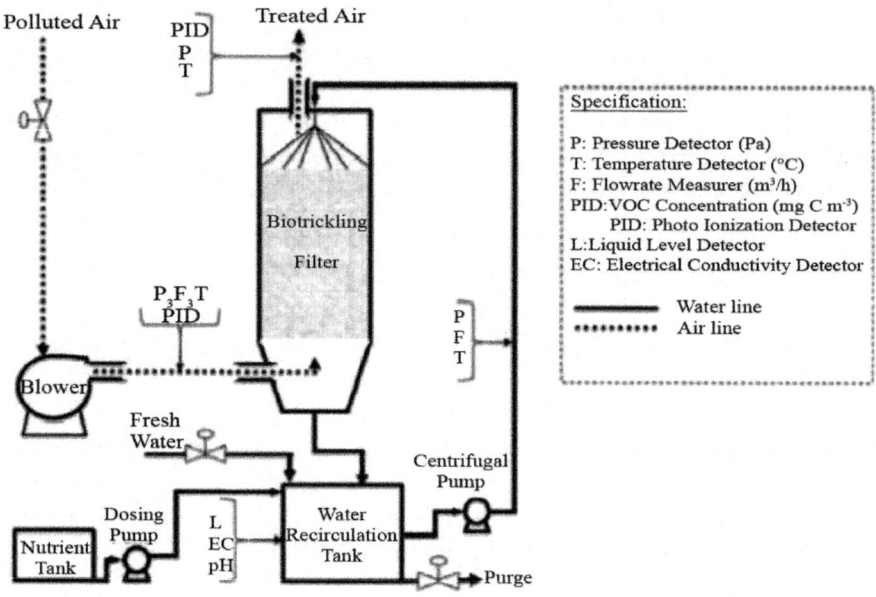

Figure 5. Schematic of the pilot-scalebiotrickling filter unit (VOCUS™).

The pilot-scale biotricklingsystem was inoculated with activated sludge from the municipal Wastewater Treatment Plant (located in Soria, Spain) without further acclimation, in order to simulate operational protocols at an industrial site. An inoculum volume of 100 L was added to the recirculation tank and continuously flowed through the bed for 72 hours.

At the plastic coating facility, rear view mirrors for cars are coated in three serial robotic spray booths (primer, base coat and clear coat layers). From the three existing sources of VOC emissions, the clear coating spray booth was selected as the focus for this work. The gaseous emissions from this booth were piped to the pilot unit, and contained a mixture of oxygenated and aromatic compounds coming from the applied product formulation (a mixture of a specific clear coat with its thinner) in the coating process. The pilot-scale unit was operated at EBRTs between 30 and 80 seconds (air flows between 34 and 90 m³ h⁻¹) over a period of 3 months, in order to determine the minimum EBRT value that enables legal regulations to be met (emission limit value (ELV): an average value of 75 mg C Nm⁻³for all valid readings, with none of the hourly averages exceeding the ELV by more than a factor of 1.5).

ANALYTICAL TECHNIQUES

Inlet and outlet gas temperatures, inlet and outlet total VOC concentrations (measured using two photo ionisation detectors, or PIDs), air flow, the pressure drop between the gas inlet and the outlet of the media bed, tank levels, and the conductivity, pH and temperature of the trickling solution were continuously monitored. A total hydrocarbon analyzer (Nira Mercury 901 model, Spirax-Sarco, Spain) was periodically used to check and calibrate the PIDs sensors. Samples of the recirculated water were collected to analyze COD, suspended solids and nitrogen and phosphorus content.

Parameters for Characterization of Biodegradation Performance

In general, the performance of the above described biotechnologies is evaluated using the parameters defined below:

Empty bed residence time (s):

$$EBRT = V_f / Q \tag{1}$$

where Q = air flow rate ($m^3 s^{-1}$) and V_f = filter bed volume (m^3).

Removal efficiency (%):

$$RE = 100 \cdot \left(1 - C_o / C_i\right) \tag{2}$$

where C_i and C_o = inlet and outlet pollutant concentration, respectively (g Nm^{-3}).

Inlet load (g m^{-3} h^{-1}):

$$IL = C_i Q / V_f \tag{3}$$

Elimination capacity (g m^{-3} h^{-1}):

$$EC = \left(C_i - C_0\right) Q / V_f \tag{4}$$

RESULTS AND DISCUSSION

Laboratory-Scale Systems

The performance parameters of both biotechnologies are summarised in Table 3 for the different stages of the experimental plan. The performance of each unit during the entire operation period is shown inFig. 6,illustrating the different stages described in Table 3. The evolution of inlet and outlet total VOC concentrations and the total removal efficiency (RE) for the biofilter and biotrickling filter are plotted in Fig. 6 (a) and 6 (b), respectively. The average temperature in the reactors was 25.3 ± 1.6 and 21.2 ± 2.3 C for BF and BTF, respectively. Biofilter leachate pH was maintained at approximately 5.8 ± 0.2, whilethe pH and conductivity of the BTF recirculation solution were 8.7 ± 0.3 and 4.3 ± 1.4 mS cm^{-1}, respectively. Pressure drop ranged from 330 to 500 Pa m^{-1} for the BF unit, and from 40 to 280 Pa m^{-1}for the BTF system, indicating the absence of clogging. Soluble COD values for the BTF recirculation solution were periodically analysed, and werestableat approximately 154 ± 56 mg L^{-1}, representing less than 2% of the weekly inlet load fed to the system. Therefore, the organic carbon quantity removed in the purge was considered negligible. For the BF unit, the moisture content of the media (a key parameter for optimum biofilm development), was maintained at appropriate values and a slight stratification was observed: values (wet basis, %wt) varied between $78.4\% \pm 2.9\%$ in the upper zone and $80.7\% \pm 1.7\%$ in the lower zone of the bed.

Table 3. Operational and performance parameters for both systems on the different stages.

	Days	EBRT, s	IL, g C m-3 h-1	EC, g C m-3 h-1	RE, %
Biofilter					
Stage BF1	0 – 112	60.0 ± 1.3	18.8 ± 1.5	17.7 ± 1.4	94.1 ± 2.7
Biotrickling filter					
Stage BTF1	0 – 12	59.1 ± 0.1	11.2 ± 0.8	8.5 ± 0.8	74.6 ± 6.4
Stage BTF2	12 – 36	59.1 ± 0.1	21.1 ± 1.9	11.5 ± 1.7	54.8 ± 7.6
Stage BTF3	36 – 50	30.0 ± 0.3	19.6 ± 1.4	13.0 ± 1.4	66.0 ± 2.5
Stage BTF4	50 – 63	30.0 ± 0.3	31.7 ± 2.9	21.2 ± 3.3	66.6 ± 4.4
Stage BTF5	63 – 79	15.1 ± 0.7	39.6 ± 1.7	21.2 ± 2.4	53.5 ± 5.4
Stage BTF6	79 – 92	15.1 ± 0.7	72.4 ± 1.2	32.7 ± 1.3	45.3 ± 2.2

After inoculation, both systems were operated at an approximate EBRT of 60 s, with moderate ILs (18.8 and 11.2 g C m^{-3} h^{-1} for BF and BTF, respectively). High and stable REs were reached in 2 – 5 days for both bioreactors, indicating appropriate development of the inoculum. For the BF unit, total REsremained at very high values (as high as 97%) throughout the entire experimental period. In the case of the BTF unit, no complete removal was obtained, even for the lower applied IL (stage BTF1), due to the low removal of toluene and m-xylene in comparison with the complete degradation of n-butyl acetate. Kinney & Moe (2004) also observed lower biodegradability of aromatic compounds in the treatment of gas emissions contaminated with 2-pentanone, n-butyl acetate, ethyl 3-ethoxypropianate, toluene and p-xylene, using a biotrickling filter.

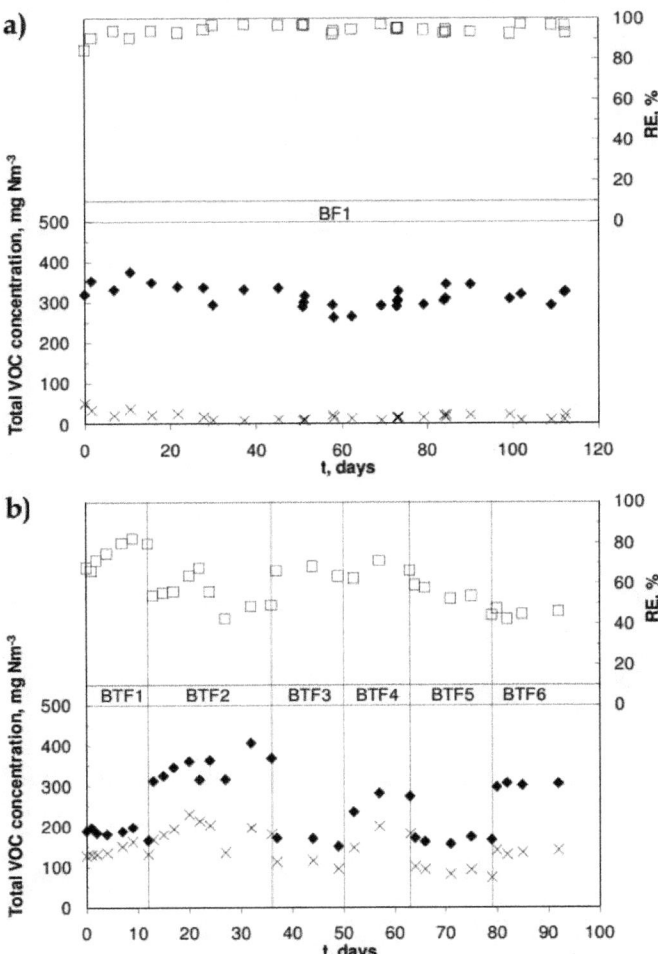

Figure 6. Evolution of inlet VOC concentrations (♦),outlet VOC concentrations (x) and total RE values (□) in (a) BF and (b) BTF with time.

The decrease in the EBRT caused a progressive drop in the total RE of the BTF system.For example, at an inlet VOC concentration of approximately 170 mg C Nm^{-3}, average RE values decreased from 75 to 54% for EBRTs rangingfrom 59 to 15 s (stages BTF1 and BTF5, respectively). At this concentration, the BTF was able to meet legal limits (ELV), even for the lowest EBRT (IL of 39.6 g C m^{-3} h^{-1}, stage BTF5). Similarly, the total RE was negatively affected when the inlet VOC concentration was duplicated. In this case, total RE values increased slightly, from 45 to 54%, when IL was dropped from 72.4 to 39.6 g C m^{-3} h^{-1}(stages BTF6 and BTF5, respectively).

The variation of EC with IL at different EBRTs for the biofilter and the biotrickling filter is plotted inFig. 7. ECswere calculated from the top of the bioreactors to each sampling port (first quarter, half, three-quarters and total bed volume for the BF; and first third, two-thirds and total bed volume for the BTF). Maximum ECs for both systems were estimated from data taken from the first section of the bed. EC values were nearly directly proportional to the total IL of the VOC mixture: up to 50 g C m^{-3}h^{-1} for the BF, and 20 g C m^{-3} h^{-1} for the BTF. No significant differences between both biotechnologies were observed for the maximum EC values. A maximum EC value of 45 g C m^{-3} h^{-1}was obtained for both systems. It is worth noting the slight influence of the applied EBRT (which ranged between 15 and 60 s) on EC values for the applied IL in the biotrickling filter.

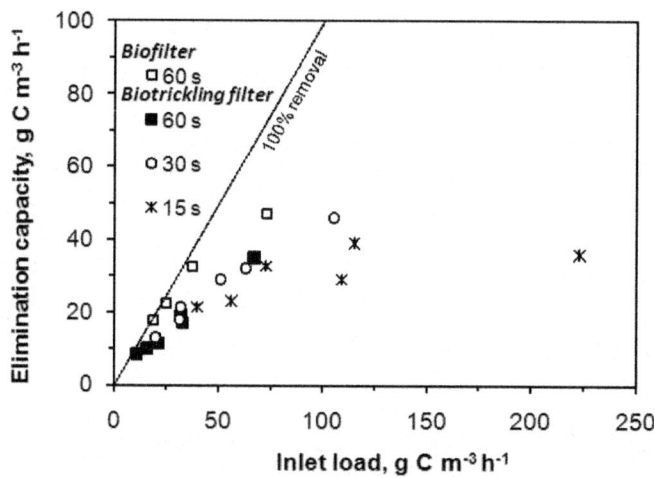

Figure 7. Variation of EC vs. IL at different EBRTs for the biofilter and biotrickling filter.

The evolution of CO_2 production as a function of elimination capacity (EC) is presented in Fig. 8 for both biotechnologies. A proportional ratio exists between EC and CO_2 production. As can be seen, the biofilterdatum follows the same trend as the biotrickling filter results. Linear regression of these data yielded a value of 1.74 g CO_2 g C^{-1}. Assuming a general biomass composition formula of $C_5H_7O_2N$, and negligible organic and inorganic carbon removalduring the periodic purges, overall yield coefficient (defined as g C of dry biomass synthesised per g C of substrate consumed) can be determined from the following biodegradation reaction balance:

$$C_xH_yO_z + aO_2 + NH_3 \longrightarrow bC_5H_7O_2N + cCO_2 + dH_2O \tag{5}$$

yielding a value of 0.53.

The degradation of each of the three compounds of the VOC mixture is shown in Fig. 9, where normalised pollutant gas concentration profiles along the bed length of the BF and BTF have been plotted. For the BTF, concentration profiles from the three highest ILs are shown. As can be seen in the Fig. 9, complete removal of n-butyl acetate was always observed in the first two-thirds of the bioreactors for all stages. Greater penetration was observed for aromatic compounds, with individual removal efficiencies of approximately 30% for each aromatic compound in the first quarter of the BF, whereas no degradation of these compounds was obtained in the first third of the BTF. Among aromatic compounds, greater emissions were observed for m-xylene, which is less biodegradable than toluene. Previous studies have reported similar phenomena (Álvarez-Hornos et al., 2007; Paca et al., 2006).

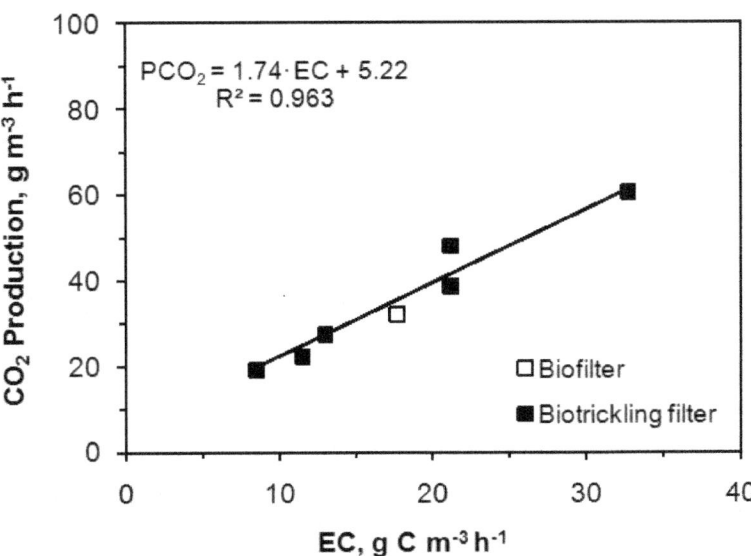

Figure 8. The relationship between EC and CO_2 production for both biotechnologies.

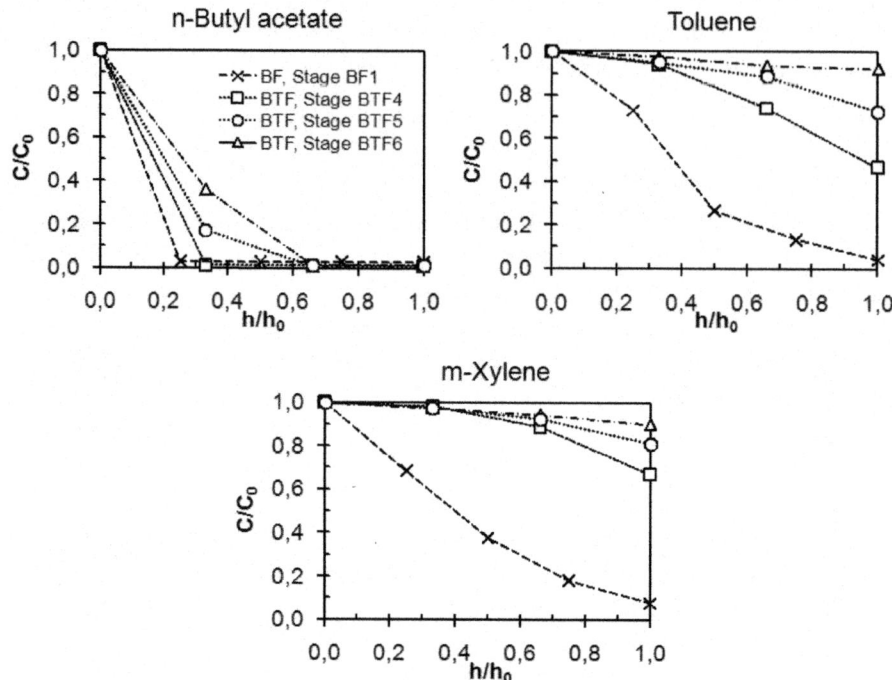

Figure 9. Pollutant concentration profiles along the length of the BF and BTF.

Pilot Unit

Characterization of Emissions from the Spray-paint Booth

The industrial site operates full-time, 24-hours a day, from 6:00 am on Monday to 6:00 am on Saturday, with shut down periods during weekends. Gas flow rate for the emissions coming from the clear coating spray booth was approximately 35 000 Nm^3 h^{-1}, with daily average emission temperatures ranging between 8 and 22 °C throughout the experimental period. The VOC concentration was relatively stable during working hours, with hourly average values ranging between 100 and 450 mg C Nm^{-3}, and with a daily average concentration of 235 ± 57 mg C Nm^{-3}. A typical VOC concentration emission pattern from the booth system over one working day is shown in Fig. 10. As can be seen, because the inlet hourly average VOC concentration exceeded the ELV by a factor greater than 1.5 (maximum hourly legal limits), treatment of booth emissions may be required.

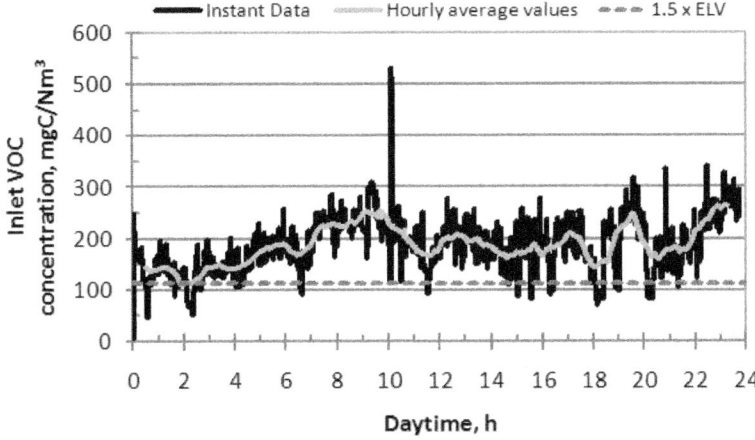

Figure 10. Typical instant and hourly VOC concentration emission patterns from the booth system. The discontinuous line represents a factor of 1.5 of the ELV.

GC-MS analyses of both emission samples and samples of the solvents used in the coating process detected the following compounds: oxygenated compoundssuch as butyl acetate, butyl glycol acetate and methyl acetate(comprised > 70%); and solvent naphtha including small proportions of aromatics compounds. The two product formulations used by the facility during the experimental period are summarised in Table 4.

Table 4. Product formulations used for the clear coating process during the experimental period.

	Product A	Product B
Clear coat (70% wt)	xylene 10-25% n-butyl acetate 10-25% 2-methoxy-1-methylethyl acetate 10-25% Isobutyl acetate 10-25%	butyl glycol acetate 25-50% naphtha, light aromatic 10-25% naphtha, heavy aromatic 10-25% 1,2,4-trimethyl benzene 2.5-10%

Thinner (30% wt)	n-butyl acetate 100%	n-butyl acetate 25-50% butyl glycol acetate 10-25% 1,2,4-trimethyl benzene 2.5-10% naphtha, light aromatic 2.5-10% xylene 2.5-10%

Pilot Unit Performance

After inoculation, the pilot unit was operated at an approximate EBRT of 20 s. During the first month, a progressive increase in removal efficiency was observed until stable performance was achieved. Once the system was stable, the air flow rate through the system was periodically adjusted to values of 33.8, 60, 41.5 and 90 Nm³ h⁻¹, corresponding to EBRTs of 80, 65, 45 and 30, respectively. In Fig. 11, (a) to (d), the inlet and outlet hourly average VOC concentrations obtained during a typical day of operation are plotted for the four EBRTs applied.

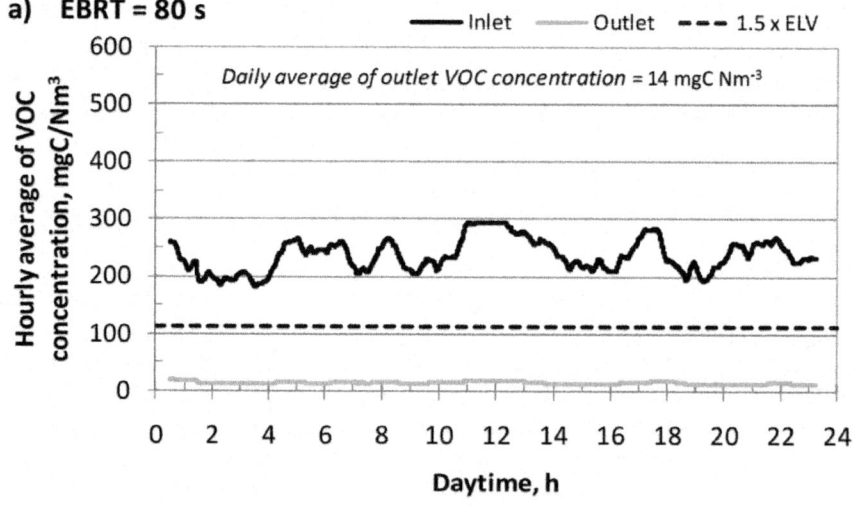

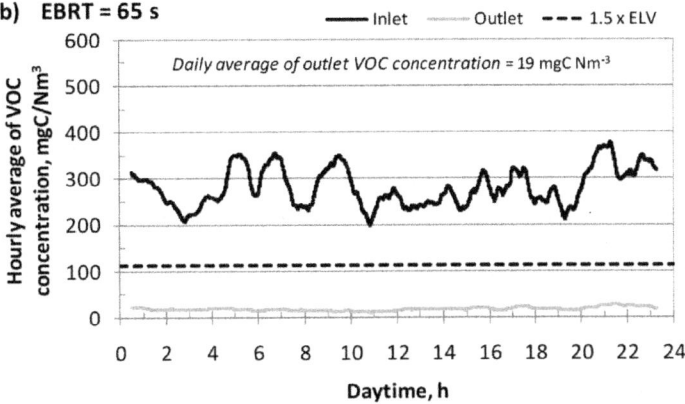

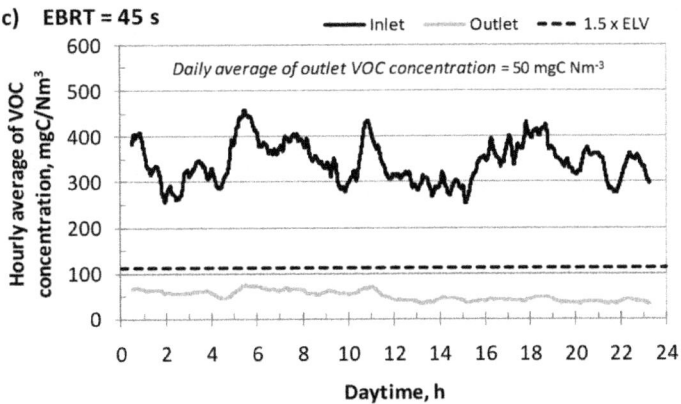

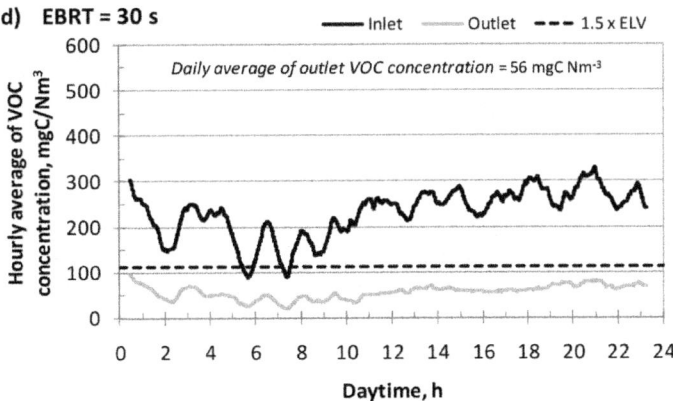

Figure 11. Daily pilot unit monitoring at four EBRTs: hourly average values working at an EBRT of (a) 80 s, (b) 65 s, (c) 45 s and (d) 30 s. The discontinuous line represents a factor of 1.5 times the ELV.

As can be seen, outlet emissions met legal regulations for all of the tested EBRT, with daily average values of 14, 19, 50 and 56 mg C Nm^{-3}, for 80, 65, 45 and 30 s of EBRT, respectively. Therefore, with respect to legal compliance,bioreactor performance was not significantly affected by the decrease in the EBRT. It is worth noting that outlet emissions did not exceeded the legal limit after the facility started the coating process on Monday mornings after weekend shutdown periods (data not shown here).

Fig. 12 summarises the performance of the pilot unit during the test period. The variation of overall RE versus the four applied EBRTs is presented in Fig. 12 (a). The variation of the daily average outlet concentration versus the daily average inlet concentration as a function of the tested EBRT is shown inFig. 12 (b). As can be seen in Fig. 12 (a), the overall removal efficiency was not significantly affected (RE values were between 85 and 95%) when the pilot unit was operated with EBRTs ranging between 45 and 80 s. However, RE dropped to approximately 70% when the EBRTdecreased to 30 s. Still, the lowest EBRT (30 s) was enough to maintain outlet VOC concentrations below legal limits (75 mg C Nm^{-3}), despite the observed decreasein RE (see Fig. 12 (b)). Furthermore, when the pilot unit was operated at an EBRT >45 s, the outlet concentration was always in compliance with the ELV over the entire inlet concentration range. Therefore, the minimum EBRT value that would allow legal regulations to be met within an adequate safety margin is between 35 and 40 s.

The pressure drop in the pilot unit was low (< 60 Pa m^{-1}) during the entire test period, indicating that the short-term starvation periods during weekend closures, combined with the relatively low VOC inlet load (~ 25 g C m^{-3} h^{-1}) avoided clogging episodes in the reactor. Soluble COD values measured in the recirculation liquid varied between 200 and 300 mg COD L^{-1}, representing < 1% of the inlet organic carbon fed to the pilot unit. The pH and conductivity average values of the recirculation liquid were 7.2 and 1.0 mS cm^{-1}, respectively.

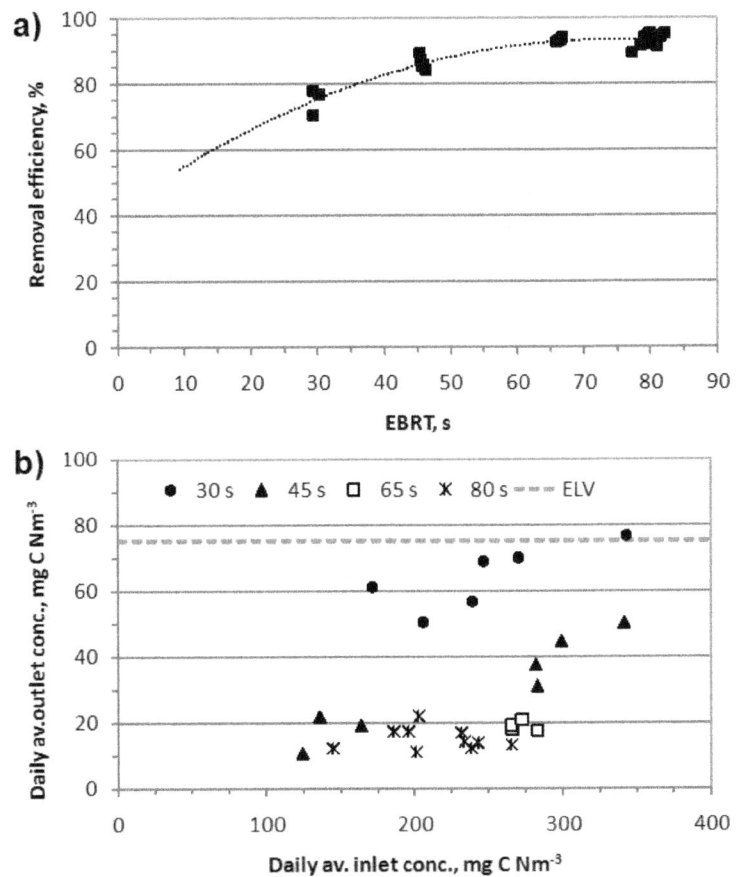

Figure 12. Summary of pilot unit performance during the entire experimental period: (a) Overall RE vs. EBRT and (b) Daily average outlet concentration vs. daily average inlet concentration.

Comparison with Laboratory-Scale Biotechnologies

A comparison of the pilot-scale unit performance with results obtained from the laboratory-scale biofilter and biotrickling filter (used to treat a continuous VOC emission polluted with a 2:1:1 (wt) n-butyl acetate:toluene:m-xylene mixture) is shown in Fig. 13. The variation in RE versus EBRT for the pilot unit, biofilter and biotrickling filter, for every section of the bed height, is presented in Fig. 13 a. Results from the biotrickling filter correspond to an inlet VOC concentration of approximately 200 mg C Nm^{-3} (stage BTF1, BTF3 and BTF5), similar in magnitude to the industrial VOC emission at the pilot unit. The variation of EC with IL for all data obtained from the three systems is plotted in Fig. 13 (b).

As can be seen, results obtained from the pilot-scale unit match those from the laboratory-scale experiments. The slightly better performance of the pilot unit versus the lab-scale BTF may be due to the following: (1) the oxygenated compound composition tested in the pilot-scale industrial emission (> 70%) was higher than that tested in the laboratory-scale BTF unit (50%); (2) greater variability in the organic composition of the industrial emission could have derived from a more complex microbial ecosystem; and (3) the short-term starvation periods due to weekend closures at moderate operational conditions could help to improve the activity of the biological system (Wright et al., 2005). Importantly, comparison between the laboratory-scale and pilot-scale units indicate that laboratory studies could be a timesaving tool for obtaining valuable data for scale-up.

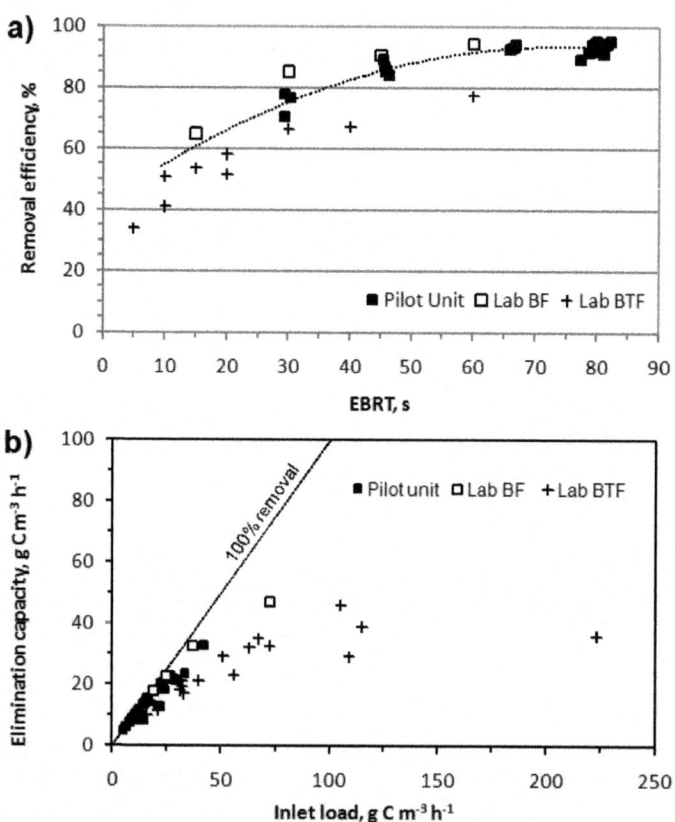

Figure 13. Comparison of the performance of the pilot unit with results obtained from

the laboratory-scale biofilter and biotrickling filter: (a) Overall RE vs. EBRT and (b) EC vs. IL.

CONCLUSIONS

Biological removal of VOCs from exhaust gases from a robotic spray paint booth at an industrial plastic coating facility has been studied. The present work evaluated the performance of environmentally friendly biotechnologies, such as biofilters and biotrickling filters, on a laboratory-scale; for comparison with a biotrickling filter pilot-scale unit connected to an industrial facility. Treatment of air contaminated with three compounds typically found in air emissions from paint and coating processes (n-butyl acetate, toluene and m-xylene) was conducted using the laboratory-scale systems. Although the biofilter displayed a higher removal efficiency than the biotrickling filter during the test period, a similar maximum elimination capacity of 45 g C m^{-3} $h^{-1}$was estimated for both laboratory biotechnologies. Results from the laboratory-scale experiments demonstrate the slight influence of the applied EBRT (which ranged between 15 and 60 s) on the EC for the applied IL in the biotrickling filter. However, application of EBRTs lower than 60 s to the BF would cause greater drying of the organic packing material, resulting in a decrease in the removal efficiency. Because of the higher EBRTs required, industrial scale biofilter applications would require greater sizes, and therefore higher investment costs. Moreover, BTF units are compatible with 8 – 10 m high bioreactors, enabling construction of units with smaller footprints than comparable BF units. In addition, it is also worth noting that the pressure drop in the BTF unit (< 60 Pa m^{-1}) was lower than that observed in the BF (around 400 Pa m^{-1}), due to the larger void fraction of the inert packing material: this difference in pressure drop is significant, because lower values require less power consumption from the blower equipment. Although complete removal of n-butyl acetate was achieved in both cases, aromatic compounds penetrated throughout the bed, indicating that mechanisms of gas removal for hydrophilic and hydrophobic compounds may be significantly different. Our evaluation of the operation of the pilot-scale plant for 3 months demonstrates the suitability and robustness of the biological process for controlling VOC emissions from a robotic spray paint booth. Importantly, legal emission limits were always achieved for the four tested empty bed residence times (EBRTs). The minimum EBRT required to meet legal regulations appears to be between 35 and 40 s.From an engineering perspective, comparison of the laboratory-scale and pilot-scale units indicates that laboratory-scale studies could be a timesaving tool for obtaining valuable data to establish safe operation limits which should allow legal requirements to be met.

ACKNOWLEDGEMENTS

Financial support from the Ministerio de Ciencia e Innovación (Spain, research projects CTM2010–15031 and TRA2009_0135) is acknowledged. F. J. Álvarez-Hornos acknowledges Ministerio de Educación, Spain, forproviding a post-doctoral contractbymeans of Programa Nacional de Movilidad de Recursos Humanos del Plan Nacional de I+D+i 2008-2011. The authors would like to give special thanks to Pure Air Solutions and FICOMIRROR, SA (Ficosa International Group) for their consistent collaboration and support.

REFERENCES

1. F. J. Álvarez-Hornos, C. Gabaldón, V. Martínez-Soria, P. Marzal, J. M. Penya-roja, M. Izquierdo, 2007 Long-term performance of peat biofilters treating ethyl acetate, toluene, and its mixture in air.Biotechnology & Bioengineering, 96 651660 , 0006-3592

2. J. T. Boswell, P. C. John, B. Stewart, S. Forrest, R. Branchik, S. Morgan, 2001 Biofiltration of VOCs from Paint Manufacturing, Proceedings of A&WMA's 94th Annual Meeting & Exhibition, Orlando, USA, June 24-28, 2001

3. J. Boswell, 2009 Clearing the air: the use of bio-oxidation for industrial air emissions control.Engineered Wood Journal, Fall, 2628

4. H. H. J. Cox, M. A. Deshusses, 2001 Biotrickling filters, In: Bioreactors for waste gas treatment, C. Kennes& M.C. Veiga (Ed.), 99131 , Kluwer Academic Publisher, 978-0-79237-190-8 Dordrecht, The Netherlands

5. European Commission 2003 IPPC Reference document on Best Available Techniques in common waste water and waste gas treatment/management systems in the chemical sector, Sevilla, Spain

6. European Environment Agency (2010).European Union emission inventory report 19902008 under the UNECE Convention on Long-range Transboundary Air Pollution (LRTAP). EEA Technical report 72010 978-9-29213-102-9 Copenhagen, Denmark

7. European Pollutant Release and Transfer Register 2008 Available from http://prtr.ec.europa.eu/

8. T. Higuchi, Y. Morita, R. Minato, 2010 Biofiltration of the exhaust gas contaminated by six VOC compounds for testing the feasibility of its application to a painting process, Proceedingsof the 2010 Duke-UAM Conference on Biofiltration for Air Polllution Control, 273278 , Washington, USA, October 28-29, 2010

9. J. W. Hwang, C. Y. Choi, S. Park, 2008 Biodegradation of gaseous

styrene by Brevibacillus sp. using a novel agitating biotrickling filter. Biotechnology Letters, 30 12071212 , 0141-5492

10. C. Kennes, H. H. J. Cox, H. J. Doddema, W. Harder, 1996 Design and performance of biofilters for the removal of alkylbenzenevapors. Journal of Chemical Technology and Biotechnology, 66 300304 , 1097-4660

11. D. J. Kim, H. Kim, 2005 Degradation of toluene vapor in a hydrophobic polyethylene hollow fiber membrane bioreactor with Pseudomonas putida.Process Biochemistry, 40 20152020 , 1359-5113

12. K. A. Kinney, W. M. Moe, 2004 Optimization of an innovative biofiltration system as a VOC control technology for aircraft painting facilities, Final report- SERDP Project CP 1104. Air Force Research Laboratory, Tyndal AFB, EEUU.

13. M. Koutinas, L. G. Peeva, A. G. Livingston, 2005 An attempt to compare the performance of bioscrubbers and biotrickling filters for degradation of ethyl acetate in gas streams. Journal of Chemical Technology and Biotechnology, 80 12521260 , 1097-4660

14. V. Martínez-Soria, C. Gabaldón, J. M. Penya-roja, J. Palau, F. J. Álvarez-Hornos, F. Sempere, C. Soriano, 2009 Performance of a pilot-scale biotrickling filter in controlling the volatile organic compound emissions in a furniture manufacturing facility.Journal of the Air & Waste Management Association, 59 9981006 , 1047-3289

15. A. K. Mathur, C. B. Majumder, 2008 Biofiltration and kinetic aspects of a biotrickling filter for the removal of paint solvent mixture laden air stream. Journal Hazardous Materials, 152 10271036 , 0304-3894

16. J. Paca, E. Klapkova, M. Halecky, K. Jones, T. S. Webster, 2006 Interactions of hydrophobic and hydrophilic solvent component degradation in an air-phase biotrickling filter reactor. Environmental Progress, 25 365372 , 1944-7450

17. P. . Popescu, I. Ionel, 2010 Anthropogenic air pollution sources, In: Air quality,K. Ashok, (Ed.), 122 , Sclyo, 978-9-53307-131-2 Rijeka, Croatia

18. F. Sempere, C. Gabaldón, V. Martínez-Soria, P. Marzal, J. M. Penya-roja, F. J. Álvarez-Hornos, 2008 Performance evaluation of a biotrickling filter treating a mixture of oxygenated VOCs during intermittent loading. Chemosphere, 73 15331539 , 0045-6535

19. Z. Shareefdeen, A. Singh, 2005 Biotechnology for odor and air pollution control. Springer, 978-3-54074-049-0 Heidelberg, Germany

20. J. A. Steele, F. Ozis, J. A. Fuhrman, J. S. Devinny, 2005 Structure of microbial communities in ethanol biofilters.Chemical Engineering

Journal,113 135143 , 1385-8947

21. T. S. Webster, A. P. Togna, W. J. Guarini, L. Mc Knight, 1999 Application of a biological trickling filter reactor to treat volatile organic compound emissions from a spray paint booth operation, Metal Finishing, 97 2026 , 0026-0576

22. World Health Organization 2004 Health aspects of air pollution, WHO Regional Office for Europe, Copenhagen, Denmark

23. W. F. Wright, E. D. Schroeder, D. P. Y. Chang, 2005 Regular transient loading response in a vapor-phase flow-direction-switching biofilter. Journal Environmental Engineering, 131 16491658 , 0733-9372

Chapter 7

ONE-WAY ANOVA METHOD TO RELATE MICROBIAL AIR CONTENT AND ENVIRONMENTAL CONDITIONS

José A. Orosa

University of A Coruña, Spain

INTRODUCTION

During the past few years, people passed most of their lives indoors than ever before. While one part was spent in their homes, the other part was spent in their working environments, such as factories and offices.

As a consequence of spending their lives in these two environments, we found that people developed certain health-related symptoms, such as headache, fatigue, nausea and getting irritated with other people. When these symptoms detected in indoor environments, it is called sick building syndrome (SBS). On the other hand, when these symptoms are stronger and related to their workplace, it is called work risk hazard.

To identify these symptoms, sampling apparatus is employed. There is a growing interest in employing bioindicators that, after tests in the laboratory, can now be employed in real case studies.

In this chapter, a new methodology based on the statistical study of One-Way ANOVA was developed to test bioindicators, such as fungi, in real case studies. With this statistical study it was possible to relate bioindicators with indoor conditions like pets' presence, limited space and presence of localized humidity problems.

PROBLEMATIC INDOOR AMBIENCES

Sick Building Syndrome

The SBS is defined by the World Health Organization (WHO, 1983) as the occurrence of an increased prevalence of no specific symptoms among populations in determined buildings (Thörn, 1998).

Between its more important and common symptoms, we find eyes, nose and throat irritation, mental fatigue, headaches, nausea, skin irritation, irritability and lack of concentration (Hedge et al., 1996;Raw et al., 1996; Gupta et al., 2007).

It is difficult to detect SBS due to its non-specific symptoms that appear with a higher prevalence than the expected value. Consequently, there is no clear consensus at the moment to define SBS and to determine if it is diagnosed by the exclusion of other causes.

On the other hand, subjective detection of SBS is being analyzed in recent research works. This detection is based on surveys and questionnaires about the perception of occupants of an indoor environment, such as that developed previously to analyze indoor thermal comfort (ASHRAE, 2003). All the subjective and objective parameters are analyzed in depth below.

As stated above, there are some parameters that can define SBS. To control these parameters like indoor air temperature, relative humidity and dust, amongst others, are some of the questions that must be asked.

The first question is if there is a need to employ any natural or mechanical ventilation. After a review of recent research works on this topic, it was concluded that there is a prevalence of some symptoms when the mechanical ventilation is working.

These symptoms are, for example, some upper respiratory problems and fatigue. The results showed that these effects can be related to the dust deposition in ducts, humidifiers and chillers of the HVAC system. On the other hand, the air remains clean when it is introduced into the house through natural ventilation, such as doors and windows.

The corrective actuation way to relax these symptoms is by HVAC system corrections and building design improvements. For example, recent research works like that of Kolari et al. (2005) showed that duct cleaning will not imply an improvement of indoor air quality, but is related to an improvement of indoor perception of air quality and its relation to nasal symptoms. Furthermore, a clear reduction, in the building, of the volatile organic compounds (VOC), carbon dioxide (CO_2) and fungal spore concentrations reach values below the actual after a duct cleaning process.

At the same time, the percentage of outdoor air that can be introduced into the indoor ambience must be considered. In this sense, it is very important to remember that, in the last few years, to reduce the energy consumption in HVAC system the air changes of indoor air was reduced (Osayintola & Simonson, 2006); consequently, a percentage of indoor air returns to the same indoor ambience mixed with outdoor air with a corresponding increment in CO_2 and moist air humidity.

We must consider that air changes are due to mechanical ventilation and infiltrations of outdoor air. In this sense, the mechanical ventilation must be improved and buildings must be less airtight. What is more, an increment in air changes of indoor air implies an improvement of indoor air quality, as reported by Haghighatt & Donnini, 1999.

Finally, outdoor air is not the only source of contaminants of indoor environments. Other parameters, such as indoor materials contamination must be considered as a source of pollution.

Once the objective parameters to detect the SBS (WHO, 1989) are defined, it is the right moment to define the subjective parameters. Among the main subjective parameters, we must consider job stress and perception of the indoor air quality.

Studies by Hedge et al. (1996) showed a clear relationship between SBS and working conditions than with environmental parameters. Furthermore, parameters such as type of job can influence the perception of SBS.

The other parameter, which is perception of indoor air quality, must be considered. In this sense, we find that the perception of indoor air quality depends not only on combustion gases, VOC, dust, particles and but also on parameters like mucous membranes temperature and humidity that can alter these perceptions (Salonvaara & Simonson, 2000 and Simonson et al., 2001). Other parameters such as air movement perception can alter this situation too.

Finally, to control the subjective parameters, we can employ questionnaires that are recognized as a more important tool to define the relationship between objective and subjective perception.

These surveys must cover an area for objective parameters that must be sampled, such as indoor air temperature and relative humidity and, at the same time, the same survey must be present in another region for the subjective parameters that must question the occupants of a building, such as indoor air perception and job stress.

The SBS detected by symptoms or by some bioindicators must be corrected. In other words, we must consider that SBS is related with building construction and not necessarily with the occupants. Consequently, parameters like air temperature, relative humidity and dust are sampled to define the SBS.

Objective Detection of Indoor Air Quality

Despite the fact that indoor air quality can be sampled with different apparatus like multi-gas samplers, it was learned that, in the past few years, there is an increasing interest in employing some natural indicators and monitors because they present some advantages with respect to most of loggers.

The main advantages of these indicators are based on the fact that they do not need any kind of calibration or energy source. It is the work of fungi to evaluate indoor environments and mosses and lichens to evaluate outdoor ambiences.

From these concepts, we do an initial definition of bioindicator and biomonitor. A bioindicator is an organism that can be used for identification and qualitative determination of human-generated environmental factors.

At the same time, we can define accumulative bioindicators which have the ability to store contaminants in their tissues and are used for the integrated measurement of the concentration of such contaminants in the environment, as a result of the equilibrium process of biota compound intake/discharge from and into the surrounding environment.

Biomonitors are organisms used for the quantity determination of contaminants and can be sub-classified as sensitive and accumulative. The methodology based in biomonitors present the problem of the need of a background level of this contaminant in the environment objective of study.

DETECTION OF INDOOR AND OUTDOOR AIR QUALITY

Fungi

Owing to its feasibility to be employed in a real case study, once defined, a few examples of biomonitors and bioindicators are explained below. To sample outdoor ambiences, we can employ mosses and lichens, and to sample indoor ambiences we can employ fungi.

Despite the fact that fungi develop in nature some functions, such as recycled energy and nutrients, most of these tasks are not adequate if it is to be developed in indoor environments.

It is due to fungi being related to spores, fungal fragments, mycotoxins and VOCs emissions is the reason for the failing health of the occupants. In this sense, nowadays we find indoor environments that present higher VOCs concentration than in years before and related to the fact that building designs were modified in accordance with energy saving, and consequently, buildings are more airtight and present a low number of air changes. When most researchers tried to find where fungi were located in indoor environments, the results showed that it developed on walls, roofs and in materials wherever it can be find dust to develop a growing media.

This fungi development increased with higher air temperatures and relative humidity values of over 75%. This is the higher value of relative humidity that must never be passed to prevent fungi developing in indoor environments.

At the same time, fungi emit mycotoxins that are low molecular weight compounds and toxic for animals and men (Cabral , 2010).

Mosses and Lichens

To analyze outdoor air quality, bioindicators, such as lichens and mosses, are employed due to they present some advantages respect traditional sampling methods. For example, these bioindicators were selected as they do not present any seasonal variation and their longevity. Consequently, these bioindicators let us sample indoor conditions for long periods without calibration and without any kind of energy source, which is a clear advantage respect traditional loggers.

Lichens are defined as a symbiotic association (Newbound *et al.*, 2010) of a fungus and an alga, and can be employed to develop a map of all the species detected in a sampling area. Another method is basically in the sampling process of pollutants in the thallium of the lichen.

Other methods are based in the transplantation of native lichens to a place where it will be killed by pollutants after a reduced period of time. Finally, new methods are being developed to define the climate change with different lichens sampling processes over calcareous rocks.

As a function of previous sampling processes, two indexes can be defined based in lichens measurements. The first index is called the index of atmospheric purity and the second is the index of poleotolerance.

Mosses are employed to define, at the same time as lichens, outdoor air quality (Szczepaniak & Biziuk, 2003). Mosses present some advantages as bioindicator as, for example, it can be employed in different regions due to their growth in different environmental conditions, such as industrial and urban areas.

Another advantage of mosses is based on the fact that sampling process is cheap and simple. It will allow a very large number of sites to be sampled, obtaining a better sampling map. At the same time, these natural indicators present some disadvantages. For example, to develop a comparative study with mosses and lichens, the same species of lichens and trees are needed to obtain adequate results to do a comparative study.

On the other hand, fungi growth in indoor environments depends on indoor temperature and relative humidity in all zones of the building. Consequently, one of the methods to control fungi growth in indoor ambiences is to control occupants' habits. For example, most researchers proposed an increment in the air changes during cooking in the kitchens, in the bathroom during mornings and in bedrooms during the nights. This increment in air changes can be obtained by natural ventilation through open windows.

PRACTICAL CASE STUDY IN FLATS

Recent research works is related to airborne microorganisms with some infections or allergic disorders (Parat et al., 1997), and some epidemiologic studies have been related to dust mite exposure with some degree of asthma. In general, however, we can say that in ambiences with higher allergen exposure we find higher asthma prevalence (Liu, 2004).

Despite this, there is not always a relationship between fungi development and asthma in children. Consequently, some researchers reached the conclusion that, with actual methodologies, it cannot be done (Jovanovic et al., 2004).

Nowadays, there is very little information on how to prevent allergies, such as environmental hygiene, avoidance of some foods and prevention of contact with some kind of pets.

However, things are more complicated and not the same for all allergens. Although increased exposure to house dust mite allergen is paralleled by increased sensitisation rates, the same is not true for cat allergen.

One possible explanation for the different effects of different allergens may be their biochemical properties: mite allergens, in contrast to cat and dog allergens, contain proteolytic enzymes. It has been shown that, concerning house dust mites, a low-allergen environment can be achieved (Lauener, 2003).

To summarize, we can say that new standards are needed to show the better methodology to sample indoor ambiences and define the effect of indoor pollutants over health to reduce sensitization to these parameters (Lauener, 2003).

Other factors like distance of the building from the source (a nearby park) and supermicrometre particle concentrations will be associated with the concentration levels of fungi in indoor ambiences of occupied buildings (Hargreaves et al., 2003).

At the present time, the only way to guarantee lower mite allergen levels in modern homes in the western world is to remove the carpets and to encase the mattresses and beddings. Furthermore, to reduce this exposure, we must improve IAQ.

The three primary considerations in improving IAQ are (1) evaluation of construction failures that allow moisture into the walls and roofs, (2) poor ventilation, causing excessive humidity and accumulation of gaseous and/or chemical exposure from materials in the living space, and (3) poorly designed or failing HVAC systems that contribute to poor air calculation.

About the two last points, some authors (Parat *et al.*, 1997) have analyzed that massive proliferation of microorganisms may take place in HVAC unit with certain risk factors, such as low efficiency filters, cold mist humidifiers using water recycling, areas in which condensation water remains stagnant, large recirculation of air and faulty or deficient maintenance conditions.

They demonstrated that compared to a naturally ventilated building, a HVAC system which is well designed and well maintained improves the microbiological quality of indoor air and, in consequence, can reduce health hazards for its occupants.

Finally, not all indoor allergens are necessarily equal in their propensity to cause asthma and its related health effects like shortness of breath and coughing. For example, using dust allergen concentration as a proxy for exposure, recent studies have revealed that indoor cockroach allergen exposure, but not mite or cat allergen exposure, is a significant risk factor for asthma (Hens, 2007).

Despite the fact that the data collected on household characteristics varied greatly between the studies and that building materials techniques are very different in different parts of the world, some common themes have emerged.

For example, concentrations of moulds varied hardly between areas (Jovanovic *et al.*, 2004), and neither climatologically conditions nor differences between urban and rural regions exhibited a systematic influence.

In another example (Perfetti *et al.*, 2004), no association was found between the concentration of mite allergens and the environmental characteristics (geographic location, floor above ground, type of ventilation) and no correlation was found between indoor humidity and allergen levels.

We must consider the fact that house dust mites live in an environment where there is no liquid water, and they are dependent on the ambient humidity to absorb water from the atmosphere. To get this, water dust mites can gain it by diffusion through the body or extract the water vapor from air via hygroscopic crystals in their supracoxal glands, located at the base of their first pair of legs.

The optimum relative humidity for mite growth is 75–95%, at temperatures of 15–30°C, whereas above 70% relative humidity conditions may be optimal for fungal growth (Liao *et al.*, 2004). Relative humidity has a major influence on the survival of mite colonies and therefore levels of mite allergens.

Although laboratory and early field studies suggested that there was a strong relationship between relative humidity and mite allergen levels, this had not been conformed by more recent large-scale studies when other factors have been considered in a multivariate analysis.

For example, freezing and/or dry weather can damage fungi and reduce the spore counts on outdoor samples, but the conditions indoors may be very hospitable to fungal growth non-seasonally (Zhou *et al.*, 2000).

From these studies, we can conclude that outdoor relative humidity influences indoor relative humidity, but other household factors can influence mite allergen levels and that means allergen levels in different geographical areas tend to be influenced by the local climate.

As a result of this, novel techniques have been developed recently, which allows measurements of relative humidity to be made within the mite microhabitat, that is, where it matters, in the depth of the carpet or mattress.

This has revealed that the relative humidity in the carpet may be higher than that in the room air, and that with different types of construction, the differences between room RH and floor RH will vary. This suggests that the relative humidity in the room air does not necessarily reflect the RH in the micro-habitat of the mite—in the depth of the carpet pile.

The object of this chapter was to get information about Spanish apartments' microbial levels and relate it with their characteristics. Results will be useful to get a healthy home, taking into account costs versus energy saving, and improve health outcomes (Bernstein *et al.*, 2008).

MATERIALS AND METHODS

Apartments

In our case study, different apartments located in the northwest of Spain were selected, in accordance with different criteria defined to obtain realistic comparisons between indoor ambiences.

The first criterion is that, in all apartments, the residents presented some kind of health problems related with the relative humidity, which is typical in this area. The reason for this work is to relate health problems with indoor conditions, in building constructions' humid areas and occupants' habits.

In particular, fungi and bacterial growth were sampled in these indoor environments.

All apartments present natural ventilation to remove all indoor air. Despite this, other mechanical ventilation system was located in toilets to reduce the humidity released during bath.

The heating system consists of heat water radiators and is employed only for a few months in the winter season. On the other hand, there is no cooling system, as the temperatures during the summer season are not too high. Consequently, during summer ventilation is enough to reduce indoor temperature.

To obtain adequate comparative results in apartments, sampling process was developed in accordance with the daily life conditions indicated by occupants. Furthermore, all buildings present the same construction and located in the same city. Consequently, outdoor weather conditions were the same for all the buildings.

Outdoor weather conditions were sampled by some weather stations located in a representative zone of the area where the buildings are located. This sampling process was developed by weather stations from MeteoGalicia (2002), with a sampling and frequency of 10 minutes.

On the other hand, humidity and temperature were measured by a 1221 Datalogger, with sensors of temperature and relative humidity, and tinytag Plus 2 dataloggers were employed.

These loggers were located in each apartment, in accordance with the ISO Standard indications. In particular, each sampling point was separated from heat sources conditions and as near as possible of center of gravity each room to obtain representative values of indoor

As explained earlier, a microbiological analysis off indoor ambiences was done. Consequently, two culture media were employed: Trypticase Soy Agar to find the total number of bacteria and Malt Agar was used to define fungi growth.

One-Way ANOVA

To compare sampled mean values of temperature and relative humidity for fungi and bacteria, a statistical study of one-way ANOVA was done. This statistical study consists in an analysis of the variance of one factor for a significance level of 0.05.

Furthermore, different statistical studies, such as Duncan and Student-Newman-Keuls *post hoc*analyses, let us define groups of apartments that present the same condition for this level of significance. In this study, two assumptions were defined. The first is based on the fact that the dependent variable is normally distributed. The second is that the two groups have approximately equal variance on the dependent variable.

In the same study, two hypotheses were considered. The hypothesis null is that there are no significant differences between the groups' mean scores. The alternate hypothesis is that there is a significant difference between the groups' mean scores. Finally, to develop this task, the statistical software SPSS 11.0 was employed. More information on how to employ this software SPSS can be found in their website (SPSS).

RESULTS AND DISCUSSION

At this point, the main results obtained are shown. In particular, Table 1 shows us the main indoor and outdoor air temperature and relative humidity of 25 apartments during the sampling process.

Despite this, we must consider the fact that this table shows the main value of sampled data, and consequently, conclusions about instantaneous values at different hours cannot be obtained from this table.

From Figs. 1, 2 and 3, we can conclude that Coruña, located in the northwest coast of Spain, presents a mild climate. In this sense, we can see that outdoor temperature is not too high in summer and too low in winter season. Mean temperature values of 11°C during winter and 16°C during summer, respectively, can be expected.

Table 1. Indoor/outdoor sampled variables.

Flat	tindoor (°C)	RHindoor	toutdoor (°C)	RHoutdoor
A	20.97	63.45	16.72	74.6
B	24.09	65.09	17.76	88.8
C	19.42	62.1	14.12	82
D	20.38	64.87	17.7	59.6
E	21.46	63.59	17.7	62
F	23.43	65.03	22.28	66.8
G	22.2	70.92	16.24	94
H	19.92	63.73	15.64	75
I	21.24	49.62	18.08	45

J	23.78	55.7	17.76	47
K	25.11	48.09	17.6	58.8
L	23.63	65.58	20.04	73.2
M	22.37	67.19	19.28	78.4
N	21.74	63.65	15.6	74.76
O	24.05	50.31	15.2	74.2
P	20.29	59.22	12.08	88
Q	20.32	62.23	11.84	88
R	20.1	62.66	14.4	89.2
S	17.4	69.88	14.4	87.4
T	19.42	64.79	16.6	72
U	21.13	61.1	15.8	76
V	22.63	65.22	21.4	59.4
W	25.12	61.18	21.64	64.4
X	19.14	64.56	14.84	78
Y	23.83	65.65	21.5	60.8

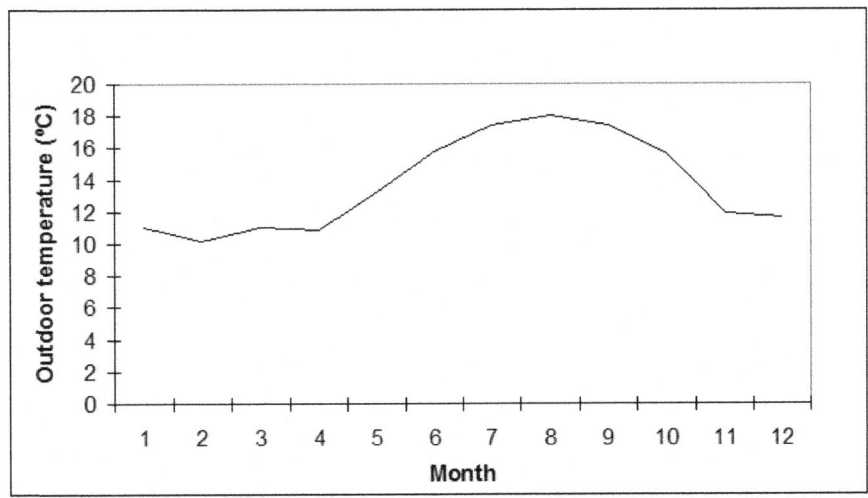

Figure 1. Outdoor temperature.

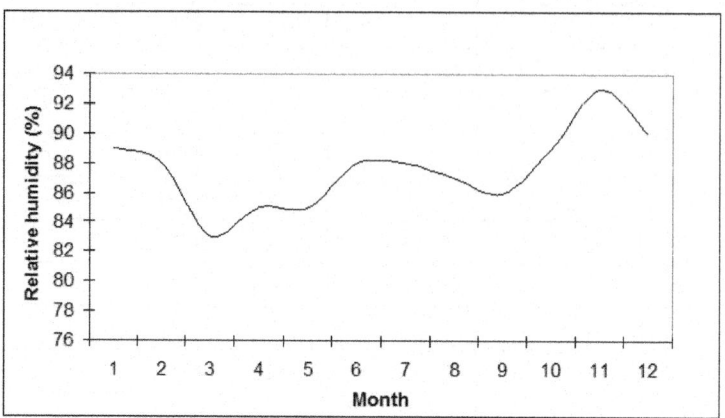

Figure 2. Outdoor relative humidity.

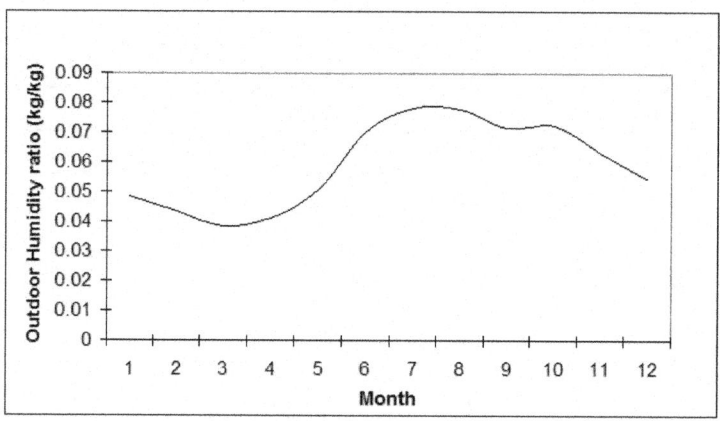

Figure 3. Outdoor humidity ratio.

On the other hand, outdoor relative humidity showed mean values between 83% in March and 93% in November. Consequently, mean outdoor relative humidity of 86% is obtained throughout the year, as we can see in Fig. 2.

Finally, another way to show the relationship between temperature and relative humidity was to express the outdoor air humidity ratio, as we can see in Fig. 3. This humidity has shown yearly values between 0.04 and 0.08 kg of water per kilogram of dry air.

This high humidity is incremented with different indoor moisture sources. This increment of humidity ratio under temperatures of 21°C will imply relative humidity values over 75% in some daily life periods, as we can see in Fig. 3.

From Table 1, we can conclude that the northwest of Spain present apartments with an indoor mean relative humidity about 62% and temperature of about 21.7°C. Thus, value is relatively high, but not excessively high for a coastal area. However, due to an indoor relative humidity of not more than 75% with an adequate cleaning procedure, development of fungi can be prevented.

Table 2. Observed characteristic.

Flat	Characteristics
A	Pet
B	Normal
C	Normal
D	Limited space
E	Normal
F	Humidity problems
G	Humidity problems
H	Limited space
I	Normal
J	Pet
K	Humidity problems
L	Limited space
M	Normal
N	Humidity problems
O	Normal
P	Normal
Q	Pet
R	Normal
S	Normal
T	Humidity problems
U	Normal
V	Normal
W	Humidity problems
X	Normal
Y	Normal

At the same time, temperature and relative humidity were sampled and different characteristics of each apartment flood were considered (Table 2). This Table shows us parameters like pets' presence, limited space and presence of localized humidity problems in the walls and roofs that were considered.

Finally, if none of previous commented parameter was detected, then the apartment was considered normal.

Once Tables 1 and 2 have shown us the main value of temperature, relative humidity and apartment characteristics, it is the right moment to show the results of fungi and bacteria growth, as we can see inFigs. 4 and 5.

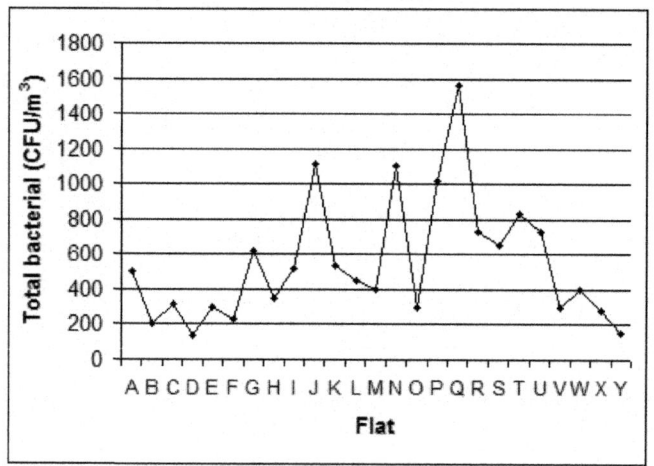

Figure 4. Total bacteria sampled (CFU/m³).

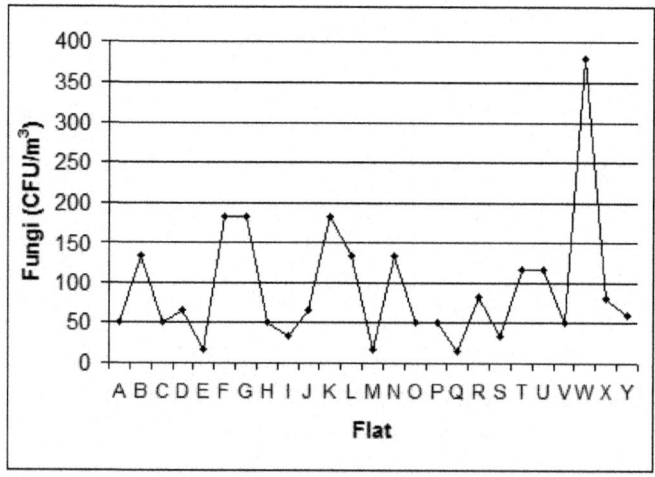

Figure 5. Total fungi sampled (CFU/m³).

After fungi and bacteria growth in these indoor environments are sampled to relate these pollutants with indoor relative humidity, like in most of laboratory studies (NTP 335, 2008), it is necessary that this is applied to real case studies. In particular, in this chapter a relationship between fungi and bacteria with the particular parameters detected in each building, reflected in Table 2, was proposed.

From this study, we see that there is no possibility to obtain an adequate linear regression between humidity and fungi reflected by a correlation factor below 0.9, see Figure 6.

It is related with the fact that, in real buildings, there are other parameters that does not influence in most laboratory studies, but in real case studies it can alter the situation. It is the case of the presence of pets, moisture sources and moisture-damaged walls and roofs.

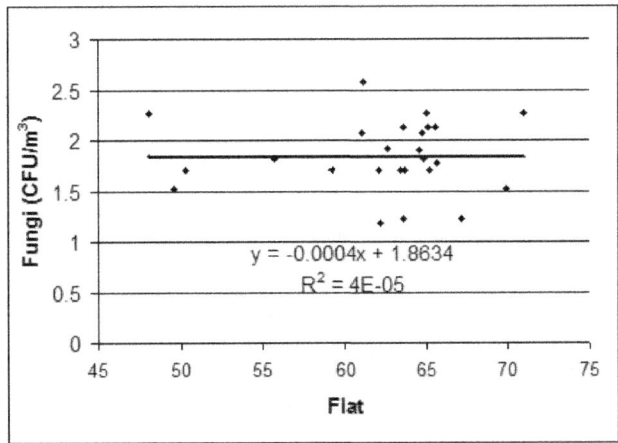

Figure 6. Fungi linear regression.

To relate these indoor conditions with sampled indoor parameters, it was proposed to develop one-way ANOVA analyses with different *post hoc* studies to define which groups of indoor environments experience the same evolution with time.

In particular, the Duncan *Post hoc* analysis with mean total bacteria and fungi was developed as we can see in Tables 3 and 4.

From the one-way ANOVA analysis, we can concluded that apartments having pets showed the same indoor air bacteria evolution with time and, consequently, can be separated as an independent group, as we can see in Table 3.

Table 3. One-way ANOVA and Duncan *post hoc* with mean total bacterial (CFU/m³).

A	J	Q	B	C	E	I	M	O	P	R	S	U	V	X	Y	D	H	L	F	G	K	N	T	W
Group 1: Pets			Group 2: Normal, limited space and humidity problems																					

Table 4. One-way ANOVA and Duncan *post hoc* with mean fungi (CFU/m³).

A	J	Q	B	C	E	I	M	O	P	R	S	U	V	X	Y	D	H	L	F	G	K	N	T	W
Group 1: Normal, limited space and pets			Group 2: Humidity problems																					

On the other hand, it was concluded that there exists a clear different indoor air fungi developed in apartments that present some humidity problems on walls and roofs with respect to others, as we can see in Table 4.

CONCLUSIONS AND FUTURE RESEARCH WORKS

This research work tried to relate indoor air conditions with fungi and bacteria growth. In this sense, objective and subjective parameters were considered. So, parameters like indoor and outdoor temperature and relative humidity were sampled in relation to bacteria and fungi growth. At the same time, parameters such as presence of pets and humidity problems in walls and roofs were considered too.

The results showed us that it is not easy to relate fungi growth with indoor air relative humidity like in laboratory studies. It is owing to the fact that there are some factors that can alter this situation. Furthermore, pets' presence was related to the increment in bacteria in indoor air, and humidity problems were related with fungi developed in a statistical way.

In conclusion, we can say that one-way ANOVA is an interesting tool to be employed by engineers to approach real case studies with laboratory conclusions.

ACKNOWLEDGEMENTS

I express my gratitude to INEGA, the University of A Coruña, Xunta de Galicia and all individuals and institutions that collaborated during the writing of this chapter.

REFERENCES

1. ASHRAE Standard 55P 2003 Thermal Environmental Conditions for Human Occupancy,. American Society of Heating, Refrigerating and Air-conditioning Engineers, Atlanta, USA.

2. J. A. Bernstein, N. Alexis, H. Bacchus, L. Bernstein, p. Fritz, E. Horner, N. Li, S. Mason, A. Nel, J. Oullette, K. Reijula, T. Reponen, J. Seltzer, A. Smith, S. Tarlo, 2008 The health effects of nonindustrial indoor air pollution. American Academy of Allergy, Asthma & Immunology. Article in press.

3. J. P. S. Cabral, 2010 Can we use indoor fungi as bioindicator of indoor air quality? Historical perspectives and open questions. Science of the Total Environment. 408 42854295 .

4. S. Gupta, M. Khare, R. Goyal, 2007 Sick building syndrome- A case study in a multi-storey centrally air-conditioned building in the Delhi City. Building and Environment, 42 27972809 .

5. F. Haghighat, G. Donnini, 1999 Impact of psycho-social factors on perception of the indoor air environment studies in 12 office buildings. Building and Environment 34 479503 .

6. M. Hargreaves, S. Parappukkaran, L. Morawska, J. Hitchins, C. He, D. Gilbert, 2003 A pilot investigation into associations between indoor airborne fungal and non-biological particle concentrations in residential houses in Brisbane, Australia. The Sci Total Environ. 312 89101 .

7. A. Hedge, W. A. Erickson, G. Rubin, 1996 Predicting sick building syndrome at the individual and aggregate levels. Environmental International 22 1 319 .

8. H. Hens, 2007 Indoor climate in student rooms: measured values. IEAEXCO energy conservation in buildings and community systems annex 41 "moist-eng" Glasgow meeting.

9. S. Jovanovic, A. Felder-Kennel, T. Gabrio, B. Kouros, B. Link, V. Maisner, I. Piechotowski, K. Schick, M. Schrimpf, U. Weidner, I. Zöllner, M. Schwenk, 2004 Indoor fungi levels in homes of children with and without allergy history. International Journal of Hygiene and environmental Health. 207 369378 .

10. S. Kolari, U. Heikkilä-Kallio, M. Luoma, P. Pasanen, P. Coronen, E. Nykyri, K. Reijula, 2005 The effect of Duct clearing on perceived work environment and symptoms of office employees in non-problem buildings. Building and Environment 40 16651671 .

11. Lauener R.P. 2003 Primary prevention of allergies. Revue française d'allergologie et d'immunologie clinique. 43 423426 .

12. C. M. Liao, W. C. Luo, S. C. Chen, J. W. Chen, H. M. Liang, 2004 Temporal/seasonal variation of size-dependent airborne fungi indoor/outdoor relationship for a wind-induced naturally ventilated airspace. Atmos Environ. 38 44154419 .

13. Liu A.H. 2004 Something Old, something New: Indoor endotoxin, allergens and asthma. Paedriatic Respiratory Reviews. 5 Suppl A), 6571 .

14. MeteoGalicia. Anuario climatolóxico de Galicia 2002 Consellería de Medio Ambiente. Xunta de Galicia. 8-44533-520-0

15. National Institute of Safety and Hygiene at Work; 2008 NTP 335: Indoor air quality: pollen grains and fungi spores evaluation.http://www.mtas.es/insht/ntp/ntp_335.htm.

16. M. Newbound, Mccarthy, T. Lebel, 2010 Fungi and the urban environment: a review. Landscape and urban planning, 96 138145 .

17. O. F. Osayintola, C. J. Simonson, 2006 Moisture buffering capacity of hygroscopic buildings materials: Experimental facilities and energy impact. Energy and Buildings. 38 12701282 .

18. S. Parat, A. Perdrix, H. Fricker-hidalgo, I. Saude, R. Grillot, P. Baconniers, 1997 Multivariate analysis comparing microbial air content of an air-conditioned building and a naturally ventilated building over one year. Atmospheric environment. 31 441449 .

19. L. Perfetti, M. Ferrari, E. Galdi, V. Pozzi, D. Cottica, E. Grignani, C. Minoia, C. Moscato, 2004 House dust mites (Der p 1, Der f 1), cat (Fel d 1) and cockroach (Bla g 2) allergens in indoor work-places (offices and archives). Science of the Total Environment. 328 1521 .

20. G. J. Raw, M. A. S. Roys, C. Whitehead, D. Tong, 1996 Questionnaire design for sick building syndrome: an empirical comparison of options. Environmental International. 22 1 6172 .

21. M. H. Salonvaara, C. J. Simonson, 2000 Mass transfer between indoor air and a porous building envelope: Part II. Validation and numerical studies. Proceedings of Healthy Buildings, 3

22. C. J. Simonson, M. Salonvaara, T. Ojalen, 2001 Improving indoor climate and comfort with wooden structures. Espoo 2001.Technical Research

Centre of Finland, VTT Publications 431.200p.+ app 91 p.

23. K. Szczepaniak, M. Biziuk, 2003 Aspects of biomonitoring studies using mosses and lichens as indicators of metal pollution. Environmental Research. 93 221230 .

24. A. Thörn, 1998 The sick building syndrome: A diagnostic dilemma. Social Science & Medicine 47 9 13071312 .

25. World Health Organization 1983 Indoor air pollutants: exposure and health effects: report on a WHO meeting. WHO Regional Office for Europe. (EURO reports and studies 78). Copenhagen:

26. World Health Organization 1989 Indoor air quality: organic pollutants. WHO Regional Office for Europe (EURO Report and Studies I 111), Copenhagen.

27. G. Zhou, W. Z. Whong, T. Ong, B. Chen, 2000 Development of a fungus-specific PCR assay for detecting low-level fungi in a indoor environment. Molecular and Cellular Probes. 14 339348 .

Chapter 8

INDOOR AIR QUALITY - VOLATILE ORGANIC COMPOUNDS: SOURCES, SAMPLING AND ANALYSIS

Alessandro Bacaloni, Susanna Insogna and Lelio Zoccolillo

Department of Chemistry, University of Rome "La Sapienza", Italy

INTRODUCTION

Since the 70s, research has found in Europe and in the United States that individuals spend between 70 and 90% of their time indoors. Health studies have found that exposures to a variety of air pollutants indoors can be substantially higher than outdoors, even in urban environment. Volatile Organic Compounds (VOCs) are often considered among the more important indoor pollutants, because of by their continue emission from many sources and their diffusion properties. With the aim to evaluate the occupants' discomfort and health effects and in order to develop guidelines and standards, Indoor Air Quality (IAQ) assessment and control is an essential step. IAQ assessment will complain:

- Sources: Identification and characterization of sources, as emissions from materials, products or activities, is best done under laboratory conditions; so it is possible measuring rates of emissions (especially chemicals such as VOCs). Exposure characterization is the second level of source identification; after the measurement of contaminants' concentrations in controlled environment, characterised by known sources, adsorbing and absorbing surfaces, these data can be used in validation of current exposure models.

- Sampling methods: In order to determine concentrations of VOCs and exposures of building occupants via inhalation, field studies can be carried out by sampling methods (and analysis) in accordance with existing official methods (EC, NIOSH, OSHA, ACGIH, etc.). This way may be expensive and cumbersome; in addition, it can be not exhaustive in predicting the discomfort and health impact. A greater number of perspectives are offered by using some "descriptors" that can be more adequate in characterizing anthropogenic pollution. Specific sampling methods may be reserved for contaminants with specific toxic effects (e.g., formaldehyde, benzene, monomers, etc.). Measurements of specific contaminants' can be necessary for sources that cause high room concentrations for relatively short periods (e.g., in case of freshly applied coatings on walls, etc.). Currently, diffusion (passive) samplers are mainly used in order to evaluate long term exposures (days to weeks and more)

• Analysis and data meaning: existing analytical methods are validated and generally show adequate limits of detection (LODs) and limits of quantification (LOQs) even in measuring subtoxic contaminants' concentrations. Analytical methods for the more frequent indoor contaminants are presented and commented. More difficulties lie in the interpretation of results, due to the limited indications suggested by European and International Standards (such as EN, EN ISO, etc.) that will be discussed. The difficulty in data evaluation is growing depending on the simultaneous occurrence of contaminants at subtoxic concentration.

In this chapter, there will be presented the way to handle the problem in IAQ assessment, and some practical applications, in order to provide the logical pathway to face the majority of actual cases.

INDOOR AIR QUALITY AND VOCS

Research on pollutant sources is needed to identify pollutants and emission levels from building materials and other products. Work in this area will serve two purposes: (1) providing measurement protocols and data to employ in exposure reduction actions, and (2) providing correlations between research on health effects and pollutant sources of contaminant critical levels. Finally, research is required to determine definition, causes of, and solutions to multiple chemical sensitivity (MCS). Identifying the physiologic nature of MCS is the first step in understanding whether and how IAQ eventually contributes to the syndrome. Maroni et al. (1995) recommended the following definitions that should be used in the description of indoor related complaints and illnesses:

• Building-Related Environmental Complaints (BREC): Complaints of poor IAQ or poor indoor air environment. BREC are usually registered in the complaint (annoyance) part of questionnaire studies.

• Building-Related Symptoms (BRS) or Building Related Health Complaints (BRC): The health complaints (subjective symptoms) reported by the single individual as occurring inside a building and usually subsiding shortly after leaving it.

- Sick Building Syndrome (SBS): denotes a situation where a significant number of the occupants of a building complain of a typical group of general, unspecific, and irritating symptoms, including particularly headache, lethargy, dry eyes, blocked nose, and sore throat. The symptoms usually fade after the person has left the indoor environment but the specific cause is unidentified.

- Building-Related Illness (BRI): clinical condition with defined symptoms and signs in which the cause (aetiology) is building-related and identifiable.

The difference between BRI and SBS is that the building problems are identified in the former and undiagnosed in the latter. The term «sick building» probably should not be used on its own, but it might be replaced with the expression «building with indoor climate problems» or «problem building.»

VOCs are often more important in assessing IAQ because of their ubiquity; VOCs may be used like "descriptors" that can be more adequate in characterizing anthropogenic pollution. Specific sampling methods may be reserved for contaminants with specific toxic effects (e.g., formaldehyde, benzene, monomers, etc.). In many cases, investigations of the indoor air are carried out because of complaints about poor IAQ, which are made by persons living or working indoors. Such complaints may be perception of unknown or unpleasant odours, headache, irritation of the eyes, nose and throat, dryness of the skin or symptoms like tiredness, lack of concentration and unspecific hypersensitivity reactions. Investigations resulting from observed or suspected health problems of occupants are quite similar and require the same sampling strategy.

SOURCES

Identification of indoor pollutant sources may be very important; emission from point (or surface) sources generates a pronounced concentration gradient near the source. The existence of such concentration gradients can be used to trace the source in an indoor environment by a careful selection of sampling sites. This strategy can lead to recommendations on how to improve indoor air quality, e.g. removing identified materials and strong emitters.

Maroni et al (1995) proposed these recommendations concerning source control:

Material Selection

- Manufacturers should be required to provide sufficient information for the evaluation of their products' safety. Protection of legitimate confidentiality must not impede evaluation.

- Standardization and harmonization of emission testing procedures must continue at international level. Meanwhile, research and application of different techniques and methodological intercomparison should be encouraged.

- Labelling of products and their ranking for safety and emission properties should be encouraged to facilitate appropriate selection and use by consumers. The choice of process may vary according to the nature of the risk involved and the local priorities.

Biological Contaminants

- The growth of molds and other microorganisms in the indoor environment should be avoided. This can be achieved effectively by elimination of moisture.

- Good housekeeping and sanitation rather than the use of biocides are appropriate control measures also for mites.

Volatile Organic Compounds (VOCs)

- Overall exposure to VOCs should be kept at the lowest possible level primarily by proper control of the emission sources.

- Pending improved methods, the total volatile organic compounds concept is a useful tool for practical assessment of the overall quality of materials and of indoor air. However, at least the type of major VOC species must be known.

- More research is recommended for the sensory effects of VOC mixtures and for some individual compounds typically present in indoor air mixtures for which little toxicological knowledge is available; this recommendation is strongly current, due to the growing interest in persistent organic pollutants (POPs) as endocrine disruptors, also in classic subtoxic concentrations.

SAMPLING METHODS

The primary objective of taking indoor air is to determine the quality of indoor air with the aim of assessing any risk to the health of the population and of individuals due to indoor air pollution. Monitoring is useful also in testing effectiveness of remedial actions, such as modifications to a building, its systems or equipment, aimed at reducing indoor air pollution. The measurements for testing their effectiveness are comparison measurements before and after the remedial actions. It is also important to record intermediate and long-term trends of indoor air pollution concentration. Respective analysis of trends will help to maintain, improve and establish abatement or risk management procedures. Finally, beside screening measurements, it can be very helpful to evaluate indoor air concentrations at abnormal or "worst-case" conditions of the indoor climate (temperature, humidity, ventilation etc.) and during particular activities. Sampling and analytical methods may be also used in validation of indoor pollution models.

To determine reference values of indoor air pollutants, the following elements of sampling strategy shall be used in addition to the basic information characterizing every sampling method (EN 14412, 2004):

- Time of sampling: to rule out any seasonal effects the individual sampling events shall be evenly distributed over the year. The time of individual sampling shall be fixed in a way that representative concentration values can be assessed. However, for sampling for ≥ one week, the choice of sampling time tends to be less important.

- Sampling duration and sampling frequency: the duration of sampling has to be set in a way that the limits of determination for all compounds of interest are exceeded and that representative readings are obtained. It is necessary to maintain a sampling duration of one week or a multiple to achieve a maximum of representativity per sampling site. This is to assess all different pollution loads that occur on the different days of the week at least once. One sampling per site is adequate.

- Sampling site and determinations with resolution in space. The sampling site shall be representative, e.g. the center of the room most inhabited by all occupants shall be used.

General methods (like NIOSH, OSHA or similar) developed for industrial workplaces may be adapted for VOC determination in air; as an example, we report the synthesis of NIOSH Method 2549 (Table 1).

Applicability

This method has been used for the characterization of environments containing mixtures of volatile organic compounds (see Table 2). The sampling has been conducted using multi-bed thermal desorption tubes. The analysis procedure has been able to identify a wide range of organic compounds, based on operator expertise and library searching.

Interferences

Compounds which coelute on the chromatographic column may present an interference in the identification of each compound. By appropriate use of background subtraction, the mass spectrometrist may be able to obtain more representative spectra of each compound and provide a tentative identity (see Table 2).

Other Methods

Other methods have been published for the determination of specific compounds in air by thermal desorption/gas chromatography [1-3]. One of the primary differences in these methods is the sorbents used in the thermal desorption tubes.

Reagents

- Air, dry
- Helium, high purity
- Organic compounds of interest for mass spectra verification (See Table 2).*
- Solvents for preparing spiking solutions: carbon disulfide (low benzene chromatographic grade), methanol, etc. (99+% purity)

    * See SPECIAL PRECAUTIONS

Equipment

- Sampler: Thermal sampling tube, ¼" s.s. tube, multi-bed sorbents capable of trapping organic compounds in the C_3-C_{16} range. Exact sampler configuration depends on thermal desorber system used. See Figure 1 as an example.
- Personal sampling pump, 0.01 to 0.05 L/min, with flexible tubing.
- Shipping containers for thermal desorber sampling tubes.
- Instrumentation: thermal desorption system, focusing capability, desorption temperature appropriate to sorbents in tube (~300° C), and interfaced directly to a GC-MS system.
- Gas chromatograph with injector fitted with 1/4» column adapter, 1/4» Swagelok nuts and Teflon ferrules (or equivalent).
- Syringes: 1-μL, 10-μL (liquid); 100-μL, 500-μL (gas tight).
- Volumetric Flasks, 10-mL.

- Gas bulb, 2 L.

Table 1. NIOSH method 2549: volatile organic compounds (modified).

SAMPLING AND MEASUREMENT		
SAMPLER:		THERMAL DESORPTION TUBE (multi-bed sorbent tubes containing graphitized carbons and carbon molecular sieve sorbents [See Appendix])
FLOW RATE:		0.01 to 0.05 L/min
VOL	-MIN: -MAX:	1L 6L *(even more for non industrial environments)*
SHIPMENT:		Ambient in storage containers
SAMPLE STABILITY:		Compound dependent (store @ -10 °C)
BLANKS:		1 to 3 per set
TECHNIQUE:		THERMAL DESORPTION, GAS CHROMATOGRAPHY, MASS SPECTROMETRY
ANALYTE:		See Table 1
DESORPTION:		Thermal desorption
INJECTION VOLUME:		Defined by desorption split flows (See Appendix)
TEMPERATURE	-DESORPTION:	300 °C for 10 min
	-DETECTOR (MS):	280 °C
	-COLUMN:	35 °C for 4 min; 8 °C/min to 150 °C, 15 °C/min to 300 °C
CARRIER GAS:		Helium
COLUMN:		30 meter DB-1, 0.25-mm ID, 1.0-μm film (or equivalent)
CALIBRATION:		Identification based on mass spectra interpretation and computerized library searches.
RANGE:		not applicable
ESTIMATED LOD:		100 ng per tube or less
PRECISION (S$_r$):		not applicable

Special Precautions

Some solvents are flammable and should be handled with caution in a fume hood. Precautions should be taken to avoid inhalation of the vapors from solvents as well. Skin contact should be avoided.

Sampling

NOTE: Prior to field use, clean all thermal desorption tubes thoroughly by heating at or above the intended tube desorption temperature for 1-2 hours with carrier gas flowing at a rate of at least 50 mL/min. Always store tubes with long-term storage caps attached, or in containers that prevent contamination. Identify each tube uniquely with a permanent number on either the tube or tube container. Under no circumstances should tape or labels be applied directly to the thermal desorption tubes.

- Calibrate each personal sampling pump with a representative sampler in line.
- Remove the caps of the sampler immediately before sampling. Attach sampler to personal sampling pump with flexible tubing.
- NOTE: With a multi-bed sorbent tube, it is extremely important to sample in the correct direction, from least to maximum strength sorbent.
- For general screening, sample at 0.01 to 0.05 L/min for a maximum sample volume of 6 L. Replace caps immediately after sampling. Keep field blanks capped at all times. Tubes can act as diffusive samplers if left uncapped in a contaminated environment.
- Collect a "humidity test" sample to determine if the thermal adsorption tubes have a high water background.

NOTE: At higher sample volumes, additional analyte and water (from humidity) may be collected on the sampling tube. At sufficiently high levels of analyte or water in the sample, the mass spectrometer may malfunction during analysis resulting in loss of data for a given sample.

- Collect a "control" sample. For indoor air samples this could be either an outside sample at the same location or an indoor sample taken in a non-complaint area.
- Ship in sample storage containers at ambient temperature. Store at -10° C.

Sample Preparation

- Allow samples to equilibrate to room temperature prior to analysis. Remove each sampler from its storage container.
- Analyze "humidity test" sampler first to determine if humidity was high during sampling (step 10).

- If high humidity, dry purge the tubes with purified helium at 50 to 100 mL/min for a maximum of 3 L at ambient temperature prior to analysis..
- Place the sampler into the thermal desorber. Desorb in reverse direction to sampling flow.

Calibration And Quality Control

- Tune the mass spectrometer according to manufacturer's directions to calibrate.
- Make at least one blank run prior to analyzing any field samples to ensure that the TD-GC-MS system produces a clean chromatographic background. Also make a blank run after analysis of heavily concentrated samples to prevent any carryover in the system. If carryover is observed, make additional blank runs until the contamination is flushed from the thermal desorber system.
- Maintain a log of thermal desorber tube use to record the number of times used and compounds found. If unexpected analytes are found in samples, the log can be checked to verify if the tube may have been exposed to these analytes during a previous sampling use.
- Run spiked samples along with the screening samples to confirm the compounds of interest. To prepare spiked samples, use the procedure outlined in the Appendix.

Measurement

- See Appendix for conditions. MS scan range should cover the ions of interest, typically from 20 to 300 atomic mass units (amu). Mass spectra can either be identified by library searching or by manual interpretation (see Table 2). In all cases, library matches should also be checked for accurate identification and verified with standard spikes if necessary.

Evaluation of Method

The method has been used for a number of field screening evaluations to detect volatile organic compounds. Estimate of the limit of detection for the method is based on the analysis of spiked samples for a number of different types of organic compounds. For the compounds studied, reliable mass spectra were collected at a level of 100 ng per compound or less. In situations where high levels of humidity may be present on the sample, some of the polar volatile compounds may not be efficiently collected on the internal trap of the thermal desorber. In these situations, purging of the samples with 3 L of helium at 100 mL/min removed the excess water and did not appreciably affect the recovery of the analytes on the sample.

Table 2. Common volatile organic compounds with mass spectral data.

Compound/Synonyms	CAS# RTECS	Empirical Formula	MWa	BPb (°C)	VPc @ 25°C mm Hg kPa		Characteristic Ions, m/z
Aromatic Hydrocarbons							
Benzene /benzol	71-43-2 CY1400000	C6H6	78.11	80.1	95.2	12.7	78*
Xylene /dimethyl benzene	1330-20-7 ZE2100000	C8H10	106.7				91, 106*, 105
o-xylene				144.4	6.7	0.9	
m-xylene				139.1	8.4	1.1	
p-xylene				138.4	8.8	1.2	
Toluene /toluol	108-88-3 XS5250000	C7H8	92.14	110.6	28.4	3.8	91, 92*
Aliphatic Hydrocarbons							
n-Pentane	109-66-0 RZ9450000	C5H12	72.15	36.1	512.5	68.3	43, 72*, 57
n-Hexane /hexyl-hydride	110-54-3 MN9275000	C6H14	86.18	68.7	151.3	20.2	57, 43, 86*, 41
n-Heptane	142-82-5 MI7700000	C7H16	100.21	98.4	45.8	6.1	43, 71, 57, 100*,41
n-Octane	111-65-9 RG8400000	C8H18	114.23	125.7	14.0	1.9	43, 85, 114*, 57
n-Decane /decyl hydride	124-18-5 HD6500000	C10H22	142.29	174	1.4	0.2	43, 57, 71, 41, 142*
Ketones							
Acetone /2-propanone	67-64-1 AL3150000	C3H6O	58.08	56	266	35.5	43, 58*
2-Butanone / methyl ethyl ketone	78-93-3 EL6475000	C4H8O	72.11	79.6	100	13	43, 72*
Methyl isobutyl ketone /MIBK, hexone	108-10-1 SA9275000	C6H12O	100.16	117	15	2	43, 100*, 58
Cyclohexanone / cyclohexyl ketone	108-94-1 GW1050000	C6H10O	98.15	155	2	0.3	55, 42, 98*, 69
Alcohols							
Methanol /methyl alcohol	67-56-1 PC1400000	CH3OH	32.04	64.5	115	15.3	31, 29, 32*
Ethanol /ethyl alcohol	64-17-5 KQ6300000	C2H50H	46.07	78.5	42	5.6	31, 45, 46*

Isopropanol /1-methyl ethanol	67-63-0 NT8050000	C3H7OH	60.09	82.5	33	4.4	45, 59, 43
Butanol /butyl alcohol	71-36-3 E01400000	C4H9OH	74.12	117	4.2	0.56	56, 31, 41, 43
Glycol Ethers							
Butyl cellosolve /2-butoxyethanol	111-76-2 KJ8575000	C6H14O2	118.17	171	0.8	0.11	57, 41, 45, 75, 87
Diethylene glycol ethyl ether / Carbitol	111-90-0 KK8750000	C6H14O3	134.17	202	0.08	0.01	45, 59, 72, 73, 75, 104
Phenolics							
Phenol /hydroxy-benzene	108-95-2 SJ3325000	C6H5OH	94.11	182	47	0.35	94*, 65, 66, 39
Cresol	1319-77-3 G05950000	C7H7OH	108.14				108*, 107, 77, 79
2-methylphenol	95-48-7			190.9	1.9	0.25	
3-methylphenol	108-39-4			202.2	1.0	0.15	
4-methylphenol	106-44-5			201.9	0.8	0.11	
Chlorinated Hy-drocarbons							
Methylene chlo-ride /dichloro-methane	75-09-2 PA8050000	CH2Cl2	84.94	40	349	47	86*, 84, 49, 51
1,1,1-Trichloro-ethane /methyl chloroform	71-55-6 KJ2975000	CCl3CH3	133.42	75	100	13.5	97, 99, 117, 119
Perchloroethylene /hexachloroethane	127-18-4 KX3850000	CCl3CCl3	236.74	187 (subl)	0.2	<0.1	164*, 166, 168, 129, 131, 133, 94, 96
o-,p-Dichloroben-zenes		C6H4Cl2	147.0				146*, 148, 111, 113, 75
/1,2-dichloroben-zene	95-50-1 CZ4500000			172-9	1.2	0.2	
/1,4-dichlorobenzene	106-46-7 CZ4550000			173.7	1.7	0.2	
1,1,2-Trichlo-ro-1,2,2- trifluoro-ethane /Freon 113	76-13-1 KJ4000000	CCl2FC-ClF2	187.38	47.6	384	38	101, 103, 151, 153, 85, 87
Terpenes							
d-Limonene	5989-27-5 OS8100000	C10H16	136.23	176	1.2		68, 67, 93, 121, 136*
Turpentine (Pi-nenes)	8006-64-2	C10H16	136.23	156 to 170	4 @ 20"		93, 121, 136*, 91
α-pinene	80-56-8			156			
β-pinene	127-91-3			165			

Aldehydes							
Hexanal /caproal-dehyde	66-25-1 MN7175000	C6H12O	100.16	131	10	1.3	44, 56, 72, 82, 41
Benzaldehyde / benzoic aldehyde	100-52-7 CU4375000	C7H12O	106.12	179	1.0	0.1	
Nonanal /pelar-gonic aldehyde	124-19-6 RA5700000	C9H18O	142.24	93	23	3	43, 44, 57, 98, 114
Acetates							
Ethyl acetate / acetic ether	141-78-6 AH5425000	C4H8O2	88.1	77	73	9.7	43, 88*, 61, 70, 73, 45
Butyl acetate / acetic acid butyl ester	123-86-4 AF7350000	C6H12O2	116.16	126	10	1.3	43, 56, 73, 61
Amyl acetate / banana oil	628-63-7 AJ1925000	C7H14O2	130.18	149	4	0.5	43, 70, 55, 61
Other							
Octamethylcyclo-tetra- siloxane	556-67-2 GZ4397000	C8H24O-4Si4	296.62	175			281, 282, 283
a Molecular Weight							
b Boiling Point							
c Vapour Pressure							
*Indicates molecular ion							

APPENDIX

Multi-bed Sorbent Tubes

Other sorbent combinations and instrumentation/conditions shown to be equivalent may be substituted for those listed. In particular, if the compounds of interest are known, specific sorbents and conditions can be chosen that work best for that particular compound(s). The tubes that have been used in NIOSH studies with the Perkin Elmer ATD system are ¼ " stainless steel tubes, and are shown in the figure in the next page:

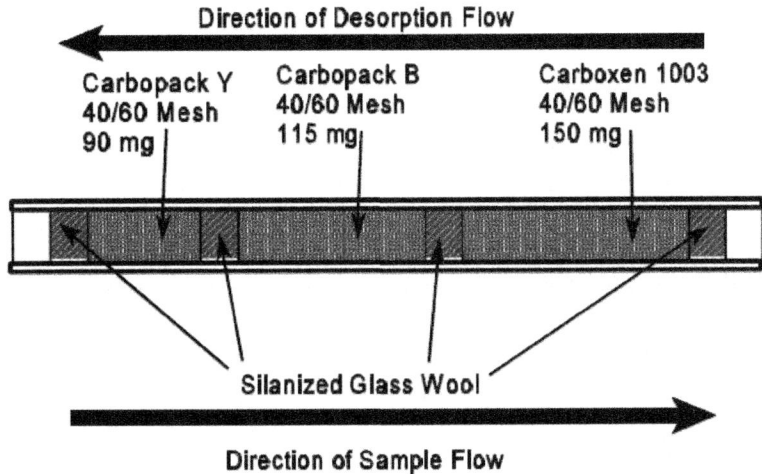

Figure 1. Carbopack™ and Carboxen™ adsorbents are available from Supelco, Inc.

Preparation of Spiked Samples

Spiked tubes can be prepared from either liquid or gas bulb standards.

Liquid Standards

Stock solutions are prepared by adding known amounts of analytes to 10-mL volumetric flasks containing high purity solvent (carbon disulfide, methanol, toluene). Solvents are chosen based on solubility for the analytes of interest and ability to be separated from the analytes when chromatographed. Highly volatile compounds should be dissolved in a less volatile solvent. For most compounds, carbon disulfide is a good general purpose solvent, although this will interfere with early eluting compounds.

Gas Bulb Standards

Inject known amounts of organic analytes of interest into a gas bulb of known volume filled with clean air. Prior to closing the bulb, a magnetic stirrer and several glass beads are placed in the bulb to assist in agitation after introduction of the analytes. After injection of all of the analytes of interest into the bulb, warm the bulb to 50 °C and place it on a magnetic stirring plate and stir for several minutes to ensure complete vaporization of the analytes. After the bulb has been stirred and cooled to room temperature, remove aliquots from the bulb with a gas syringe and inject into a sample tube as described below.

Tube Spiking

Fit a GC injector with a ¼" column adapter. Maintain the injector at 120°C to assist in vaporization of the injected sample. Attach cleaned thermal desorption tubes to injector with ¼" Swagelok nuts and Teflon ferrules, and adjust helium flow though the injector to 50 mL/min. Attach the sampling tube so that flow direction is the same as for sampling. Take an aliquot of standard solution (gas standards 100 to 500 µL; liquid standards, 0.1 to 2 µL) and inject into the GC injector. Allow to equilibrate for 10 minutes. Remove tube and analyze by thermal desorption using the same conditions as for field samples.

Instrumentation

Actual media, instrumentation, and conditions used for general screening of unknown environments are as follows: Perkin-Elmer ATD 400 (automated thermal desorption system) interfaced directly to a Hewlett-Packard 5980 gas chromatograph/HP5970 mass selective detector and data system.

ATD conditions

Tube desorption temperature: 300°C

Tube desorption time: 10 min.

Valve/transfer line temperatures: 150°C

Focusing trap: Carbopack B/Carboxen 1000, 60/80 mesh, held at 27°C during tube desorption

Focusing trap desorption temperature: 300°C

Desorption flow: 50-60 mL/min.

Inlet split: off

Outlet split: 20 mL/min.

Helium: 10 PSI

GC conditions

DB-1 fused silica capillary column, 30 meter, 1-µm film thickness, 0.25-mm I.D. Temperature program: Initial 35°C for 4 min, ramp to 100°C at 8 /min, then ramp to 300°C at 15 /min, hold 1-5 min.

Run time: 27 min.

MSD conditions

Transfer line: 280°C

Scan 20-300 amus, EI mode

EMV: set at tuning value

Solvent delay: 0 min for field samples; if a solvent-spiked tube is analyzed, a solvent delay may be necessary to prevent MS shutdown caused by excessive pressure.

Alternative methods may be obtained adapting, for instance, NIOSH Method 1500 (Hydrocarbons b.p. 36 – 216°C) or NIOSH Method 1501 (Hydrocarbons, aromatic).

Main differences – not fundamentals – may consist in: sorbent choice (charcoal tube), desorption technique (elution with CS_2), sampling time (longer than in workplaces) and other. More important is the possibility of using diffusive (passive) samplers.

PASSIVE SAMPLING

The original purpose of the development on passive sampling (based on Fick's laws) is to provide technology at low cost, enabling air quality surveys to be routinely executed at multiple locations within urban and rural areas, industrial sites and forests. This requires the examination of the performance characteristics of the diffusive sampler over long sampling periods, in comparison with established methods, and in practical applications of urban monitoring (Brown et al., 1994; Brown et al., 1999). Several authors proposed the use of different models of diffusive (passive) devices (Brown, 1993; Harper, 2000; Bertoni et al., 2001; Brown, 2002; Kot-Wasik et al., 2007) conceived for the determination of long-term averaged concentration of some airborne volatile and semi-volatile organic compounds, relevant on the human health protection (VOCs, polycyclic aromatic hydrocarbons and nicotine). The diffusive sampling technique is known to be the cheapest and easiest way to perform extensive sampling campaigns, both in temporal and spatial terms. Moreover, this is the only collection technique allowing the true separation of the vapour phase species from the particle bound fraction (Bertoni et al., 2004; Namiesik et al., 2005).

A lot of different models of passive sampler have been proposed by many researchers; in this chapter we can show only a "classic" sampler (Analyst, developed by the Italian National Research Council, IIA-CNR, Figg. 2 and 3) and a more recent sampler, based on radial diffusion principle (Ring, IIA-CNR, Fig. 4).

All the diffusive samplers have to respect the requirements and test methods according to EN 13528-1, EN 13528-2, EN 13528-3, EN 14412 and EN-ISO 16017-2.

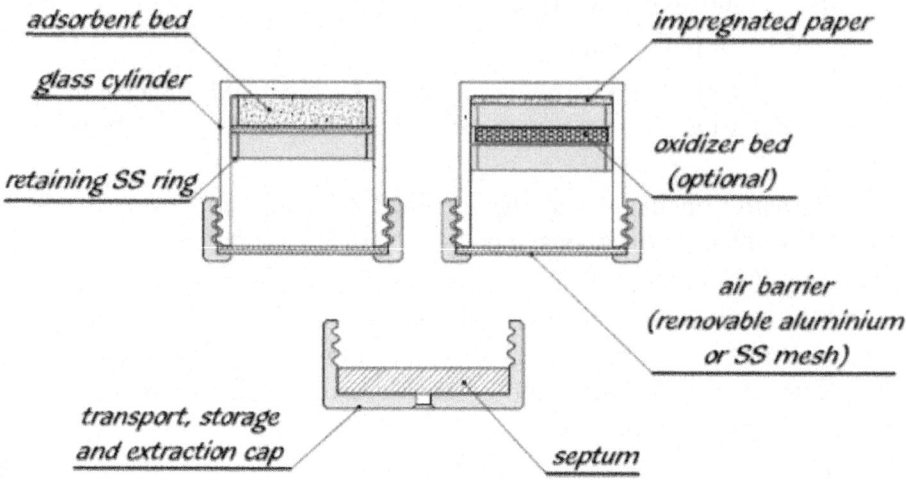

Figure 2. Analyst sampler scheme

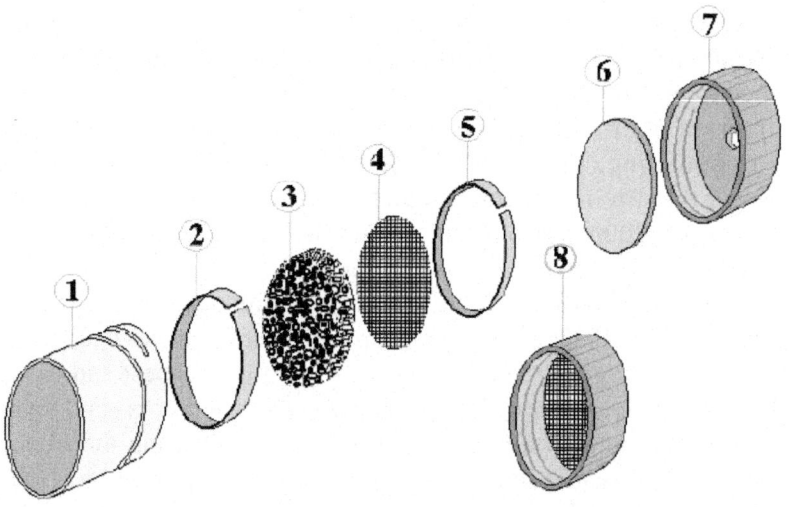

Figure 3. Scheme of the Analyst : 1: glass cylinder (i.d. = 20 mm); 2 and 5: retaining S.S. rings; 3 adsorbent bed; 4: viewing S.S. ring; 6: Teflon seal; 7: cap; 8: aluminium diffusion cap.

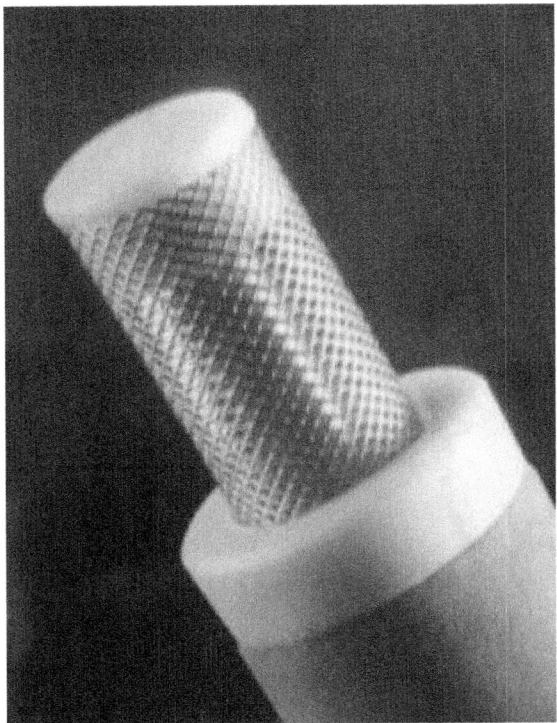

Figure 4. Diffusive sampler "Ring" (IIA CNR).

DATA EVALUATION AND EXPOSURE RISK ASSESSMENT

Studies have found that home levels of several organics average 2 to 5 times higher indoors than outdoors. During and for several hours immediately after certain activities, such as paint stripping, levels may be 1,000 folds background outdoor levels. Despite direct reading instruments can be used (like the IR used in indoor monitoring shown in Figure 5), as above mentioned, currently it is better to try to obtain very low LODs and LOQs by employing sampling methods as described.

Sometimes they are used Threshold Limit Values (TLVs) that are the guideline values set by the American Conference of Governmental Industrial Hygienists (ACGIH) to minimize workers exposure to hazardous concentrations as much as possible. The TLVs are published yearly for more than 700 chemical substances and physical agents (ACGIH, 2010). This use is incorrect and unauthorized by ACGIH, because the TLVs respect is not a warranty of no health effects, that is on the contrary a fundamental requirement in nonindustrial environment.

No standards have been set for VOCs in non-industrial settings. OSHA regulates formaldehyde, a specific VOC, as a carcinogen. OSHA has adopted a Permissible Exposure Level (PEL) of 0.75 ppm, and an action level of 0.5 ppm. Based upon current information, it is advisable to mitigate formaldehyde that is present at levels higher than 0.1 ppm.

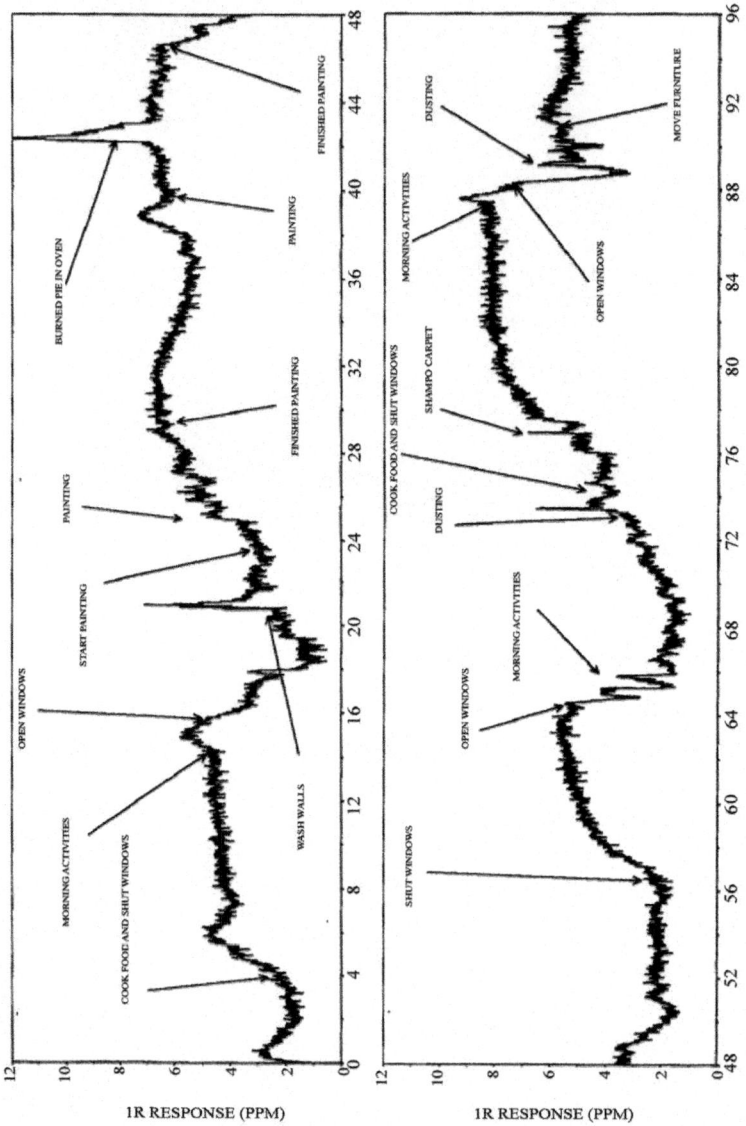

Figure 5. Variation in indoor VOC levels in house house as detected by an IR instrument. VOC levels are expressed in ppm of equivalent 33% propane/67% butane. Human activities at different times are shown (Clobes et al., 1992).

Recently the World Health Organization (WHO, 2010) provided health-based guidelines for 55 airborne inorganic and organic compounds for carcinogenic and non-carcinogenic health endpoints. The non-carcinogenic endpoints include development toxicity, reproduction toxicity, respiratory toxicity, neurotoxicy, hepatoxicity, hematoxicity, eye/nose/throat irritation, and odor annoyance. The lowest concentration at which effects are observed in humans, animals, and plants was used as a staring point for the non-carcinogenic endpoints. Uncertainty factors determined through scientific judgment in consensus and averaging time were also taken into account in determining the health endpoint for non-carcinogenic compounds. The classification by the international Agency for Research on Cancer (IARC) was used to determine a chemical as a carcinogen. The endpoint of carcinogen was determined by linear extrapolation from the high dose level, which is characteristic of animal experiments or occupation exposure with cancer responses.

This objective requires measurements for checking whether specified limit or guideline values are being exceeded. Examples of limit and guideline values (except workplace atmospheres and ambient air) for indoor environments are given in Table 3 and 4, established by the ad hoc committee of IRK/AOLG of Germany for organic chemicals.

Table 3. Examples of limit values.

Pollutant	Country	Limit value	Pollutant
Tetrachloroethene	Germany	0.1 mg/m3	7 days

Table 4. Examples of guideline values.

Pollutant	Country/Organisation	Guideline Value	Averaging Time
Formaldehyde	Germany	0.12 mg/m3	not specified
Nitrogen dioxide	Germany	60 µg/m3 a	7 days
Toluene	Germany	0.3 mg/m3 a	not specified
Styrene	Germany	0.03 mg/m3 a	7 days
Dichloromethane	Germany	0.2 mg/m3 a	24 h
TVOC	Germany	0.2 – 0.3 mg/m3	not specified

a Guideline value (RW I-value, "Richtwert I"= guideline value I) aimed at hygienic prevention.			

REFERENCES

1. American Conference of Governmental Industrial Hygienists (ACGIH) 2010 Threshold Limit Values and Biological Exposure Indices, Cincinnati OH- USA.

2. G. Bertoni, R. Tappa, I. Allegrini, 2001 "The internal consistency of the 'Analyst' Diffusive sampler- a long-term field test"- Chromatographia 54, November (9 10) 653- 657

3. G. Bertoni, R. Tappa, A. Cecinato, 2001 "Environmental Monitoring of semi-volatile polyciclic aromatic hydrocarbons by means of diffusive sampling devices and GC-MS analysis" Chromatographia, 53, Suppl., S312 - S316.

4. G. Bertoni, C. Ciuchini, R. Tappa, 2004 "Long-term diffusive samplers for the indoor air quality evaluation",- Ann. Chim. 94 (Rome), 637644 .

5. R. H. Brown, 2002 Monitoring volatile organic compounds in air- the development of ISO standards and a critical appraisal of the methods- J Environ Mon 4, 112N-118N

6. R. H. Brown, M. D. Wright, N. T. Plant, 1999 The Use of Diffusive Sampling for Monitoring of Benzene, Toluene and Xylene in Ambient Air (IUPAC Technical Report) Pure Appl. Chem., 71 10 19932008 .

7. R. H. Brown, 1993 The use of diffusive samplers for monitoring of ambient air- Pure & Appl. Chem. 65 18591874

8. R. H. Brown, R. P. Harvey, C. J. Purnall, K. J. Saunders, 1984 A Diffusive Sampler Evaluation Protocol- Am. Ind. Hyg. Assoc. J. 45, 67 EOF75 EOF .A. L. Clobes, G. P. Ananth, A. L. Hood, J. A. Schroeder, K. A. Lee, 1992 Human Activities as Source of Volatile Organic Compounds in Residential Environments; in "Sources of Indoor Air Contaminants", Tucker W.G., Leaderer B.P., Molhave L. and Cain W.S. editors. Annals of the New York Academy of Sciences, 641 79-86.

9. M. Harper, trapping. Review-Sorbent, volatile. of, compounds. organic, air. from-J, Chrom, 885 885 2000 129151

10. A. Kot-Wasik, B. Zabiegala, M. Urbanovicz, E. Dominiak, A. Wasik,

J. Namiesnik, 2007 Advances in passive sampling in environmental studies- Anal Chim Acta 602 141164

11. M. Maroni, R. Axelrad, A. Bacaloni, 1995 "NATO's Efforts to set IAQ Standards" Am. Ind. Hygiene Ass. J., 56 499508

12. J. Namiesnik, Z. Bozena, A. Kot-Wasik, M.. Partyka, A. Wasik, 2005 Passive sampling and/or extraction techniques in environmental analysis: a review- Anal Bioanal Chem 381 279301

13. EN 13528- 1 (Sept. 2002) Ambient air quality: Diffusive samplers for the determination of concentrations of gases and vapours- Requirements and test methods. Part 1: General requirements

14. EN 13528- 2 (Sept. 2002) Ambient air quality: Diffusive samplers for the determination of concentrations of gases and vapours- Requirements and test methods. Part 3: Specific requirements and test methods

15. EN 13528- 3 (Dec. 2003) Ambient air quality: Diffusive samplers for the determination of concentrations of gases and vapours- Requirements and test methods. Part 3: Guide to selection, use and maintenance

16. EN 14412 (Sept. 2004) Indoor air quality- Diffusive samplers for the determination of concentrations of gases and vapours- Guide for selection, use and maintenance

17. EN ISO 16017-2 (May 2003) Indoor, ambient and workplace air- Sampling and analysis of volatile organic compounds by sorbent tube/ thermal desorption/capillary gas chromatography. Part 2: Diffusive sampling

18. NIOSH Manual of Analytical Methods, 1996 Method 2549 Volatile Organic Compounds (screening) 4th rev., 1-ed . Cincinnati, OH- USA.

19. NIOSH Manual of Analytical Methods, 2003 Method 1500 Hydrocarbons b.36216 C, 3-ed . Cincinnati, OH- USA.

20. NIOSH Manual of Analytical Methods, 2003 Method 1501 Hydrocarbons, aromatic, 3-ed . Cincinnati, OH- USA.

21. WHO guidelines for indoor air quality: selected pollutants 2010 978-9-28900-213-4

22. http://www.epa.gov/iaq/atozindex.html

23. http://www.umweltbundesamt.de/uba-info-daten-e/daten-e/irk.htm#4

24. http://www.umweltbundesamt.de/uba-info-daten/daten/irk.htm

25. http://www.osha.gov/dts/osta/otm/otm_iii/otm_iii_2.html

Chapter 9

INDOOR AIR POLLUTANTS AND THE IMPACT ON HUMAN HEALTH

Marios. P. Tsakas, Apostolos P. Siskos, and Panayotis A. Siskos

[1]Laboratory of Environmental Chemistry, National and Kapodestrian University of Athens, Greece

INTRODUCTION

The major area of public concern and government policy, in terms of the impact of air pollution on human health, continues to be outdoor air. However, over the last two decades, indoor air quality (IAQ) has caused increasing concern due to the adverse effects that it may have on human health. The term "indoors" is used in relative literature to refer to a variety of environments, including homes, workplaces, and buildings used as offices or for recreational purposes. In addition, a number of studies have been carried out to measure various compounds inside vehicles during commuting activities. Most people in the developed world spend up to 90% of their time in an indoor environment and up to 60% of the workforce work in an office. (Tsakas & Siskos, 2010; McCurdy et al., 2000; Ashford & Caldart, 2008; Andersson & Klevard Setterwall, 1996) Decreased ventilation rates for energy conservation, along with increased use of synthetic materials in buildings, have resulted in increased health complaints from building occupants (Siskos, 2003). Many indoor pollutants are either known, or suspected to be, allergens, carcinogens, neurotoxins, immunotoxins or irritants, while all may contribute to sick building syndrome (SBS). The set of health symptoms associated with SBS includes nasal, ocular and generalised diseases. According to various studies performed in public buildings by the National Institute of Occupational Safety and Health (Soldatos et al., 2003), the three most significant symptoms that were experienced in more than 70% of the buildings are dry eyes, dry throat and headaches.

The IAQ and the presence of air pollutants in indoor environment is a worldwide issue, since many governments and environmental institutes have faced this serious phenomenon. Starting in the 1990s in Japan, tightly sealed buildings with low ventilation rates have been constructed. This, combined with the use of some new types of building materials has often resulted in IAQ problems. Many inhabitants suffering with SBS and multiple chemical sensitivity (MCS), have been reported (McCurdy et al., 2000; Zhang & Niu, 2004). As a result the Japanese Ministry of Health, Labour and Welfare have introduced indoor air guidelines for a range of VOCs including HCHO based on hazard assessments (Shinohara et al., 2009).

The importance of IAQ has also been recognised in Europe and has been identified as an important element within the European Collaborative Action (ECA) (ECA, 1998) and the European Environment and Health Action Plan (Dimitroulopoulou et al., 2006). In America, the State of California has adopted an active programme for the last two decades aiming to the reduction of indoor air pollution, which has led to a range of policy instruments (Waldman & Jenkins, 2004). Over recent years, important steps have been made towards setting IAQ standards and guidelines in the UK (Dimitroulopoulou et al., 2006) and recently, the UK Department of Health Committee on the Medical Effects of Air Pollutants (Short, 2001), launched a guidance document on the effects of indoor air pollutants. In many developing countries, exposure to indoor air pollution causes a major health burden (Committee on the Medical Effects of Air Pollutants, 2004). Increased concern regarding indoor air quality especially in the last two decades has led to a number of studies and meetings on the subject. For example, in Greece many researchers have conducted significant studies about IAQ issue (Siskos & co-workers, 2001, 2003, 2005, 2010; Helmis & co-workers, 2007, 2009; Santamouris et al., 2001). With increasing concern in relation to health effects, in recent years the problem has come into sharper focus. Additional new sources of contaminants are being introduced, which haven't been measured before. In several countries, studies have been undertaken, in some cases involving comprehensive investigations of the factors governing air quality, so that effective control measures ranging from the setting of minimum ventilation standards, to controlling, or even banning, certain products such as urea-formaldehyde foam insulation or unvented paraffin or gas heaters. It is nevertheless recognized that some of the responsibility for maintaining acceptable and healthy indoor air quality will continue to rest with building owners and occupants of buildings.

Some studies have revealed a variety of contaminants of indoor air including odorous, non-odorous gases and vapours, and particles, and although there were suggestions that some of these contaminants could be responsible for health effects, proving causal relationships is exceedingly difficult even where elevated levels of potentially toxic substances exist (World Health

Organization [WHO], 1989; Perry & Kirk, 1986; WHO, 1986; Priorities for Indoor Air Research and Action, 1991).

INDOOR AIR POLLUTANTS AND THEIR SOURCES

There are many indoor air contaminants, which can be separated based on their effects on human health, the frequency of their appearance, their usual concentration levels, their sources etc. This chapter is focused, primarily, on those species common to indoor and outdoor air environments and those who are measured more often in indoor environments.

RADON

The main source of indoor radon is its immediate parent radium-226 in the ground of the site and in the building materials (Nero, 1988, 1989). Outdoor air also contributes to the radon concentration indoors, via the ventilation air. Tap-water and the domestic gas supply are usually radon sources of minor importance, with a few exceptions. In most situations it appears that elevated indoor radon levels originate from radon in the underlying rocks and soils (Castren et al., 1985). This radon may enter living spaces in dwellings by diffusion or pressure driven flow if suitable pathways between the soil and living spaces are present. It should be noted, however, that in a minority of cases elevated indoor radon levels may arise due to the use of building materials containing high levels of radium-226. Examples of such materials, used in some buildings, are by-product gypsum, alum shale and volcanic tuffs.

The United Nation Scientific Committee on the Effects of Atomic Radiations (UNSCEAR) has made a very simple model to try to estimate the relative contribution of these sources: for a "typical" house, with a radon concentration of 50 Bq/m^3 at ground floor, the contributions of soil, building materials and outdoor air are, respectively, 60%, 20% and 20%, while for the upper floors in high rise buildings, where the radon concentration-is estimated to be "typically" 20 Bq/m^3, these values become 0%, 50% and 50% (UNSCEAR, 1993).

Soil

For those who live close to the ground, e.g. in detached houses or on the ground floor of apartment buildings without cellars, the most important radon source is radium in the ground.

The radium concentration in soil usually lies in the range 10 Bq/kg to 50 Bq/kg, but it can reach values of hundreds Bq/kg, with an estimated average of 40 Bq/kg (UNSCEAR, 1993). Typical radon concentrations in soil gas range from 10000 Bq/m³ into 50000 Bq/m³. The potential for radon entry from the ground depends mainly on the activity level of radium-226 in the subsoil and its permeability with regard to air flow. Example of terrains with a high radon potential are alum shales, some granites and volcanic rocks, due to high concentrations of radium-226 and the presence of eskers (gravel, sand and rounded stone deposited from subglacial streams during the ice ages), all these being characterised by high permeability. The ground could also be contaminated with waste tailings from uranium or phosphate mining operations with enhanced activity levels (Tyson et al., 1993).

The ingress of radon from the soil is predominantly one of pressure-driven flow, with diffusion playing a minor role (de Meijer et al., 1992). The magnitude of the inflow varies with several parameters, the most important being the air pressure difference between soil air and indoor air, the tightness of the surfaces in contact with the soil on the site, and the radon exhalation rate of the underlying soil. If there is no airtight layer between the basement and the ground, the underpressure indoors causes radon to be drawn in from the ground under the building. Underpressure occurs in most houses if either the adjustment of inlet and outlet of air in forced ventilation systems or the outdoor air supply for vented combustion appliances is inappropriate. The underpressure may be considerable for all types of ventilation systems when the inlet air is restricted too much. The tightness of the structures has to do with e building regulations and techniques and is very dependent on cracks, openings and joints. Structures are hardly ever so airtight that radon inflow is completely prevented. For example, to get a radon daughter concentration of less than 100 Bq/m³ EER in a house with a volume of 500 m³and a ventilation rate of 0.5 air changes per hour, not more than 1 m³ per hour must be allowed to leak into the house if the radon gas concentration in soil air is about 50000 Bq/m³. Such values are quite typical.

Building Materials

Building materials are generally the second main source of radon indoors, while in the Seventies they were considered the principal one (UNSCEAR, 1977; Meyer et al., 1986). Radon exhalation from building materials depends not only on the radium concentration, but also on factors such as the fraction of radon produced through material release, the porosity of the material and the surface preparation and finish of the walls. In general, no action needs to be taken concerning traditional building materials. Typical values for radium and thorium content in building materials are 50 Bq/kg or less (Nuclear Energy Agency Organisation for Economic Co-operation and Development - NENOECD, 1979). Building materials containing by-product gypsum

(UNSCEAR, 1982) and concrete containing alum shale (Swedjemark & Mjones, 1984) may have much higher radium concentrations. The activity concentrations in brick and concrete may also be high if the raw materials have been taken from locations with high levels of natural radioactivity. Examples of such natural materials, used in some buildings, are volcanic tuffs and pozzolana (Sciocchetti et al., 1983; Campos Venuti et al., 1984; Battaglia et al., 1990), where radium and thorium content can reach some hundreds of Bqlkg. Other measurements of radioactivity content and exhalation of building materials are reported in NENOECD (1979).

Building materials are the main sources of radon-220 (also called "thoron") in indoor air. Due to its short half life (55 s), thoron originating in soil in effect is usually prevented from entering buildings and therefore makes negligible contribution to indoor thoron levels. For this reason and due to the greater difficulties of measurement, thoron concentration measurements are very much fewer than those for radon. Although the indoor thoron concentrations are usually low (Cliff, 1992; UNSCEAR, 1993), in some cases the doses due to this isotope and its daughters are significant and comparable to those due to radon- 222 (Sciocchetti et al., 1983, 1992; Guo et al., 1992; Bochicchio et al., 1993; Doi & Kobayashi, 1994).

Outdoor Air

Outdoor air usually acts as a diluting factor, due to its normally low radon concentration, but in some cases, as in high rise apartments built with materials having very low radium content, it can act as a real source. The radon concentration in outdoor air is mainly related to atmospheric pressure, and (in case of non-perturbative weather) it shows a typical oscillating time pattern, with higher values during the night.

Until a few years ago the average level of radon gas concentrations in the atmosphere at ground level was, in most cases, assumed to be of the order of few Bq/m^3 -e.g. in the range of 4 to 15 Bq/m^3 in USA (Gesell, 1983), but more recent measurements seem to indicate higher values, reaching some tens of Bq/m^3 (Hopper et al., 1991; Robé et al., 1992; Bochicchio et al., 1993; Deyuan, 1993; Grasty 1994;Price et al., 1994). Quite high radon concentrations in the outdoor air have been reported near substantial radon sources, such as mine tailings (Tyson et al., 1993), or in the case of particular weather conditions, such as thermal inversion or very low precipitation (Grasty, 1994).

Ambient air over oceans has very low values ($\sim$ 0.1 Bq/m^3) of radon concentrations, due to the minimum presence of radium in the sea water and the high solubility of radon in water at low temperatures. Therefore radon concentration in outdoor air of islands and coastal regions is generally lower than in continental countries, e.g. United Kingdom and Japan have an average outdoor air value of ~4 Bq/m³.

Taking into account recent measurements, the mean value of outdoor radon concentrations adopted by UNSCEAR in its last report has been changed from 5 to 10 Bq/m³ for continental areas and somewhat less in coastal regions (UNSCEAR, 1993).

Tap Water

In wells drilled in rock the radon concentrations of water may be high. When such water is used in the household, radon can be partially released into the indoor air, causing an increase in the average radon concentrations. In a few regions, such as Finland and Maine (USA), the tap water from wells drilled in rock has been shown to contribute significantly to radon concentrations indoors. Radon concentrations in tap-water from deep wells can range from 100 kBq/m³ to 100 MBq/m³ (UNSCEAR, 1988). The indoor radon concentrations in these regions may already be high due to high rates of radon entry from the ground. The world average radon concentration in all types of water supplies is assumed to be 10 kBq/m³ (UNSCEAR, 1993).

Domestic Gas

In some regions, natural gas used for cooking and heating contains elevated concentrations of radon, which is released on combustion. Normally this source is insignificant, and can be monitored at transmission and distribution points. Typically the radon level in natural gas is about 1000 Bq/m³. Natural gas, as it is usually supplied, contains gas from a number of wells and fields and thus can vary over time, depending on the proportions supplied by different sources (UNSCEAR, 1993).

Oxides of Nitrogen

NOx

A large number of studies of NO and NO_2 have been carried out in many different indoor air environments (Finlayson-Pitts, 1999; Pitts et al., 1985). Because of air exchange, indoor levels are generally higher when outdoor levels increase (Hoek et al., 1989; Rowe et al., 1991; Hisham & Grosjean, 1991; Spengler et al., 1994; Weschler et al., 1994; Baek et al., 1997). However, enhanced indoor levels can be found when combustion sources are present. These include gas stoves, paraffin heaters, water heaters, and cigarette smoke (Wade et al., 1975; Marbury et al., 1988; Ryan et al., 1988;Petreas et al., 1988;

Hoek et al., 1989; Pitts et al., 1989; Spengler et al., 1994; Levy et al., 1998). While combustion generates primarily NO, the focus indoors has been on NO_2 because of its health impact. Again, the use of gas stoves was highly correlated with indoor NO_2, with an indoor/outdoor concentration ratio of 1.19 for homes with a gas range compared to 0.69 for those without a gas stove. The ratio was even higher for homes with a paraffin space heater, 2.3 compared to 0.85 without such a heater (Levy et al., 1998). Both the indoor and outdoor concentrations of NO_2 were higher in cities where at least 75% of the homes had gas stoves; for example, the mean outdoor NO_2 concentration in such gas-intensive cities was 38 ± 20 ppb, compared to 14 ± 6 ppb in cities where fewer than 25% of the households had gas stoves installed. High concentrations of NO_2 have also been measured in indoor skating rinks where the use of ice resurfacing machines powered by propane, gasoline, or diesel fuel results in significant emissions (e.g., Brauer & Spengler, 1994; Brauer et al., 1997; Pennanen et al., 1997). Mean concentrations of NO_2 of ~200 ppb have been reported, with some rinks having concentrations up to 3 ppm! The indoor-to-outdoor ratios of the arithmetic mean concentrations varied from about 1 to 41, with an overall mean of 20. In the absence of such sources of NOx, indoor and outdoor concentrations are quite similar (Weschler et al., 1994), since removal of NO and NO_2 indoors, e.g., on surfaces, is relatively slow. However, as it has been discussed shortly, although the surface reaction of NO_2 is relatively slow, it is still of interest since it generates nitrous acid (HONO). Different surfaces found inside homes have been found to have different removal rates for NO_2. In short, there is a variety of evidence that there are higher levels of NO_2 indoors when combustion sources are present and that the concentrations generated indoors can be quite substantial in some circumstances. One word of caution is in order, however, particularly in regards to earlier measurements of NO_2.

HONO and HNO_3

HONO is formed by the reaction of NO_2 with water on surfaces. The reaction is usually represented as

$$2NO_2 + H_2O \rightarrow HONO + HNO_3 \tag{1}$$

Although the detailed mechanism is not known; gaseous HNO_3 is not generated in equivalent amounts, something which has been attributed to its remaining being adsorbed on the surface. This overall reaction occurs on a variety of surfaces in the laboratory and hence might be expected to also occur on surfaces in other environments, such as homes. This, indeed, is the case. (Pitts et al., 1985) first used differential optical absorption spectrometry (DOAS) to establish unequivocally that NO_2 injected into a mobile home forms HONO. Interestingly, the dependence of the rate of HONO generation on the NO_2 concentration was similar to that measured in laboratory systems, consistent with production in, or on, a thin film of water adsorbed on surfaces. A number of studies have confirmed that the behaviour is similar to that in laboratory systems; i.e., the rate of production of HONO increases with NO_2 and with relative humidity. Indoor levels of HONO as high as 8 ppb as a 24-h average and 40 ppb as a 6-h average have been reported in normal, in-use buildings and homes (Febo & Perrino, 1991; Spengler et al., 1993; Weschler et al., 1994). The ratio of HONO to NO_2 indoors can be quite large, up to ~0.15 (e.g., Febo & Perrino, 1991; Brauer et al., 1990, 1993; Spengler et al., 1993). This can be compared to typical values of a few percent outdoors. High levels of HONO (up to ~ 30 ppb) have also been measured in automobiles in use in polluted urban areas, and again, the ratio of HONO to NO2 was quite large, ~0.4, compared to 0.02-0.03 measured outdoors in the same study (Febo & Perrino, 1995). The generation of NO was attributed by Spicer and co-workers to a reaction of gaseous NO_2 with adsorbed HONO:

$$NO_2(g) + HONO(ad) \rightarrow H^+ + NO_3^- + NO_3(g) \tag{2}$$

The same process was hypothesised to explain some time periods in a commercial office building when indoor NO actually exceeded outdoor NO (Weschler et al., 1994). As is the case in laboratory systems, equivalent amounts of HNO_3 are not observed as might be expected from the stoichiometry of reaction (1), likely due to HNO_3 remaining on the surface after formation and/ or being taken up by surfaces. The accumulation of nitrate on indoor surfaces in a commercial building has been reported by Weschler and Shields (1996) and attributed to the formation and uptake of HNO_3 via reactions of NO_3 and/ or oxidation of nitrite (i.e., adsorbed HONO) in an aqueous surface film. Subsequently, it was shown that HONO is also directly emitted by gas stoves (Pitts et al., 1989). In a house used for investigating indoor air pollution that had natural gas fueled appliances (a convective heater, a radiant heater, and a range with four burners), both the surface reaction of NO_2 and the direct combustion emissions contributed significantly to the measured indoor HONO. When an appliance was operational, the contribution of direct emissions was the more important source (Spicer et al., 1993). In short, the "dark reaction" of NO_2 with water on surfaces is ubiquitous and occurs not only in laboratory systems but also indoors. The combination of this heterogeneous reaction with combustion sources of HONO can produce significant concentrations of HONO indoors. As a result, there is a concern regarding the health impacts of nitrous acid, not only because it is an inhalable nitrite but also because it is likely the airborne acid present in the highest concentrations indoors.

CO and SO$_2$

As for NOx, combustion sources such as gas stoves and paraffin heaters can be significant sources of indoor CO. The ratio of indoor to outdoor concentrations of CO in homes using gas stoves has been measured to be 1.2-3.8 (Wade *et al.*, 1975), with the highest ratios found close to the source. Similarly, higher CO levels indoors compared to outdoors have been reported for restaurants in Korea, with those using charcoal burners as well as gas giving much higher concentrations (Baek *et al.*, 1997). In buildings where motor vehicle exhaust can be entrained from outdoors or attached parking garages, elevated indoor CO levels may also result (Hodgson *et al.*, 1991). On the other hand, in homes and offices where there was no direct indoor source of CO, the indoor to-outdoor ratio was about one, and sometimes less. For example, in Riyadh, Saudi Arabia, CO concentrations were measured indoors and outdoors; the indoor to- outdoor ratio varied from 0 to 2, but was typically below one (Rowe *et al.*,1989). There have been a number of measurements of CO in the "indoor environment" of automobiles. Given that cars are major CO sources in urban areas, one might expect higher concentrations of CO during commutes and this is indeed the case. Typical CO concentrations of ~9-56 ppm have been measured inside automobiles during commutes in major urban areas (Flachsbart *et al.*, 1987; Koushki*et al.*, 1992; Ott *et al.*, 1994, 1995; Dor *et al.*, 1995; Fernandez-Bremauntz & Ashmore, 1995). This can be compared to peak outdoor levels of ~ 10 ppm in highly polluted urban areas. Thus, a significant enhancement of CO inside automobiles during commutes is common. For example, Chan *et al.* (1991) report a ratio of the in-vehicle CO concentration to that outdoors of ~ 4.5 in Raleigh, North Carolina. As is the case for CO, SO$_2$ levels indoors and outdoors tend to be similar if there are no combustion sources indoors.

Volatile Organic Compounds (VOCs)

Volatile organic compounds (VOC) are ubiquitous components not only of ambient air but also of indoor air environments, including offices, commercial and retail buildings, and homes (Shah & Singh, 1988; Finlayson-Pitts, 1999). There are three sources/categories for VOC: (1) entrainment of air from outside the building, (2) emissions from building materials, and (3) human activities inside buildings. As might be expected, given the nature of the sources, a very large variety of organic compounds have been identified and measured indoors (e.g., Brown *et al.*, 1994; Crump, 1995;Kostiainen, 1995). These numbers in the hundreds of different compounds, with the particular species and their concentrations depending on the particular sources present as well as the air exchange rates. Some of the compounds associated with the three sources: entrainment from outdoors, emissions from building materials, and anthropogenic activities - are now briefly reviewed.

Entrainment of Air from Outdoor Sources

Entrainment of outdoor air through ventilation systems brings with it the species found in ambient air. Some of them, such as HNO_3, can be removed on surfaces such as those in air conditioning systems, and hence the indoor concentrations tend to be lower than those outdoors. Others such as NO tend to have similar concentrations indoors and outdoors if there are no significant combustion sources indoors (e.g., Weschler *et al.*, 1994). In the case of hydrocarbons, the concentrations of compounds that do not have significant indoor sources tend to be about the same as the outdoor concentrations. For example, Lewis and Zweidinger (1992) measured VOC in 10 homes in winter and showed that the concentrations of ethene, benzene, 2-methylpentane, methylcyclopentane, 2,2,4-trimethylpentane, and 2,3-dimethylbutane indoors were within experimental error of those outdoors. There are, however, some specific outdoor sources that can lead to higher concentrations of certain VOCs indoors than in the general outdoor air environment. For example, gases generated in landfills or from petroleum contamination can migrate through the soil and groundwater to adjacent buildings and homes to give larger indoor concentrations, particularly in basements and crawl spaces, than otherwise expected (Moseley & Meyer, 1992; Hodgson *et al.*, 1992;Fischer *et al.*, 1996). In one such case, the total hydrocarbon concentration was measured to be 120 ppm in a crawl space beneath the floor of a school where petroleum contamination was present from adjacent sources, compared to < 80 ppb outdoors (Moseley & Meyer, 1992). Although concentrations in various rooms were lower, they were still elevated compared to outdoors, ranging from 0.13 to 3.4 ppm. The use of pesticides *outside* buildings can also lead to enhanced concentrations of these compounds indoors. For example, Anderson and Hites (1988) measured the concentrations of chlorinated pesticides indoors and found elevated levels inside, e.g., a factor of 7 times higher for y-chlordane compared to outdoor levels. One home that had the highest indoor concentrations had been treated with chlordane about a decade earlier, presumably by subsurface injection from which the pesticide migrated into the house through cracks in the basement walls. Enhanced levels of chlorpyrifos were observed indoors in homes where soil surrounding the home had been treated on a regular basis. Another source of VOC is motor vehicle emissions, which can be drawn into buildings from outdoors or parking garages (e.g., Perry & Gee, 1994; Daisey et al., 1994). For example, motor vehicles were major sources (responsible for > 75%) of 12 of 39 individual compounds measured in a dozen buildings by Daisey *et al.* (1994). Of the 12 compounds, 5 were alkanes and 7 were aromatics. Similarly, Baek *et al.* (1997) report that vehicle emissions are important VOC sources indoors in Korea during the summer in homes and offices, as has been reported in the United States (e.g.,Hodgson *et al.*, 1991; Daisey *et al.*, 1994).

Building Materials

Emissions associated with building materials are major contributors to indoor levels of VOC. New buildings often have higher concentrations of certain compounds compared to older buildings. For example, enhanced levels of n-dodecane, n-decane, and n-undecane, the xylenes, and 2-propanol have been measured in new buildings, and the total VOC concentration is generally larger (by factors of 4-23) compared to established buildings (Brown *et al.*, 1994). Kostiainen (1995) identified more than 200 individual VOCs indoors in 26 houses. In addition, they compared the VOC concentrations in normal houses to those where complaints of odours or illness had been registered. A number of different VOCs were present at increased concentrations in the houses with complaints compared to the normal houses; these included a variety of aromatic hydrocarbons, methylcyclohexane, n-propylcyclohexane, terpenes, and chlorinated compounds such as 1,1,1-trichloroethane and tetrachlorethene. Carpets are a major source of VOCs in homes. For example,Sollinger *et al.* (1993, 1994) have identified 99 different VOCs emitted from a group of 10 carpet samples, and Schaeffer *et al.* (1996) identified more than 100 different VOCs emitted from the carpet cushion alone. Emissions come not only from the carpet fibres but also from the backing materials and the adhesives used to bind the carpet to the backing. As a result, the individual compounds emitted by carpets can vary substantially, depending on the carpet construction. Many of the compounds emitted are known to be used in the manufacturing processes (e.g., e-caprolactam is used in Nylon-6 production) and /or are common solvents Emissions of VOC from carpets tend to decrease with time and increase with temperature.

The dependence of VOC emissions from building materials on relative humidity is more complex, with some emissions increasing with relative humidity, but others not. For example, Sollinger *et al.*(1994) report that the VOC emissions from carpets did not change with relative humidity over the range from 0 to 45% RH. On the other hand, the emissions of formic and acetic acids from latex paints have been reported to increase dramatically with relative humidity; for example, for one paint sample the emission rate for acetic acid almost tripled when the relative humidity was changed from 4-5% to 5-23% (Reiss *et al.*, 1995b). A number of different aldehydes have been measured indoors (Crump & Gardiner, 1989; Lewis & Zweidinger, 1992; Zhang *et al.,* 1994; Daisey *et al.,* 1994; and Reiss *et al.,*1995a), some of which are directly emitted and some of which are formed by chemical reactions indoors of VOCs such as styrene. Of these, there is an enormous amount of evidence for direct emissions of HCHO from building materials. Interest in formaldehyde emissions and levels in homes and other buildings stems from its well-known health effects, which include possible human carcinogenicity and eye, skin, and respiratory tract irritation (Feinman, 1988). Formaldehyde is emitted from urea-formaldehyde foam insulation as well as from resins used in reconstituted wood products such as particleboard and plywood (Meyer and Reinhardt, 1986); urea-formaldehyde resins comprise about 6-8% of the weight of particleboard and 8-10% of mediumdensity fiberboard (Meyer and Hermanns, 1986). Other sources include permanent press fabrics (such as draperies and clothing), floor finishing materials, furniture, wallpaper, latex paint, varnishes, some cosmetics such as fingernail hardener and nail polish, and paper products (Kelly, 1996; Howard *et al.,* 1998a, 1998b).

Many measurements of HCHO have been made in indoor air environments. In conventional homes, average concentrations are typically about 10-50 ppb (Stock, 1987; Zhang *et al.,* 1994; Reiss *et al.,*1995a). Sexton *et al.* (1989) measured concentrations of HCHO in 470 mobile homes in California and found geometric mean concentrations of 60-90 ppb, although maximum values of over 300 ppb were recorded in some cases. In a similar study in Wisconsin, levels up to 2.8 ppm were measured (Hanrahan *et al.,* 1985). Higher levels are typically found in mobile homes because of the reconstituted wood products (e.g., particleboard and plywood) used in their construction. Interestingly, HCHO does not appear to be a significant product of natural gas combustion, as levels in dwellings with and without gas stoves turned on are not significantly different (e.g., Pitts *et al.,* 1989; Zhang *et al.,* 1994). Temperature is again an important determinant of HCHO levels.

Human Activities

There are many sources of VOCs associated with human activities in buildings. For example, mixtures of C_{10} and $C_1\sim$ isoparaffinic hydrocarbons, which are characteristic of liquid process copiers and plotters, have been identified in office buildings in which these instruments were in use (Hodgson *et al.,* 1991). Emissions of a number of hydrocarbons and aldehydes and ketones have been observed during operation of dry-process copiers; these include significant emissions of ethylbenzene, o-, m-, and p-xylenes, styrene, 2-ethyl-l-hexanol, acetone, nnonanal, and benzaldehyde (Leovic *et al.,* 1996). Enhanced levels of acetaldehyde in an office building in Brazil were attributed to the oxidation of ethanol used as a cleaning agent (Brickus *et al.*, 1998), although levels outdoors were also enhanced due to the use of ethanol as a fuel. Pyrocatechol has been measured in an occupational environment where meteorological charts are mapped on paper impregnated with this compound (Ekinja *et al.*, 1995), and p-dichlorobenzene is observed when mothballs containing this compound are in use (e.g., Tichenor *et al.*, 1990; Chang and Krebs, 1992). Elevated concentrations of the n-C_{13} to n-C_{18} alkanes and branched-chain and cyclic analogs were measured in a building having a history of air quality complaints; the source was found to be volatilisation from hydraulic fluids used in the building elevators (Weschler *et al.*, 1990). Enhanced levels of chlorinated compounds have been observed indoors due to human activity as well. For example, increased levels of perchloroethylene have been observed from unvented dry-cleaning units (Moschandreas & O'Dea, 1995) and volatilisation of chlorinated organics such as chloroform from treated tap water can occur (McKone, 1987). Other sources include the use of household products. For example, chloroform emissions have been observed from washing machines when bleach containing hypochlorite was used (Shepherd *et al.*,1996). It is interesting that emissions of organics associated with the use of washing machines are decreased when the machine is operated with clothes inside (Howard and Corsi, 1998). Of course, activities such as smoking result in enhanced levels not only of nicotine (e.g., Thompson *et al.*, 1989) but also of a variety of other gases associated with cigarette smoke (e.g., California Environmental Protection Agency, 1997; Nelson *et al.*, 1998). For example, using 3-ethenylpyridine as a marker for cigarette smoke, Heavner *et al.* (1992) estimated that 0.2-39% of the benzene and 2-49% of the styrene measured in the homes of smokers was from cigarette smoke. Humans emit a variety of VOCs such as pentane and isoprene (e.g., Gelmont *et al.*, 1981; Mendis *et al.*, 1994; Phillips *et al.*, 1994;Jones *et al.*, 1995; Foster *et al.*, 1996). In addition, emissions from personal care products have been observed. Decamethylcyclopentasiloxane (D5), a cyclic dimethylsiloxane with five Si-O units in the ring, and the smaller D4 analog, octamethylcyclotetrasiloxane, are used in such products as underarm deodorant and antiperspirants at concentrations up to 40-60% by weight (Shields and Weschler, 1992; Shields *et al.*, 1996). Increased concentrations of D5 have been measured in offices and are correlated to human activity, as expected if personal care products were the major source (Shields and Weschler, 1994). In some cases, increased concentrations attributable to emissions from silicone-based caulking materials were also observed (Shields *et al.*, 1996). The use of pesticides indoors can lead to very large concentrations not only of the pesticide but of the additional VOCs used as a matrix for the pesticide, which represent most (>95%) of the mass of the material as purchased. For example, Bukowski and Meyer (1995) predict that VOC concentrations immediately after the application of a fogger could reach levels of more than 300 mg m^{-3}!

OZONE

Because O$_3$ decomposes on surfaces, indoor levels are usually lower than those outdoors due to the decomposition that occurs as the air passes through air conditioning systems and impacts building surfaces (Reiss et al., 1994; Finlayson-Pitts, 1999). The measured ratio of indoor-to-outdoor concentrations of ozone vary from 0.1 to 1, but are typically around 0.3-0.5 (e.g., Druzik et al., 1990;Hisham & Grosjean, 1991; Liu et al., 1993; Weschler et al., 1989, 1994; Gold et al., 1996; Jakobi & Fabian, 1997; Avol et al., 1998; Drakou et al., 1998; Romieu et al., 1998). Buildings with low air exchange with outside air tend to have lower ratios, ~0.1-0.3 (Druzik et al., 1990; Weschler et al.,1994; Romieu et al., 1998). For example, Gold et al. (1996) estimate that at outdoor ozone concentrations of 170 ppb in Mexico City, the indoor-to-outdoor ratio of O$_3$ at a school was 0.71 ± 0.03 with the windows and doors open, which maximised the exchange with outside air, 0.18 ± 0.02 with the windows and doors closed and the air cleaner off, and 0.15 ± 0.02 with the windows and doors closed and the air cleaner on. There are some additional sources of O$_3$ indoors. These include dry-process photocopying machines, laser printers, and electrostatic precipitators (e.g., Leovic et al.,1996; Wolkoff, 1999). Indeed, it is not unusual to detect O$_3$ by its odour during operation of some copy machines and laser printers. In the "indoor environment" in cars, ozone levels tend to be significantly less than in the surrounding area. For example, Chan et al. (1991) report that in-vehicle O$_3$ concentrations during commutes in Raleigh, North Carolina, were only about 20% of those measured in the local area at a fixed station. There are several contributing factors to these low concentrations. One is that NO concentrations are higher near roadways, so that O$_3$ is titrated to NO$_2$by its rapid reaction with NO. A second is that O$_3$ can decompose on the surfaces of the automobile air conditioning system. A similar titration effect has been observed inside homes where there are combustion sources of NO.

Particles

With the epidemiological studies suggesting increased mortality associated with particles, there has been increasing interest in indoor particle concentrations compared to outdoor levels (Finlayson-Pitts, 1999). A number of studies have examined this over the years and are summarized in a review by Wallace (1996). In general, if there are no indoor sources of particles, the levels indoors tend to reflect those outdoors. For example, application of a mass balance model to measurements of indoor and outdoor particle concentrations in Riverside, California, indicated that 75% of $PM_{2.5}$ and 65% of PM_{10} in a typical home were from outdoors (Wallace, 1996). Similar conclusions were reached by Koutrakis *et al.* (1991, 1992) for homes in two counties in New York. For example, they report that 60% of the mass of particles in homes is due to outdoor sources. However, the contribution to various individual elements in the particles varies from 22% for copper to 100% for cadmium. There are some differences in indoor levels of particulate matter in areas with low outdoor compared to high outdoor levels. In the case of high outdoor levels, the indoor concentrations tend to be somewhat lower than those outdoors; for example, Colome *et al.* (1992) report that the ratio of indoor-to-outdoor median concentrations of PM_{10} is 0.7 in residences in southern California. On the other hand, when outdoor levels are low, indoor levels tend to be higher Night time mass concentrations indoors tend to be smaller than those during the day, probably because of the decreased activity. Interestingly, when individuals wear personal exposure monitors to measure their actual exposure to particles, the measured mass concentrations tend to be higher than those measured with fixed monitors located indoors. A major source of increased particles indoors is cigarette smoking. (e.g. Spengler *et al.,* 1981;Quackenboss *et al.,* 1989; Neas *et al.,* 1994). In addition to the contribution to the *mass* concentrations of indoor particles, cigarette smoke is of concern because of the mutagens, carcinogens, and toxic air contaminants that are emitted (Löfroth *et al.,* 1991; Chuang *et al.,* 1991; California Environmental Protection Agency, 1997; Nelson *et al.,* 1998). Thus, a variety of both gaseous and particulate polycyclic aromatic hydrocarbons (PAH) and compounds (PAC) have been identified in buildings with cigarette smoke (Offermann *et al.,* 1991; Mitra & Ray, 1995). Indeed, in the homes of smokers, almost 90% of the total PAH was from tobacco smoke (Mitra and Ray, 1995). Higher levels of mutagenic particles have also been shown to be associated with indoor air containing cigarette smoke (e.g., Lewtas *et al.,* 1987; Löfroth *et al.,* 1988, 1991; Georgiou *et al.,* 1991). Other significant sources identified in a number of studies are cooking, the use of paraffin heaters, wood burning, and humidifiers. For example, a study carried out under the auspices of the U.S. Environmental Protection Agency, the TEAM study (Total Exposure Assessment Methodology), indicated that an increase in PM m of ~10-20 / μg m $^{-3}$ could be attributed to cooking (Wallace, 1996). This source will obviously depend on the amount of cooking, the types of cooking, and the ventilation. For example, Löfroth *et al.* (1991) measured emissions of particles ranging from 0.07 to 3.5 mg per gram of food cooked, depending on the particular food. Baek *et al.* (1997) measured indoor and outdoor concentrations of particles in homes, offices, and restaurants in Korea and report ratios of 1.3, 1.3, and 2.4, respectively. The higher value in restaurants, even those using only gas and not charcoal, suggests a significant contribution from cooking. Paraffin heaters can be significant sources of particles under some circumstances. For example, paraffin heaters were reported to contribute to indoor $PM_{2.5}$ in homes in Suffolk County, New York, but not Onondaga County; wood stoves and fireplaces and gas stoves did not contribute in either case (Koutrakis *et al.,* 1992; Wallace, 1996). A similar conclusion was reached in a study of eight mobile homes in North Carolina (Mumford *et al.,* 1990). However, it should be noted that even where paraffin heaters do not contribute significantly to particle *mass* concentrations, they may still be important in terms of health effects. This is because of the composition of the particles emitted, which include polycyclic aromatic compounds and other mutagenic species, as well as sulfate (Traynor *et al.,* 1990). For example, Traynor *et al.* (1990) studied the emissions from unvented paraffin space heaters and identified a number of PAHs (naphthalene, phenanthrene, fluoranthene, anthracene, chrysene, and indeno[c,d]pyrene) and nitro-PAHs (1-nitronaphthalene, 9-nitroanthracene, 3-nitrofluoranthene, and 1-nitropyrene), in addition to a host of other gaseous species.

Baek *et al.* (1997) also reported increased levels of a number of gases indoors in homes and offices in Korea due to the use of paraffin heaters. In studies of indoor air in eight mobile homes, Mumford *et al.* (1991) identified the PAHs and nitro- PAHs measured in emissions from paraffin heaters byTraynor *et al.* (1990), as well as a number of compounds that may be animal carcinogens, such as cyclopenta[c,d]pyrene, benz[a]anthracene, benzofluoranthenes, benzo-[a]pyrene, and*benzo[ghi]perylene.* While the mass concentrations of PM10 did not increase with the paraffin heater on in six of the eight homes studied, the particles in five of the homes had increased mutagenicity using TA98 with or without $9 added. In short, not only the mass emissions but also the nature of the compounds emitted must be taken into account in assessing the health effects of indoor particles.

Where indoor heating and cooking involves the use of coal or biomass, indoor particle concentrations can be extremely large. For example, Florig (1997) and Ando *et al.* (1996) report that in China typical indoor total suspended particle (TSP) concentrations can be in the range from 250 to $900/\mu g$ m^{-3} in homes using coal and 950-3500 / μg m^{-3} in those using biomass fuels. These levels can be compared to annual average outdoor concentrations of 250-410 / μg m^{-3}. The high concentrations associated with coal burning combined with the mutagenic nature of the emissions have been suggested to be responsible for enhanced lung cancer in China (Mumford *et al.,* 1987). Similarly, Davidson *et al.*(1986) measured TSP concentrations of 2900-42,000/ μg m^{-3} in homes in Nepal that used biomass fuels, compared to outdoor levels of 280

/ μg m^{-3}. For particles with diameters less than 4 /μm, the levels ranged from 870 to 14,000/ μg m^{-3}. Similar conclusions regarding the relative indoor and outdoor concentrations have been reached in studies of office and commercial buildings. For example,Ligocki *et al.* (1993) measured indoor and outdoor concentrations of particles and their components at five museums in southern California. The indoor-to-outdoor ratios of particle mass varied over a wide range, depending to a large extent on the ventilation and filtration systems in use. Ratios varied from 0.16 to 0.96 for particles with diameters less than 2.1/ μm and from 0.06 to 0.3 for coarse particles with diameters greater than this.

Microbial Pollutants

Microbial pollution is a risk to health and is associated with allergic illnesses. Published results indicate that 20% of the population can be sensitised by airborne fungal spores in the UK, while 40% of the inspected houses in Germany suffer from mould-related problems (Waubke & Kusterle, 1990). The medical consequences of immune response, allergic reactions, endotoxins, mycotoxins, and epidemiology have been extensively studied by Miller (1990), Morey (1990), Gravensen et al. (1990) and Burge et al. (1990). Similarly, Legionnaires ' disease and Pontiac fever are associated with wet cooling towers and domestic hot-water systems in complex buildings.

Accordingly to the official published figures, some 560,000 people need treatment because of indoor pollution due to mites and mould in damp houses (House of Commons Environment Committee, 1991). Indoor airborne allergic components come from two sources: outdoor air-borne spores moving inside and allergic components originating inside the dwelling. The source of biological growth within buildings is associated with moisture and the formation of microclimates; it also depends upon the type of the buildings and their ventilation. Mould fungi thrive on surfaces on which there is nourishment and suitable humidity, for example on damp water pipes, windows and walls in kitchens and bathrooms, in central air-conditioning systems, circulation pumps, blowers, ventilation ductwork and air filters, central dehumidifiers, and inside damp structures. Allergenic substances can be airborne and inhaled, such as pollen, fungus and dust, digested, such as mouldy food or drink. Investigations suggest that airborne allergies cause more problems throughout the world than all other allergies combined. Additionally, cross-infection from patient to patient is of great concern in hospitals. The medical field that treats allergies recognises the following allergenic diseases: asthma, allergic rhinitis, serous otitis media, bronchopulmonary aspergillosis, and hypersensitivity pneumonitis.

Allergic Load and Cocktail Effect

For some people, an allergic reaction in the indoor environment may be triggered by non-biological factors, such as chemicals or other indoor air pollutants, emotional stress, fatigue or changes in the weather. These factors burden allergic people further if they are suffering from allergic reactions to biological contaminants. This combination is known as 'allergic load'. Microbial contaminants propagated within the health care establishment are particularly aggressive to patients due to reduced immune system resistance.

Recently, attention has been focused on the cocktail effect of chemicals present in indoor air. Volatile organic compounds may be produced from the use of wood preservatives and remedial timber treatment chemicals, moth-proof carpets, fungicides, mouldicide-treated paints, furnishing materials such as particle board and foamed insulation which may emit formaldehyde. Biological pollutants alone or in synergetic effect with any of the above-mentioned volatile organic compounds may produce symptoms such as stuffy nose, dry throat, chest tightness, lethargy, loss of concentration, blocked, runny or itchy nose, dry skin, watering or itchy eyes or headache in sensitive people. The 'sick building syndrome' (SBS) or tight building syndromes may arise from a variety of causes. Because of the uncertainties about the causes of SBS and the rising levels of health related problems in buildings there is an increasing use of the term building-related illness (BRI) to cover a range of ailments which commonly affect building occupants.

Asbestos and Manmade Mineral Fibres

Asbestos is known to cause a number of diseases after occupational exposure (Brown & Hoskins, 1993). Before the hazards associated with the inhalation of these mineral fibres were understood these exposures were often very large with frequent reports of dust clouds so great that visibility in the workplaces was considerably reduced. This type of exposure is quantitatively quite different from those in the general environment that have provoked a response which in some quarters approaches hysteria. In the USA at least there is massive expenditure on asbestos removal, management and litigation.

Asbestos is a collective, trivial, name given to a group of highly fibrous minerals that are readily separated into long, thin, strong fibres occurring on sufficient large bulk deposits for their industrial exploitation. Asbestos minerals were usually used for their insulating properties, or in a composite, where they added strength, as in cement, or increased friction, as in brake shoes. Chrysotile, or white asbestos has counted for over 90% of the world trade in asbestos minerals. It is a serpentine mineral while the others (amosite (brown asbestos); crocidolite (blue asbestos); anthophyllite; tremolite; and actinolite) are all amphibole minerals. Amphibole asbestos has grater acid and water resistance than chryusotile and was used where these properties made it more suitable. Sometimes users would be unaware of the differences between the types of asbestos and so different minerals could have been used for a single application.

Recently the concern over the health effects of asbestos has been extended to another group of fibrous materials- the man-made mineral fibres (MMMF). While this term is self-explanatory a variety of types are produced with diverse chemical compositions, properties and uses. While sometimes referred to as 'asbestos substitutes' the majority of uses for the manmade fibres are relatively novel and ones for which the natural fibres are unsuitable. For example refractory ceramic fibres are resistant to considerably higher temperatures than are any of the natural fibres. The development of synthetic fibrous insulation materials has been given a great impetus in recent years by the need for more thermally efficient buildings and industrial processes.

MMMF can be made from most types of glass, from rock such as basalt, diabase and olivine and from various types of slag. Ceramic fibres can be made from kaolin or from pure silica and other oxide starting materials. The MMMF have been classified into four broad groups based on the manufacture and use: continuous filament glass fibre made by extrusion and winding processes, insulation wool (including ceramic fibre), and special purpose fibres. The non-continuous fibres are made by dropping molten material onto spinning disks or by air or steam jet impingement on a stream of the molten material. They contain a wide range of fibre sizes and are contaminated by small glassy balls called shot which often account for 50% of the product by weight.

FACTORS THAT INFLUENCE EXPOSURE TO INDOOR AIR POLLUTANTS

General

The types and quantities of pollutants found indoors vary temporally and spatially. Depending on the type of pollutant and its sources, sinks and mixing conditions, its concentration can vary by a factor of 10 or more, even within a small area.

Human mobility constitutes an important kind of complexity in the determination of exposure to air pollutants. Human activity patterns differ between midweek and weekend, between one season and another, and between one part of one's life and another. Activity patterns determine when and how long one is exposed to both indoor and outdoor pollutants. Therefore, in reviewing the factors that influence air-pollution exposures, we have specifically separated them into two major components: time (activity) and concentration (location).

Information on the time spent in various activities is summarized first, and then the variations in concentration often encountered in different locations. Unfortunately, most of the studies discussed were not longitudinal and thus do not offer information on seasonal differences in time spent indoors and outdoors or on regional differences in activity patterns.

Outdoor concentrations of pollutants and rates of infiltration affect the concentrations to which people are exposed indoors. Building construction techniques, as they vary geographically, and their effect on pollution infiltration are particularly important. But the measurement techniques available are limited; the need for additional studies is discussed. The rates of infiltration on a neighborhood scale have been studied by only a few researchers. Although their work has focused on energy conservation, their findings can easily be applied to the study of impact on indoor pollution.

Patterns of human behaviour and activity determine the time spent in any specific location, and thus knowledge of them is essential in estimating exposures of populations to pollutants. As indicated by Ott (Ott, 1995), a large number of variety of studies in which data on human activities were collected from population samples have been completed over the past 50 ye.

When one examines the literature on human activities, the term "time budget" ("zeitbudget", "budget de temps") is encountered often. A time budget produces a systematic record of how time is spent by a person in some specified period, usually 24 h. It contains considerable detail on a person's activities; including the locations in which the activities take place (Michelson, 1973).

One way of obtaining time budget information from the populations surveyed is to ask each respondent to maintain a diary of his or her activities over a 24-h period or longer. In another approach, the so-called "yesterday" survey approach, the interviewer asks each responder about his or her activities on the preceding day.

Several summaries of the historical development of time-budget research have been published (Chapin, 1974; Converse, 1968; Ottensman, 1972). Ott (Ott, 1995) discussed the literature on activity patterns in the context of estimation of exposure to air pollution. Owing to the small number of field monitoring studies, the geographic distribution of indoor air pollutants has not been determined. However, it is instructive to review the geographic distribution of the major factors that affect variations in the concentrations of pollutants and their impact on the quality of the indoor environment. Outdoor air quality, air-infiltration rates, and sources of emission of indoor air pollutants are the major factors. Outdoor air quality has been studied with respect to some pollutants, and the geographic distribution of these few pollutants is well understood. Descriptive statistics published annually by EPA and state and local air-quality agencies furnish much scientific information useful in discerning regional and local differences in concentrations of carbon monoxide, total suspended particles, ozone, NOx, sulfur dioxide, sulfates, and others. It should be noted that the geographic distribution of some criteria pollutants has been studied and is easily accessible from the literature; information on non-criteria pollutants is sparse and often collected and analyzed by questionable methods.

Concentrations of chemically non-reactive pollutants in residences generally correlate with those outdoors. Distribution of indoor air quality is extremely difficult to describe on a geographic scale, because indoor air quality is determined by complex dynamic relationships that depend heavily on occupant activity and highly variable structural characteristics. Weather, which has a regional character, influences indoor air concentrations of some chemicals, such as formaldehyde, and biologic contaminants, such as bacteria and molds. Therefore, the influence of relative humidity and other weather-related conditions affecting indoor environmental quality needs to be studied geographically. Research specifically addressed to geographic distribution of indoor air quality is needed.

Typically, the air-infiltration rate for American residences is assumed to be 0.5- 1.5 ach. This assumption is supported by the results of several energy and air-quality studies that experimentally determined the range of ventilation rates for typical residences to be between 0.7 and 1.1 ach (Moschandreas & Morse, 1979). However, the sample that yielded the data is small, and statistical documentation for such statements is not strong.

The quality of indoor air is a function of outdoor air quality, emission from indoor sources, air-infiltration rates, and occupant activity is likely to vary within each metropolitan and suburban area, is indeed within each neighborhood. Within a metropolitan area, it has been shown that an urban complex leads to the so-called urban heat reservoir (American Society of Heating, Refrigerating and Air-Conditioning engineers. ASHRAE, 1972). Urban characteristics-- such as city size, density of buildings, and population-- correlate with such meteorological factors as temperature, pressure and wind velocity (Gibson, & Cawley, 1977; Kostiainen, 1995). The urban heat island affects both urban pollution patterns and meteorological characteristics that affect the infiltration rates of buildings. Thus, although the exact nature of the impact on indoor air quality is not known, it is fair to expect that the heat island to have an impact on the indoor environment that is likely to be adverse. Also, the variations due to mechanical ventilation, structural differences, and air infiltration may vary within a neighborhood as a function of such factors as house orientation, tree barriers, and terrain roughness.

Occupant activity, air-infiltration rates, the indoor sources of pollutants and their chemical natures are some of the factors that cause variations within a city. A study (Moschandreas et al., 1980) in the Boston metropolitan area obtained indoor air samples from 14 residences under occupied "real-life" conditions for 2 week each. The indoor air character not only was driven by outdoor concentrations, but was greatly affected by other factors, such as indoor activities.

Wind speed, temperature difference, pressure differential, terrain characteristics (roughness and barriers, such as trees and fences), building orientation, and structure characteristics may be affected by the location of one residence relative to another within a neighborhood.

The indoor air quality of an individual building is often characterized by the 24-h average for the concentration of one pollutant measured at one sampling location. Because the activity patterns of persons are such that more time is spent in some indoor areas than in others, the question arises (Moschandreas et al., 1978): "Do indoor zones (independent areas) with distinct pollutant patterns exist?" At issue here is whether sampling from one monitoring zone is sufficient to characterize the air quality of an entire building.

In an extensive analytic study of indoor air quality, Shair and Heitner (1974) assumed that there are no pollutant gradients in the indoor environment. The experimental database of Moschandreas and co-workers (1980) verified that the gradients in concentrations of several gaseous pollutants in the residential environment are negligible. J.D. Spengler, R.E. Letz, J.B. Ferris, Jr., T. Tibbets, and C. Duffy reported (at the annual meeting of the Air Pollution Control Association, 1981) on weekly nitrogen dioxide measurements in 135 homes in Portage, Wisconsin. On the average, kitchen concentrations were twice those in bedrooms in homes that had gas stoves. A study of the air quality in a scientific laboratory by West (1977) showed an almost uniform distribution of an intern tracer continuously released in the room. Similar experiments performed by Moshandreas et al. in residential environments showed that equilibrium is reached throughout a house within an hour. Episodic release of sulphur hexafluoride tracer gas also illustrates this point. The source location was the living room; adjacent locations were the kitchen and the hall. Episodic release of this inert gas in 24 residences was followed by uniform indoor distributions within 30 min (Moschandreas et al., 1978; Peterka & Cermak, 1977). The one-zone concept does not require instantaneous mixing, because it is based on the behavior of hourly average pollutant concentrations.

Moschandreas and associates (1980) used a different database derived from the monitoring of 14 indoor environments in the Boston metropolitan area. Analysis of variance was used to reach the following conclusions:

- Pollutants (ozone and sulphur dioxide) generated principally outdoors have little or no interzonal statistical difference indoors.

- Pollutants with strong indoor generation have interzonal statistical differences in residences with gas facilities and offices, but not in electric-cooking residences. In general, the observed differences are not large, and the health differences are not expected to be serious.

- Depending on indoor activity and outdoor episodic pollutant activity, the indoor arithmetic 24-h average may or may not adequately represent the variation of hourly indoor concentrations.

- Although more than one zone would be preferable, hourly pollutant concentrations obtained from one indoor zone adequately characterize the indoor environment.

The most important factors that influence exposure to indoor air pollutants are the one described under. It should be noticed that these conclusions are not applicable to short-lived pollutants. Contaminants associated with tobacco smoke, bathroom odors, allergens, and other pollutants related to dust are expected to vary considerably in a given residence. Additional documentation is needed to determine the extent of this variation.

Site Characteristics

The characteristics of a building site that influence indoor air quality are addressed as three related subjects: air flow around buildings, proximity to major sources of outdoor pollution, and type of utility service available.

The air flow around a building has been shown to be determined by the local characteristics of the geometry of surrounding buildings (Peterka & Cermak, 1977), the location and type of surrounding vegetation (White, 1995), the terrain (Geiger, 1965), and the size and shape of the building itself. Pollutants can be transferred by the air flow from the street level, over the façade of the building and onto the roof (Cermak, 1976). Field tests of isolated buildings have been used to develop scaling coefficients for both isothermal and stratified cases of surface wind pressures, turbulence, and dispersion (Davenport, 1960.). Air flow around the building creates low pressure on the leeward side and/or the sides adjacent to the windward face, as well as the roof. Air pollutants released from stacks, flues, vents, and cooling towers in the region can re-enter the building through make-up air intakes for ventilation (Cermak, 1976).

Trees and forests have been generally studied as shelter belts in an agricultural context. Shelter belts affect air flow around buildings. When an air current reaches a shelter belt, part of it is deflected upward with only a slight change in velocity, part passes through the crowns of the trees with very low velocity, and part is deflected beneath the canopy with rapidly decreasing velocity (Federer, 1971). The changes in velocity of air flow outside may change the infiltration rate and thus affect indoor air quality.

The location of a building relative to a major outdoor pollution source can affect indoor air quality. For example, buildings near major streets or highways often have high carbon monoxide and lead concentrations, owing to the infiltration of these pollutants.

The type of utility service available is also related to the site of the building and may affect the character of its indoor environment. The availability of particular fuels (e.g., natural gas and oil) influences the types and concentrations of pollutants (e.g., combustion products) emitted by space-and water- heating. Service moratoria, development timing, and development scale are institutional elements that contribute to the variability of utility services and thus can affect indoor air quality.

Occupancy

Occupancy factor that affect indoor air quality include the type and intensity of human activity, spatial characteristics of a given activity, and the operation schedule of a building.

Several human activities-such as smoking, cleaning and cooking- generate gaseous and particulate contaminants indoors. The number of occupants of a space and the degree of their physical activity (i.e., metabolic rate at rest or under intense activity) are related to the production of various pollutants, such as carbon dioxide, water vapor, and biologic agents. If the only source of indoor carbon dioxide is that caused by occupants, ventilation rates may be proportional to the number of people and their metabolic rates (McIntyre, 1980). Although studies have shown no constant relationship between carbon dioxide concentrations and the concentrations of other pollutants, carbon dioxide concentration is often used as a general indicator of the adequacy of ventilation in an occupied space.

Building occupancy is often expressed as occupant density and the ratio of building volume to floor area. The importance of occupancy in indoor air quality is illustrated by the fact that the choice of natural or mechanical ventilation is based on occupant density and the spatial characteristics of the building under consideration.

Occupancy schedule and associated building use may affect the type, concentration, and time and space distribution of indoor pollutants. Because most buildings are unoccupied for substantial portions of each day, the manipulation of "operating schedule" is a means of controlling energy use (American Institute of Architects Research Corporation. Phase Two Report for the Development of Energy Performance Standards for New Buildings, 1979). Efforts to conserve energy through the design of ventilation systems can result to the degradation of indoor air quality. However, detailed studies relating ventilation capacity, occupancy schedules, energy requirements, and indoor air quality have only recently been implemented.

Design

Elements of building design that affect the indoor environment include interior-space design (space planning), envelope design, and selection of materials.

The evolution of space planning in many building types has resulted in flexibility in assigning functions to specific locations. However, this flexibility is accompanied by a decrease in the ability to predict exposure to air pollutants. In particular, "open-plan" offices and schools have serious technical problems of redundant service distribution, limited acoustic control, incomplete air diffusion, and incomplete pollutant dispersion indoors, compared with "fixed-plan" floor layouts.

Evaluation of the success of a floor plan in achieving space efficiency, structural economy, and energy efficiency is usually in terms of net area per occupant and ratio of net usable area to total area. Explicit planning for environmental quality must be included to ensure that spatial arrangements are acceptable to the occupants.

A building's structural envelope consists of both primary elements -foundations, floors, walls, and roofs- and secondary "skin" elements -facings, claddings, and sheathing. To various degrees, the function of these is to maintain the integrity of the structure under the stresses caused by structural load, wind pressure, thermal expansion, precipitation, earth movement, and fire. The integrity of the building envelope is a major consideration in uncontrolled air movement into and out of the building —usually referred to as "infiltration". This is a major factor in indoor air quality. There has been no systematic survey of infiltration rates of buildings in the United States. The dominant factor in determining a building's infiltration rate is the total area of effective leakage, as measured with fan pressurization. Following the leakage area in importance are the terrain and shielding near the building, the mean climatic conditions during heating (or cooling) periods, and the building height (Sherman, 1981). There is much evidence (Dickerhoff et al., 1980), both in the United States and in Europe, that houses in mild climates are "very leaky", whereas houses in severe climates are "tight".

Greater height of a building increases the "stack effect", or updraught, and exposes the building to higher wind speeds. Thus, higher wind pressures drive air through existing openings, referred to as "leakage", increasing the infiltration rate.

The dominant building factors that determine infiltration have not been identified, but a catalogue of leakage openings found in typical structures is as follows:

- Walls: Leakage around sill plates (the openings at the bottom of wallboard), electric outlets, plumbing penetrations, and headers in attics for both interior and exterior walls.

- Windows and doors: Window type is more important than manufacturer in determining window leakage. This source of leakage tends to be overrated; it contributes only about 20% of the total leakage of a house

- Fireplaces: This includes dampers, glass screens, and fireplace caps.

- Heating and cooling systems: The variables include combustion air for furnaces, dampers for stack air draft, air-conditioning units, and location of ductwork.

- Vapour barrier and insulation penetrations.

- Utility accesses: This includes recessed lighting and plumbing and electric penetrations leading to attic or outside.

- Terminal devices in conditioned space: This includes leakage of dampers, especially those for large air-handling systems.

- Structural types: Examples are drop ceilings above cupboards or bathtubs, prism-shaped enclosures over staircases in two-story houses, and elevator and utility shafts that lead from basement to attic.

Wall and ceiling materials and floor finishes are the constituents of the building interior. Modular components, weight, strength, thermal insulation, thermal stability, sound insulation, fire resistance, ease and speed of installation and ease of maintenance are among the criteria considered in the selection of materials for walls, ceiling and floors. But emphasis on first cost, ease of installation, maintenance and long service life has also led to the use of materials that may be sources of indoor contaminants.

Operations

Depending on the type of ownership (owner-occupied or developer-owned), building operation may vary considerably, and this variation may have an impact on indoor air quality. "Building operation" pertains to the following elements of a building: the building envelope, service and plant, building facilities, equipment and landscaping. Cleaning, preventive maintenance, and replacement and repair of defects are also included in building operation. The staff responsible for building operation includes management, engineering, and custodial personnel. The care responsibilities are operation of the heating, ventilation, and air-conditioning systems and building services, such as hot water, lighting and power distribution. Building operation has an impact on indoor air quality in numerous ways, but the magnitude of this impact is not known.

HEALTH EFFECTS OF INDOOR AIR POLLUTION

Indoor air pollution, apart from the health impact, has socio-economic costs. The potential economic impact of poor indoor air quality is quite high, and has been estimated to be in the order of tens of billions of ECU per year in Western Europe. This includes costs of medical care, loss of income during illness, days lost due to illness, poor working performance and lower productivity. Labor costs are significantly greater per square meter of office space than energy and other environmental control costs (ECA, 1989). In the US, the loss in productivity for each employee which is attributable to IAQ problems is currently estimated to be 3% (14 minutes/day) and 0.6 added sick days annually. Other estimates have been made by calculating the impact of IAQ on productivity. For instance, in Norway, the authorities estimate that the costs to society related to poor IAQ are in the order of 1 to 1.5 billion ECU per year or about 250 - 350 ECU per inhabitant. This estimation only includes costs related to adverse health effects requiring medical attention and does not include reduced working efficiency or job-related productivity losses. Thus, from an economic consideration, remedial action to improve indoor air quality is likely to be cost effective even if an expensive retrofit is required.

As far as it concerns the health effects on IAP, it is very interesting to present the methods of studying health effects, the criteria for the assessment of the impact of IAP on the community and the diverse effects of IAP on human health (ECA, 1991).

Methods of Studying Health Effects

Methods of studying health effects of indoor pollutants can be grouped into three broad categories:

- Human studies, subdivided into observational and experimental studies. Epidemiological studies of pollutants are mostly observational, i.e. the investigator has no means of experimentally exposing humans to pollutants, or of allocating subjects to exposed and unexposed groups. Critical issues are therefore the validity and precision of exposure assessment, and the control for confounding factors in these studies. Recent developments have stressed the importance of reducing exposure misclassification, and of studying restricted, well defined, homogenous populations to address these issues. The main advantage is that humans are studied under realistic conditions of exposure. By themselves, observational epidemiological studies are not usually sufficient to support causality of an observed association, so that additional information is needed from other types of studies. Experimental studies are among these; however, these are only suitable for studying moderate, reversible, short term effects in persons who are healthy or only moderately ill. Their main advantage is that exposure conditions and subjects election are under the control of the investigator.

- Animal studies, which can be subdivided into a number of categories depending on their length (acute, subchronic, chronic) or end-point (morbidity, mortality, carcinogenicity, irritation, etc.). Here, the investigator has full control over exposure conditions and health effects studied. However, the principle limitations lie in the fact that extrapolation from the studied animal species to man is always necessary. Also, while in human populations health effects with low incidences are often of interest (e.g., specific cancers), it is not feasible to study very large groups of animals to detect these low incidences. In practice, therefore, animal experiments are often carried out using very high experimental doses to compensate for the relatively small number of animals used and as a consequence, an additional extrapolation from high to low doses is also often necessary.

- In vitro studies, in which effects of pollutants on cell or organ cultures are studied. These studies have the advantage that they are less costly than animal studies, and that results can generally be obtained in a shorter period of time. They are useful for studying mechanisms of action, but it is not usually possible to predict effects on whole organisms from their results in a quantitative way.

Criteria for the Assessment of the Impact of IAP on the Community

The process of risk characterization for indoor pollutants occurs through several phases: hazard identification, exposure assessment, dose-effect evaluation, and finally qualitative and quantitative risk assessment. The final product of this process may be an individual risk estimate per exposure unit or the evaluation of the incidence of the concerned effects in a given population. The risk characterization through a multi-stage process as described above is particularly informative because, by dividing the analysis of the scenario of each pollutant into steps, it allows the separate recognition of the importance of each variable in the scenario and the prediction of the changes of frequency or severity of effects obtainable by modifying (increasing or decreasing) exposure.

For some types of IAP, our understanding of human health risk is well defined. For most indoor air pollutants, however, the risk assessment process has its limitations.

First, it has been applied successfully only to individual pollutants for which information is available for exposure and dose-response relationships and for which the effect is clear, certain, and measurable, such as mortality and cancer. Little progress has been made in applying the risk assessment process to environmental issues involving pollutant mixtures or effects for which the causes are difficult to ascertain precisely, such as in heart disease, allergic reactions, headache, and malaise. A different approach is needed for the assessment and characterization of the risks associated with most indoor air pollutants.

A basic and simple criterion for assessing the importance of the health risk related to indoor pollution makes reference to the severity of the effect concerned and to the size of the population affected. Important issues for the community may come from severe health impacts, particularly when affecting a large segment of the population. Minor impacts, such as those related to discomfort or annoyance may, however, become important when a large number of individuals in the community are concerned.

The Impact of IAP on Humans' Health

Respiratory Health Effects Associated with Exposure to IAP

Several effects on the respiratory system have been associated with exposure to IAP. These include acute and chronic changes in pulmonary function, increased incidence and prevalence of respiratory symptoms, augmentation of pre-existing respiratory symptoms, and sensitization of the airways to allergens present in the indoor environment. Also, respiratory infections may spread in indoor environments when specific sources of infectious agents are present, or simply because the smaller indoor mixing volumes allow infectious diseases to spread more easily from one person to the next. The latter mechanism is particularly operative in schools, nursery schools, etc.

Observed changes in pulmonary function due to exposure to, e.g., tobacco smoke in the home, have mostly been due to acute or chronic airway narrowing leading to obstruction of air flow. This is measured as a reduction in the quantity of air that can be exhaled in one second after deep inspiration (FEVI), and a limitation in the various measures of air flow such as Peak Expiratory Flow (PEF), Maximum Mid Expiratory Flow (MMEF), and Maximum Expiratory Flow at x% of Forced Vital Capacity (MEFx). In growing children, it has also been suggested that lung development could be impaired by exposure to IAP.

Asthma, manifested by attacks of excessive airway narrowing leading to shortness of breath and wheezing, can be caused or aggravated by exposure to allergens at home, but it has also been associated with exposure to substances such as nitrogen dioxide and environmental tobacco smoke (ETS). Bronchitis, manifested in inflammatory changes in the airways and mucus hyper secretion has been linked to high levels of ambient air pollution in the past, and to exposure to ETS in the home in recent studies. Respiratory symptoms which have been associated with exposure to indoor air pollutants are symptoms mostly related to the lower airways such as cough, wheeze, shortness of breath and phlegm.

In contrast to the occurrence of chemical pollutants in indoor air, attention to which has grown considerably over the past two decades, the role of infectious agents in indoor air has been known for a long time. Infectious agents can be involved in the inflammatory conditions rhinitis, sinusitis, conjunctivitis and sinusitis, in pneumonia, in asthma and in alveolitis.

Allergic Diseases Associated With Exposure to IAP

Allergic asthma and extrinsic allergic alveolitis (hypersensitivity pneumonitis) are the two most serious allergic diseases caused by allergens in indoor air. Allergic rhinoconjunctivitis and humidifier fever are other important diseases; it is not clear if or how the immunological system is involved in humidifier fever.

Allergic asthma is characterized by reversible narrowing of the lower airways. Pulmonary function during an attack shows an obstructive pattern in serious cases together with reduced ventilation capacity. Allergic asthma may be caused by exposure to indoor air pollutants, either acting as allergens or as irritants. Immunological specific IgE sensitization to an airborne allergen is a major component of this disease, but non-specific hypersensitivity is also important for the asthmatic attacks occurring on exposure to irritants in the indoor air.

The prevalence of asthma varies considerably from country to country. Although asthmatic attacks seldom lead to death, the costs of medical care are considerable in terms of hospital admissions, medication, and lost work days.

Allergic rhinoconjunctivitis is also an IgE-mediated disease, but while asthma occurs in all age groups, allergic rhinoconjunctivitis is especially prevalent among children and young adults. The main symptoms are itching of the eye andlor the nose, sneezing, watery nasal secretion and some stuffiness of the nose. The severity of the symptoms varies with the exposure to the allergen. Individuals often suffer from both allergic asthma and allergic rhinoconjunctivitis and are seldom sensitive to only one allergen. Aeroallergens from house dust mites, pets, insects, moulds, and fungi in the indoor air have been shown to be associated with allergic asthma and/or rhinoconjunctivitis. Extrinsic allergic alveolitis, also called hypersensitivity pneumonitis, is characterized by recurrent bouts of pneumonitis or milder attacks of breathlessness and flu-like symptoms. Studies of the pulmonary function during an acute episode will usually show a restrictive pattern with a decreased diffusion capacity. The disease is believed to be an inflammatory reaction in the alveoli and bronchioles involving circulating antibodies and a cell-mediated immunological response to an allergen. For example it occurs in farmers as a result of handling moldy hay ("farmer's lung") and in pigeon breeders due to bird droppings. However, the disease has also in a few cases been associated with exposure to IAP, most frequently related to humidifiers in homes and offices contaminated with bacteria, fungi, or protozoans.

Allergic asthma and extrinsic allergic alveolitis resolve with cessation of exposure to the allergen, but continued exposure in sensitized patients may result in permanent lung damage and death from pulmonary insufficiency.

Humidifier fever is a flu-like illness involving the immune system, in which X-ray abnormalities are usually absent. The exact cause is not clear. The disease may occur among persons exposed to humidification systems contaminated with microbial growth. The symptoms typically occur 4-8 h after the exposure on the first day back at work after a weekend, but resolve within 24 h. Despite continuous exposure the disease does not recur until after the next weekend. Even though pulmonary changes are seen during attacks of humidifier fever, the disease does not lead to permanent lung damage.

Cancer and Effects on Reproduction Associated with Exposure to IAP

Lung cancer is the major cancer which has been associated with exposure to IAP (radon or ETS). Asbestos exposure has been linked to cancer in workers and also in workers' family members, presumably due to asbestos fibres brought into the home on workers' clothing. However, there are no studies associating asbestos exposure in homes or public buildings from asbestos used as a construction material to the development of cancer. Effects on human reproduction have been associated with exposure to chemicals in the environment, but it is as yet unclear to what extent (if any) exposure to IAP is involved.

Sensory Effects and Other Effects on the Nervous System Associated with IAP

Sensory effects are defined as the perceptual response to environmental exposures. Sensory perceptions are mediated through the sensory systems and result in a conscious experience of smell, touch, itching, etc. Sensory effects are typically observed in buildings with indoor climate, problems because many chemical compounds found in the indoor air have odorous or mucosal irritation properties. Most indoor air chemicals with a measurable vapor pressure will be odorous when the concentration is high enough.

Sensory effects are important parameters in indoor air quality control for several reasons. They may appear as: (1) adverse health effects on sensory systems (e.g., environmentally-induced sensory dysfunctions); (2) adverse environmental perceptions which may be adverse per se or constitute precursors of disease to come on a long term basis (e.g., annoyance reactions, triggering of hypersensitivity reactions); (3) sensory warnings of exposure to harmful environmental factors (e-g., odor of toxic sulfides, mucosal irritation due to formaldehyde); (4) important tools in sensory bioassays for environmental characterization (e.g.. using the odor criterion for general ventilation requirements or for screening of building materials to find those with low emissions of volatile organic compounds).

The senses responding to environmental exposure are not only hearing, vision, olfaction and taste, but also the skin and mucous membranes. As pointed out by WHO (1989), many different sensory systems that respond to irritants are situated on or near the body surface. Some of these systems tend to respond to an accumulated dose and their reactions are delayed. On the other hand, in the case of odor perception the reaction is immediate but also very much influenced by olfactory fatigue on prolonged exposures.

Responders are often unable to identify a single sensory system as the primary route of sensory irritation by airborne chemical compounds. The sensation of irritation is influenced by a number of factors such as previous exposures, skin temperature, competing sensory stimulation, etc. Since interaction and adaptation processes are characteristic of the sensory systems involved in the perception of odour and mucosal irritation, the duration of exposure influences the perception. Humans integrate different environmental signals to evaluate the total perceived air quality and assess comfort or discomfort. Comfort and discomfort by definition are psychological and for this reason the related symptoms, even when severe cannot be documented without using subjective reports. Sensory effects reported 10 be associated with IAP are in most cases multisensory and the same perceptions or sensations may originate from different sources. It is not known how different sensory perceptions are combined into perceived comfort and into the sensation of air quality. Perceived air quality is for example mainly related to stimulation of both the trigeminus and olfactorius nerves.

Several odorous compounds are also significant mucosal irritants, especially at high concentrations. The olfactory system signals the presence of odorous compounds in the air and has an important role as a warning system. In the absence of instrumentation for chemical detection of small amounts of some odorous vapors, the sense of smell remains the only sensitive indicator system. It is well known that environmental pollution can affect the nervous system. The effects of occupational exposure to organic solvents can be mentioned as an example. A wide spectrum of effects may be of importance, ranging from those at molecular level to behavioral abnormalities. Since the nerve cells of the CNS typically do not regenerate, toxic damage to them is usually irreversible. The nerve cells are highly vulnerable to any depletion in oxygen supply.

Cardiovascular Effects Associated with IAP

Increased mortality due to Cardiovascular Diseases (CVD) has been associated with exposure to ETS in some groups of non-smoking women married to smokers. Some investigators have also addressed the question whether total mortality is influenced by exposure to ETS, but results have been contradictory. As any effect on mortality would not be expected to occur until after many years of exposure, a problem in these types of study is the accuracy and reliability of the exposure classification. Attempts have also been made to relate ETS to electrocardiographic abnormalities and cardiovascular symptoms, but results have been inconclusive.

Carbon monoxide (CO) exerts its influence primarily through binding to the haemoglobin (Hb) in blood. The affinity of CO to Hb is about 200 times higher than the affinity of oxygen to Hb, so that at relatively low levels of CO in the air. Oxygen is replaced by CO. The percentage of Hb bound to CO (O/O carboxyhaemoglobin) is a measure of recent exposure to CO. Organs with a high oxygen demand, such as the heart and the brain, are particularly susceptible to a reduced oxygenation caused by CO exposure. Early effects include reduction of time to onset of chest pain in exposed, exercising heart disease patients. At higher levels of exposure, myocardial infarctions may be triggered by CO.

BASIC CONTROL STRATEGIES

There are some basic control methods for lowering concentrations of indoor air pollutants (Ashford & Caldart, 2008), which are described bellow:

Source Management includes source removal, source substitution, and source encapsulation. Source management is the most effective control method when it can be practically applied. Source removal is very effective. However, policies and actions that keep potential pollutants from entering indoor are even better than preventing IAQ problems. Source substitution includes actions such as selecting a less toxic art material or interior paint than the products which are currently in use. Source encapsulation involves placing a barrier around the source so that it releases fewer pollutants into the indoor air (e.g., asbestos abatement, pressed wood cabinetry with sealed or laminated surfaces). Local Exhaust is very effecting on removing point sources of pollutants before they can disperse into the indoor air by exhausting the contaminated air outside. Well known examples include restrooms and kitchens where local exhaust is used. Other examples of pollutants that originate at specific points and that can be easily exhausted include science lab and housekeeping storage rooms, printing and duplicating rooms, and vocational/ industrial areas such as welding booths. Ventilation through use of cleaner (outdoor) air to dilute the polluted (indoor) air that people are breathing. Generally, local building codes specify the quantity (and sometimes quality) of outdoor air that must be continuously supplied to an occupied area. For situations such as painting, pesticide application, or chemical spills, temporarily increasing the ventilation can be useful in diluting the concentration of noxious fumes in the air. Exposure Control includes adjusting the time of use and location of use. An example of time of use for school students would be to strip and wax floors on Friday after school is dismissed, so that the floor products have a chance to off-gas over the location of use deals with moving the contaminating source as far as possible from occupants, or relocating susceptible occupants. Air Cleaning primarily involves the filtration of particles from the air as the air passes through the ventilation equipment. Gaseous contaminants can also be removed, but in most cases this type of system should be engineered on a case-by-case basis.

CONCLUSIONS

As it has been clearly proven above, indoor air pollution is a major public concern issue, which can be characterized as "global environmental phenomenon". Also, it is obvious, that the causes which created this domestic environmental problem, such as modern way of living, decreased ventilation rates for energy conservation or increased use of synthetic materials in buildings, are not expected to be reduced (it is more probable that they are going to be increased). Nevertheless, the task of reducing levels of exposure to air pollutants is rather complex. It begins with an analysis to determine which chemicals are present in the air, at what levels, and whether likely levels of exposure are hazardous to human health and the environment. It must then be decided whether an unacceptable risk is present. When a problem is identified, mitigation strategies have to be developed and implemented so as to prevent excessive risk to public health in the most efficient and cost-effective way. In addition, analyses of air pollution problems are exceedingly complicated. Some are national in scope (such as the definition of actual levels of exposure of the population, the determination of acceptable risk, and the identification of the most efficient control strategies), while others are of a more basic character and are applicable in all countries (such as analysis of the relationships between chemical exposure levels, and doses and their effects). So, it is very essential for governments of all countries- especially the governments of the more developed ones- to adopt and implement these policies, in order to effectively face this worldwide issue in combination with the need of energy saving, the use of new building materials and the modern trend of living.

REFERENCES

1.   American Institute of Architects Research Corporation. 1979 Phase Two Report for the Development of Energy Performance Standards for New Buildings. Report to U.S. Department of Housing and urban Development and U.S. Department of Energy, 197

2.   American Society of Heating, Refrigerating and Air-Conditioning engineers. 1972 ASHRAE Handbook of fundamentals. New York: American Society of Heating Refrigerating and Air-Conditioning Engineers, Inc., 688

3.   D. J. Anderson, R. A. Hites, 1988 Chlorinated Pesticides in Indoor Air.

Environ. Sci. Technol., 22 6 June 1988) 717720 , 0001-3936X

4. H. E. B. Andersson, Setterwall. A. . Klevard, Eds, 1996 The energy book.
    A resume of present knowledge and research, The Swedish council for
    building research, 9-15405-744-2 Sweden

5. M. Ando, K. Katagiri, K. Tamura, S. Yamamoto, M. Matsumoto, Y. F. Li,
    S. R. Cao, R. D. Ji, C. K. Liang, 1996 Indoor and Outdoor Air Pollution
    in Tokyo and Beijing Supercities. Atmos. Environ., 30 5 1996), 695702
    , 1352-2310

6. N. A. Ashford, C. C. Caldart, 2008 Environmental Law, Policy, and
    Economics.Reclaiming the Environmental Agenda, 0-262-01238-3978-
    0-262-01238-6, MIT.

7. E. L. Avol, W. C. Navidi, S. D. Colome, 1998 Modeling Ozone Levels
    in and around Southern California Homes. Environ. Sci. Technol., 32 4
    February 1998),463468 , 0001-3936X

8. S. Baek, O. , Y. Kim, S. , R. Perry, 1997 Indoor Air Quality in Homes,
    Offices, and Restaurants in Korean Urban Areas--Indoor/Outdoor
    Relationships. Atmos. Environ., 31 529544 , 1352-2310

9. A. Battaglia, D. Capra, G. Queirazza, A. Sampaolo, 1990 Radon
    exhalation rate in building materials and fly ashes in Italy, INDOOR
    AIR '90, Proceedings of the 5th International Conference on Indoor Air
    Quality and Climate, 3 4752 . Toronto, 29 July- 3 August

10. F. Bochicchio, Venuti. G. Campos, F. Felici, A. Grisanti, G. Grisanti, F.
    Moroni, C. Nuccetelli, S. Risica, F. Tancredi, 1993 Characterisation of
    some parameters affecting the indoor radon exposure of the population,
    Proceedings of the First International Workshop on Indoor Radon
    Remedial Action, Rimini, Italy, 27th June- 2th July 1993

11. M. Brauer, P. B. Ryan, H. H. Suh, P. Koutrakis, J. D. Spengler, N. P. Leslie,
    I. H. Billick, 1990 Measurements of Nitrous Acid inside Two Research
    Houses. Environ. Sci. Technol., 24 10 October 1990), 15211527 , 0001-
    3936X

12. M. Brauer, J. D. Spengler, 1994 Nitrogen Dioxide Exposures inside Ice
    Skating Rinks. Am. J. Public Health, 84 3 March 1994), 429433 , 0090-
    0036

13. M. Brauer, K. Lee, J. D. Spengler, R. O. Salonen, A. Pennanen, O. A.
    Braathen, E. Mihalikova, P. Miskovic, A. Nozaki, T. Tsuzuki, S. Rui-Jin,
    Y. Xu, Z. Quing-Xiang, H. Drahonovska, S. Kjaergaard, 1997 Nitrogen
    Dioxide in Indoor Ice Skating Facilities: An International Survey. J. Air
    Waste Manage. Assoc., 47 (October 1997), 10951102 , 1047-3289

14. L. S. R. Brickus, J. N. Cardoso, Neto. F. R. De Quino, 1998 Distributions of Indoor and Outdoor Air Pollutants in Rio de Janeiro, Brazil: Implications to Indoor Air Quality in Bayside Offices. Environ. Sci. Technol, 32 22 November 1998) 34853490 , 0001-3936X

15. R. C. Brown, J. A. Hoskins, 1993 Asbestos and Man-made Mineral Fibres. MRC toxicology Unit, Woodmansterne Road, Surrey, 0-94841-107-4

16. S. K. Brown, M. R. Sim, M. J. Abramson, C. N. Gray, 1994 Concentrations of Volatile Organic Compounds in Indoor Air-A Review. Indoor Air, 4 2 June 1994), 123134 , 0905-6947

17. J. A. Bukowski, L. W. Meyer, 1995 Simulated Air Levels of Volatile Organic Compounds Following Different Methods of Indoor Insecticide Application. Environ. Sci. Technol, 29 3 March 1995), 673676 , 0001-3936X

18. P. S. Burge, P. Jones, A. S. Robertson, 1990 Sick Building Syndrome: environmental comparisons of sick and healthy buildings. Indoor Air, 90 1, 479484

19. California Environmental Protection Agency, Office of Environmental. 1997 Health Hazard Assessment. Health Effects of Exposure to Environmental Tobacco Smoke. Final Report, September 1997.

20. Venuti. G. Campos, S. Colilli, A. Grisanti, G. Grisanti, G. Monteleone, S. Risica, G. Gobbi, M. P. Leogrande, A. Antonini, R. Borio, 1984 Indoor exposure in a Region of Central Italy, Radiat. Prot. Dosim, 7 1-41,271-274)

21. O. Castrén, A. Voutilainen, K. Winqist, I. Mäikeläinen, 1985 Studies of high indoor radon areas in Finland. Science of the Total Environment. 45 (October 1985), 311318 , 0048-9697

22. J. E. Cermak, 1976 Nature of air flow near buildings. ASHRAE Trans, 82 Pt.1), 10441054

23. C. Chan, C. , H. Ozkaynak, J. D. Spengler, L. Sheldon, 1991 Driver Exposure to Volatile Organic Compounds, CO, Ozone, and 2 under Different Driving Conditions. Environ. Sci. Technol., 25 No 6 (June 1991), 964972 , 0001-3936X

24. J. C. S. Chang, K. A. Krebs, 1992 Evaluation of Para-Dichlorobenzene Emissions from Solid Moth Repellant as a Source of Indoor Air Pollution. J. Air Waste Manage. Assoc., 42 9 September 1992), 12141217 , 1047-3289

25. F. S. Chapin, jr , 1974 Human activity patterns in the City: Things People

Do in time and in space. New York: John Wiley & sons, Inc., 272 0-471-14563-7 13: 9780471145639

26. J. C. Chuang, G. A. Mack, M. R. Kuhlman, N. K. Wilson, 1991 Polycyclic Aromatic Hydrocarbons and Their Derivatives in Indoor and Outdoor Air in an Eight-Home Study. Atrnos. Environ., 25B 369380 , 1352-2310

27. K. D. Cliff, 1992 Thoron daughter concentrations in UK-homes, Radiat. Prot. Dosim., 45 1-4) Supplement, 361366

28. S. D. Colome, N. Y. Kado, P. Jaques, M. Kleinman, 1992 Indoor-Outdoor Air Pollution Relations: Particulate Matter Less Than 10 /zm in Aerodynamic Diameter (PM 10) in Homes of Asthmatics. Atmos. Environ., 26 1, 21732178 , 1352-2310

29. Committee on the Medical Effects of Air Pollutants: Guidance on the Effects on Health of Indoor Air Pollutants 2004 Department of Health, London

30. P. E. Converse, 1968 Time budgets. In D. L. Sills, Ed. International Encyclopedia of the Social sciences. 16 4247 New York: The Macmillan Company and the Free Press

31. D. R. Crump, D. Gardiner, 1989 Sources and Concentrations of Aldehydes and Ketones in Indoor Environments in the UK. Environ. Int., 15 455462 , 0160-4120

32. D. R. Crump, 1995 Volatile Organic Compounds in Indoor Air," in Issues in Environmental Science and Technology (R. E. Hester and R. M. Harrison, Eds.), Chap. 4, 109124 , Royal Chem. Soc., Letchworth, UK

33. J. M. Daisey, A. T. Hodgson, W. J. Fisk, M. J. Mendell, Brinke. J. Ten, 1994 Volatile Organic Compounds in Twelve California Office Buildings: Classes, Concentrations, and Sources. Atmos. Environ., 28 22 December 1994), 35573562 , 1352-2310

34. A. G. Davenport, 1960 A rationale for determination of design wirx: velocities. Proceedings of the ASCE Journal of the Structures Division, 86 3966

35. C. I. Davidson, S. Lin, F. , J. F. Osborn, M. R. Pandey, R. A. Rasmussen, M. A. K. Khalil, 1986 Indoor and Outdoor Air Pollution in the Himalayas. Environ. Sci. Technol., 20 6 June 1986), 561567 , 0001-3936X

36. R. J. De Meijer, P. Stoop, L. Put, 1992 Contribution of radon flows and radon sources to the radon concentration in a dwelling, Radiat. Prot. Dosim. 45 1-4) Supplement, 439442

37. T. Deyuan, 1993 Indoor and outdoor air radon concentration level in China", INDOOR AIR '93, Proceedings of the 6th International

Conference on Indoor Air Quality and Climate, 4 459463 , Helsinki, Finland, July 4-81993

38. D. Dickerhoff, D. T. Grinsrud, B. Shohl, 1980 Infiltration and Air-Conditioning: A Case Study. Lawrence Berkeley Laboratory Report LBL-11674. Berkeley, Cal.: Lawrence Berkeley Laboratory

39. C. Dimitroulopoulou, S. R. Ashmore, M. T. R. Hill, M. A. Byrne, R. Kinnersley, 2006 INDAIR: A probabilistic model of indoor air pollution in UK homes. Atmos Environ, 40 63626379 , 1352-2310

40. M. Doi, S. Kobayashi, 1994 Characterization of Japanese wooden houses with enhanced radon and thoron concentrations, Health Phys., 66 3 274282 , 0017-9078

41. F. Dor, Y. Le Moullec, B. Festy, 1995 Exposure of City Residents to Carbon Monoxide and Monocyclic Aromatic Hydrocarbons during Commuting Trips in the Paris Metropolitan Area. J. Air Waste Manage. Assoc., 45 February 1995), 103110 , 1047-3289

42. G. Drakou, C. Zerefos, I. Ziomas, M. Voyatzaki, 1998 Measurements and Numerical Simulations of Indoor 0 3 and NO X in Two Different Cases. Atmos. Environ., 32 595610

43. A. V. Dremetsika, P. A. Siskos, E. B. Bakeas, 2005 Determination of formic and acetic acid in the interior atmosphere of display cases and cabinets in Athens Museums by reverse phase high performance liquid chromatography. Indoor Built. Environ., 14 1 5158

44. J. R. Druzik, M. S. Adams, C. Tiller, G. R. Cass, 1990 The Measurement and Model Predictions of Indoor Ozone Concentrations in Museums. Atmos. Environ., 24A 18131823 , 1352-2310

45. I. Ekinja, Z. Grabaric, B. S. Grabaric, 1995 Monitoring of Pyrocatechol Indoor Air Pollution. Atmos. Environ., 29 11651170 , 1352-2310

46. Collaborative. European, E. C. A. Action, 1998 Indoor air quality and its impact on man. Report 15 9-28270-119-0

47. European Collaborative Action (ECA). 1991 indoor air quality & its impact on man (formerly COST Project 61 3). Environment and Quality of Life Report 10 Effects of Indoor Air Pollution on Human Health

48. A. Febo, C. Perrino, 1991 Prediction and Experimental Evidence for High Air Concentration of Nitrous Acid in Indoor Environments. Atmos. Environ., 25A 10551061 , 1352-2310

49. A. Febo, C. Perrino, 1995 Measurement of High Concentration of Nitrous Acid inside Automobiles. Atmos. Environ., 29 345351 , 1352-2310

50. C. A. Federer, 1971 Effect of trees in modifying urban microclimate, pc,

2328 . In S. Little, and J. H. Noyes, Eds. Trees and Forests Urbanizing Environment. Amherst, Mass.: University of Massachusetts Cooperative Extension Service

51.   S. E. Feinman, 1988 Formaldehyde Sensitivity and Toxicity, CRC Press, Boca Raton, FL

52.   A. A. Fernandez-Bremauntz, M. R. Ashmore, 1995 Exposure of Commuters to Carbon Monoxide in Mexico City. Measurement of In-Vehicle Concentrations. Atmos. Environ., 29 525532 , 1352-2310

53.   B. J. Finlayson-Pitts, J. N. Pitts, 1999 Chemistry of the upper and lower atmosphere Academic Press, 0-12-257060-x

54.   M. L. Fischer, A. J. Bentley, K. A. Dunkin, A. T. Hodgson, W. W. Nazaroff, R. G. Sextro, J. M. Daisey, 1996 Factors Affecting Indoor Air Concentrations of Volatile Organic Compounds at a Site of Subsurface Gasoline Contamination. Environ. Sci. Technol., 30 10 September 1996), 29482957 , 0001-3936X

55.   P. G. Flachsbart, G. A. Mack, J. E. Howes, C. E. Rodes, 1987 Carbon Monoxide Exposures of Washington Commuters. JAPCA, 37 2 135142 , 0894-0630

56.   H. K. Florig, 1997 China's Air Pollution Risks. Environ. Sci. Technol., 31 274A279A

57.   W. M. Foster, L. Jiang, P. T. Stetkiewicz, T. H. Risby, 1996 Breath Isoprene--Temporal Changes in Respiratory Output after Exposure to Ozone. J. Appl. Physiol., 80 706710

58.   R. Geiger, 1965 The climate near the ground. Cambridge, Mass.: Harvard University Press, 611pp

59.   D. Gelmont, R. A. Stein, J. F. Mead, 1981 Isoprene--The Main Hydrocarbon in Human Breath. Biochem. Biophys. Res. Commun., 99 4 April 1981), 14561460 , 0000-6291X

60.   P. E. Georghiou, P. Blagden, D. A. Snow, L. Winsor, D. T. Williams, 1991 Mutagenicity of Indoor Air Containing Environmental Tobacco Smoke: Evaluation of a Portable PM-10 Impactor Sampler. Environ. Sci Technol., 25 8 August 1991), 14961500 , 0001-3936X

61.   T. F. Gesell, 1983 Background atmospheric Rn-222 concentrations outdoors and indoors: a review, Health Phys., 45 289302 , 0017-9078

62.   U. E. Gibson, R. E. Cawley, 1977 The heat pump solar collector interface-A practical experiment. Appliance Eng. 11 4), 6871

63.   D. R. Gold, G. Allen, A. Damokosh, P. Serrano, C. Hayes, M. Castillejos, 1996 Comparison of Outdoor and Classroom Ozone Exposures for

School Children in Mexico City. J. Air Waste Manage. Assoc., P. B. 46 April 1996), 335342 , 1047-3289

64.   R. L. Grasty, 1994 Summer outdoor radon variations in Canada and their relation to soil moisture, Health Phys. 66 2 185193 , 0017-9078

65.   S. Gravesen, L. Larson, F. Gyntelberg, P. Skov, 1990 The role of potential immunogenic components of dust (MOD) in the sick building syndrome. Indoor Air, 90 1, 914 , 0905-6947

66.   Q. Guo, M. Shimo, Y. Ikebe, S. Minato, 1992 The study of thoron and radon progeny concentrations in dwellings in Japan, Radiat. Prot. Dosim., 45 1-4) Supplement, 357359

67.   C. H. Halios, C. G. Helmis, 2007 On the estimation of characteristic indoor air quality parameters using analytical and numerical methods. Science of the Total Environment, 381 222232

68.   L. P. Hanrahan, H. A. Anderson, K. A. Daily, A. D. Eckmann, M. S. Kanarek, 1985 Formaldehyde Concentrations in Wisconsin Mobile Homes. JAPCA, 35 11641167 , 0894-0630

69.   D. L. Heavner, M. W. Ogden, P. R. Nelson, 1992 Multisorbent Thermal Desorption/Gas Chromatography/Mass Selective Detection Method for the Determination of Target Volatile Organic Compounds in Indoor Air. Environ. Sci. Technol., 26 9 September 1992), 17371746 , 0001-3936X

70.   C. G. Helmis, J. Tzoutzas, H. A. Flocas, V. Panis, O. I. Stathopoulou, V. D. Assimakopoulos, C. H. Halios, M. Apostolatou, G. Sgouros, E. Adam, 2007 Indoor Air Quality in a Dentistry Clinic. Science of the Total Environment, 377 349365

71.   C. G. Helmis, V. D. Assimakopoulos, H. A. Flocas, O. I. Stathopoulou, G. Sgouros, M. Hatzaki, 2009 Indoor air quality assessment in the Air Traffic Control Tower of the Athens Airport, Greece. Environmental Monitoring and Assessment, 148 4760

72.   M. W. M. Hisham, D. Grosjean, 1991 Sulfur Dioxide, Hydrogen Sulfide, Total Reduced Sulfur, Chlorinated Hydrocarbons, and Photochemical Oxidants in Southern California Museums. Atmos. Environ., 25A 14971505 , 1352-2310

73.   A. T. Hodgson, J. M. Daisey, R. A. Grot, 1991 Sources and Source Strengths of Volatile Organic Compounds in a New Office Building. J. Air Waste Manage. Assoc., 41 11 November 1991), 14611468 , 1047-3289

74.   A. T. Hodgson, K. Garbesi, R. G. Sextro, J. M. Daisey, 1992 Soil-Gas Contamination and Entry of Volatile Organic Compounds into a House

near a Landfill. J. Air Waste Manage. Assoc., 42 3 March 1992), 277283 , 1047-3289

75. G. Hoek, B. Brunekreef, P. Hofschreuder, 1989 Indoor Exposure to Airborne Particles and Nitrogen Dioxide during an Air Pollution Episode. JAPCA, 39 13481349 , 0894-0630

76. R. D. Hopper, R. A. Levy, R. C. Rankin, M. A. Boyd, 1991 National ambient radon study, Proceedings of the International Symposium on Radon and Radon Reduction Technology, Philadelphia, PA (USA)

77. House of Commons Environment Committee. 1991 Indoor Pollution, 6th report, HMSO, London.

78. C. Howard, R. L. Corsi, 1998a Volatilization of Chemicals from Drinking Water to Indoor Air: The Role of Residential Washing Machines. J. Air Waste Manage. Assoc., 48 (October 1998), 907914 , 0000-1047- 3289

79. E. M. Howard, R. C. Mc Crillis, K. A. Krebs, R. Fortman, H. C. Lao, Z. Guo, 1998b Indoor Emissions from Conversion Varnishes. J. Air Waste Manage. Assoc., 48 (October 1998), 924930 , 0000-1047- 3289

80. G. Jakobi, P. Fabian, 1997 Indoor/Outdoor Concentrations of Ozone and Peroxyacetyl Nitrate (PAN). Int. J. Biometeorol., 40 3 May 1997), 162165 , 0020-7128

81. A. W. Jones, V. Lagesson, C. Tagesson, 1995 Origins of Breath Isoprene. J. Clin. Pathol., 48 10 October 1995), 979980 , 0021-9746

82. T. J. Kelly, 1996 Determination of Formaldehyde and Toluene Diisocyanate Emissions from Indoor Residential Sources. California Air Resources Board, Research Division, 2020 L Street, Sacramento, CA 95814, Final Report, Contract 93-315 93315 , November, 1996.

83. R. Kostiainen, 1995 Volatile Organic Compounds in the Indoor Air of Normal and Sick Houses. Atmos. Environ., 29 693702 , 1352-2310

84. P. A. Koushki, K. H. Al-Dhowalia, S. A. Niaizi, 1992 Vehicle Occupant Exposure to Carbon Monoxide. J. Air Waste Manage. Assoc., 42 12 December 1992), 16031608 , 1047-3289

85. P. Koutrakis, M. Brauer, S. L. K. Briggs, B. P. Leaderer, 1991 Indoor Exposures to Fine Aerosols and Acid Gases. Environ. Health Perspect., 95 2328

86. P. Koutrakis, S. L. K. Briggs, B. P. Leaderer, 1992 Source Apportionment of Indoor Aerosols in Suffolk and Onondaga Counties, New York. Environ. Sci. Technol., 26 3 March 1992), 521527 , 0001-3936X

87. K. W. Leovic, L. S. Sheldon, D. A. Whitaker, R. G. Hetes, J. A. Calcagni, J. N. Baskir, 1996 Measurement of Indoor Air Emissions from Dry-

Process Photocopy Machines. J. Air Waste Manage. Assoc., 46 821829 , 1047-3289

88. J. I. Levy, K. Lee, J. D. Spengler, Y. Yanagisawa, 1998 Impact of Residential Nitrogen Dioxide Exposure on Personal Exposure: An International Study. J. Air Waste Manage. Assoc., 48 553560 , 1047-3289

89. C. W. Lewis, R. B. Zweidinger, 1992 Apportionment of Residential Indoor Aerosol VOC and Aldehyde Species to Indoor and Outdoor Sources, and Their Source Strengths. Atmos. Environ., 26A 21792184 , 1352-2310

90. J. Lewtas, S. Goto, K. Williams, J. C. Chuang, B. A. Petersen, N. K. Wilson, 1987 The Mutagenicity of Indoor Air Particles in a Residential Pilot Field Study: Application and Evaluation of New Methodologies. Atmos. Environ., 21 443449 , 1352-2310

91. G. Lfroth, P. I. Ling, E. Agurell, 1988 Public Exposure to Environmental Tobacco Smoke. Mutat. Res., 202 103110 , 0027-5107

92. M. P. Ligocki, L. G. Salmon, T. Fall, M. C. Jones, W. W. Nazaroff, G. R. Cass, 1993 Characteristics of Airborne Particles inside Southern California Museums. Atmos. Environ., 27A 697711 , 1352-2310

93. L. Liu, J. S. , P. Koutrakis, H. H. Suh, J. D. Mulik, R. M. Burton, 1993 Use of Personal Measurements for Ozone Exposure Assessment: A Pilot Study. Environ. Health Perspect., 101 318324

94. G. Löfroth, C. Stensman, M. Brandhorst-Satzkorn, 1991 Indoor Sources of Mutagenic Aerosol Particulate Matter: Smoking, Cooking, and Incense Burning. Mutat. Res., 261 2128 , 0000-0027- 5107

95. M. C. Marbury, D. P. Harlos, J. M. Samet, J. D. Spengler, 1988 Indoor Residential 2 Concentrations in Albuquerque, New Mexico. JAPCA, 38 392398 , 0894-0630

96. T. Mc Curdy, G. Glen, L. Smith, 2000 The National Exposure Research Laboratory's Consolidated Human Activity Database. J Exposure Analysis Environ Epidemiol, 10 6 November 2000), 566578 , 1559-0631

97. T. E. Mc Kone, 1987 Human Exposure to Volatile Organic Compounds in Household Tap Water: The Indoor Inhalation Pathway. Environ. Sci. Technol., 21 12 December 1987), 11941201 , 0001-3936X

98. D. A. McIntyre, 1980 Indoor Climate. London: Applied Science Publishers Ltd. 443 pp, 0-85334-868-5

99. S. Mendis, P. A. Sobotka, D. E. Evler, 1994 Pentane and Isoprene in Expired Air from Humans--Gas Chromatographic Analysis of a Single Breath. Clin. Chem., 40 8 August 1994), 14851488 , 0009-9147

100. B. Meyer, K. Hermanns, 1986 Formaldehyde Release from Wood Products. An Overview. in Formaldehyde Release from Wood Products, ACS Symposium Series 316, Chapter 1, 116 , Am. Chem. Soc., Washington, DC, 9780841209824

101. W. Michelson, 1973 Time-budgets in environmental research: Some introductory considerations, 262268 . In W. F. E. Preiser, Ed. Environmental Design Research. 2 symposia and Workshops. Fourth International EDRA Conference. Stroudsburg, Pa.: Dowden, Hutchinson, and Ross, Inc.

102. J. Miller, 1990 Fungi as contaminants in indoor air. Indoor Environment, 90 5, 5164

103. S. Mitra, B. Ray, 1995 Patterns and Sources of Polycyclic Aromatic Hydrocarbons and Their Derivatives in Indoor Air. Atmos. Environ., 29 33453356 , 1352-2310

104. A. Morey, 1990 Biological contaminants in the Indoor Environment ASTM. ISBN-EB: 978-0-8031-5131-4

105. D. J. Moschandreas, S. S. Morse, 1979 The relationship between energy conservation measures and exposure to indoor toxic pollutants. Paper 27 b., presented at the American Institute of Chemical Engineers 87th National Meeting, Boston, Massachusetts, August 19-22, 1979

106. D. J. Moschandreas, Ed, 1978 Indoor Air pollution in the Residential Environment. Vol. II. Field Monitoring Protocol, Indoor Episodic Pollutant Release Experiments and Numerical Analyses. U.S. Environmental Protection Agency Report EPA-600 -7-78-229b. Research triangle Park: U.S. Environmental Protection Agency, Environmental Monitoring and Support Laboratory, 1978, 240 pp.

107. D. J. Moschandreas, J. Zabransky, D. J. Pelton, 1980 Comparison of Indoor-Outdoor concentrations of Atmospheric Pollutants. GEOMET Report ES-823 . Palo Alto, Cal.: Electric Power Research Institute

108. D. J. Moschandreas, D. S. O'Dea, 1995 Measurement of Perchloroethylene Indoor Air Levels Caused by Fugitive Emissions from Unvented Dry-to-Dry Dry Cleaning Units. J. Air Waste Manage. Assoc., 45 February 1995), 111115 , 1047-3289

109. C. L. Moseley, M. R. Meyer, 1992 Petroleum Contamination of an Elementary School: A Case History Involving Air, Soil-Gas, and Groundwater Monitoring. Environ. Sci. Technol., 26 1 January 1992), 185192 , 0001-3936X

110. J. L. Mumford, X. Z. He, R. S. Chapman, S. R. Cao, D. B. Harris, X. M. Li, Y. L. Xian, W. Z. Jiang, C. W. Xu, J. C. Chuang, W. E. Wilson, M.

Cooke, 1987 Lung Cancer and Indoor Air Pollution in Xuan Wei, China. Science, 235 4785 January 1987), 217220 , 0036-8075

111. J. L. Mumford, R. W. Williams, D. B. Walsh, R. M. Burton, D. J. Svendsgaard, J. C. Chuang, V. S. Houk, J. Lewtas, 1990 Indoor Air Pollutants from Unvented Paraffin Heater Emissions in Leaderer, B. P., P. M. Boone, and S. K. Hammond, "Total Particle, Sulfate, and Acidic Aerosol Emissions from Paraffin Space Heaters. Environ. Sci. Technol., 24 6 June 1990), 908912 , 0001-3936X

112. L. M. Neas, D. W. Dockery, J. H. Ware, J. D. Spengler, B. G. Ferris, Jr , F. E. Speizer, 1994 Concentration of Indoor Particulate Matter as a Determinant of Respiratory Health in Children. Am. J. Epidemiol., 139 11 June 1994), 10881099 , 0002-9262

113. P. R. Nelson, S. P. Kelly, F. W. Conrad, 1998 Studies of Environmental Tobacco Smoke Generated by Different Cigarettes. J. Air Waste Manage. Assoc., 48 April 1998), 336344 , 1047-3289

114. NENOECD (Nuclear Energy Agency Organisation for Economic Co-operation and Development) 1979 Exposure to Radiation from the Natural Radioactivity in Building Materials, Expert Group Report, OECD, Paris

115. A. V. Nero, Jr (1988, In, (1988).In:Radon and its decay products in indoor air: an overview, In Radon and its decay products in Indoor Air, W.W. Nazaroff and A.V. Nero Jr. edit., 156 156 Wiley Interscience, 0-47162-810-7

116. A. V. Nero, Jr , 1989 Earth, Air, Radon and Home. Physics Today. 42 4 (April 1989), 3239 , 0031-9228

117. F. J. Offermann, S. A. Loiselle, A. T. Hodgson, L. A. Gundel, J. M. Daisey, 1991 A Pilot Study to Measure Indoor Concentrations and Emission Rates of Polycyclic Aromatic Hydrocarbons. IndoorAir, 4 497512 , 0905-6947

118. W. Ott, P. Switzer, N. Willits, Carbon Monoxide Exposures inside an Automobile Traveling on an Urban Arterial Highway. 1994 J. Air Waste Manage. Assoc., 44 August 1994), 10101018 , 1047-3289

119. W. R. Ott, 1995 Human activity patterns: A review of the literature for air pollution exposure estimation. SIMS Technical report. Stanford Cal.: University, Department of statistics

120. J. R. Ottensman, 1972 Systems of Urban Activities and Time: An interpretive Review of the Literature. An Urban Studies Research Paper. Chapel Hill, N.C.: University of North Carolina, Center for Urban and Regional Studies, 1972. 45pp.

121. A. S. Pennanen, R. O. Salonen, S. Aim, M. J. Jantunen, P. Pasanen,

1997 Characterization of Air Quality Problems in Five Finnish Indoor Ice Arenas. J. Air Waste Manage. Assoc., 47 October 1997), 10791086 , 1047-3289

122. R. Perry, P. W. Kirk, 1986 Indoor and Ambient Air Quality. Selper, London

123. R. Perry, I. L. Gee, 1994 Vehicle Emissions and Effects on Air Quality: Indoors and Outdoors. Indoor Environ., 3 224236

124. J. A. Peterka, J. E. Cermak, 1977 Turbulence in Building wakes, 447463 . In K.J. Eaton, Ed. Proceedings of the fourth international conference on wind effects on buildings and structures, Heathrow, 1975. New York: Cambridge University Press, 0-52120-801-7

125. M. Petreas, K. Liu, S. , B. Chang, H. , S. B. Hayward, K. Sexton, 1988 A Survey of Nitrogen Dioxide Levels Measured inside Mobile Homes. JAPCA, 38 647651 , 0894-0630

126. M. Phillips, J. Greenberg, J. Awad, 1994 Metabolic and Environmental Origins of Volatile Organic Compounds in Breath. J. Clin. Pathol., 47 11 November 1994), 10521053 , 0021-9746

127. J. N. Pitts, T. J. Wallington, H. W. Biermann, A. M. Winer, 1985 Identification and Measurement of Nitrous Acid in an Indoor Environment. Atmos. Environ., I9 763767 , 1352-2310

128. J. N. Pitts, H. W. Biermann, E. C. Tuazon, M. Green, W. D. Long, A. M. Winer, 1989 Time-Resolved Identification and Measurement of Indoor Air Pollutants by Spectroscopic Techniques: Gaseous Nitrous Acid, Methanol, Formaldehyde, and Formic Acid. JAPCA, 39 13441347 , 0894-0630

129. J. G. Price, J. G. Rigby, L. Christensen, R. Hess, D. D. La Pointe, A. R. Ramelli, M. Desilets, R. D. Hopper, T. Kluesner, S. Marshall, 1994 Radon in outdoor air in Nevada, Health Phys. 66 4 433438 , 0017-9078

130. Priorities for Indoor Air Research and Action. 1991 International Conference, Montreux.

131. J. J. Quackenboss, M. D. Lebowitz, C. D. Crutchfield, 1989 Indoor-Outdoor Relationships for Particulate Matter: Exposure Classifications and Health Effects. Environ. Int., I5 353360 , 0160-4120

132. R. Reiss, P. B. Ryan, P. Koutrakis, 1994 Modeling Ozone Deposition onto Indoor Residential Surfaces. Environ. Sci. Technol., 28 3 March 1994), 504513 , 0001-3936X

133. R. Reiss, P. B. Ryan, S. J. Tibbetts, P. Koutrakis, 1995a Measurement of Organic Acids, Aldehydes, and Ketones in Residential Environments and

Their Relation to Ozone. J. Air Waste Manage. Assoc., 45 October 1995), 811822 , 1047-3289

134. R. Reiss, P. B. Ryan, P. Koutrakis, S. J. Tibbets, 1995b Ozone Reactive Chemistry on Interior Latex Paint. Environ. Sci. Technol., 29 8 August 1995), 19061912 , 0001-3936X

135. E. M. C. Rob, A. Rannou, J. Le Bronec, 1992 Radon measurement in the environment in France, Radiat. Prot. Dosim. 45 1-4) Supplement, 455457

136. I. Romieu, M. C. Lugo, S. Colome, A. M. Garcia, M. H. Avila, A. Geyh, S. R. Velasco, E. P. Rendon, 1998 Evaluation of Indoor Ozone Concentration and Predictors of Indoor-Outdoor Ratio in Mexico City. J. Air Waste Manage. Assoc., 48 April 1998), 327335 , 1047-3289

137. D. R. Rowe, A. , K. H. Dhowalia, M. E. Mansour, 1989 Indoor- Outdoor Carbon Monoxide Concentrations at Four Sites in Riyadh, Saudi Arabia. J. Air Waste Manage. Assoc., 39 8 August 1989), 11001102 , 1047-3289

138. D. R. Rowe, K. H. Al-Dhowalia, M. E. Mansour, 1991 Indoor- Outdoor Nitric Oxide and Nitrogen Dioxide Concentrations at Three Sites in Riyadh, Saudi Arabia. J. Air Waste Manage. Assoc., 41 7 July 1991), 973976 , 1047-3289

139. P. B. Ryan, M. L. Soczek, J. D. Spengler, I. H. Billick, 1988 The Boston Residential 2 Characterization Study: I. Preliminary Evaluation of the Survey Methodology. JAPCA, 38 2227 , 0894-0630

140. L. G. Salmon, W. W. Nazaroff, M. P. Ligocki, M. C. Jones, G. R. Cass, 1990 Nitric Acid Concentrations in Southern California Museums. Environ. Sci. Technol., 24 7 July 1990), 10041012 , 0001-3936X

141. M. Santamouris, N. Papanikolaou, I. Livada, I. Koronakis, C. Georgakis, A. Argiriou, D. Assimakopoulos, 2001 On the impact of Urban climate on the energy consumption of buildings. Solar Energy, 70 201216

142. V. H. Schaeffer, B. Bhooshan, S. Chen, B. , J. S. Sonenthal, A. T. Hodgson, 1996 Characterization of Volatile Organic Chemical Emissions from Carpet Cushions. J. Air Waste Manage. Assoc., 46 September 1996), 813820 , 1047-3289

143. G. Sciocchetti, G. F. Clemente, G. Ingrao, F. Scacco, 1983 Results of a survey on radioactivity of building materials in Italy. Health Phys. 45 2 August 1983), 385388 0017-9078

144. G. Sciocchetti, M. Bovi, G. Cotellessa, P. G. Baldassini, C. Battella, I. Porcu, 1992 Indoor radon and thoron surveys in high radioactivity areas of Italy, Radiat. Prot. Dosim., 45 1-4) Supplement, 509513

145. K. Sexton, M. X. Petreas, K. S. Liu, Formaldehyde Exposures inside

Mobile Homes 1989 Environ. Sci. Technol., 23 8 August 1989), 985988 , 0001-3936X

146. J. J. Shah, H. B. Singh, 1988 Distribution of Volatile Organic Chemicals in Outdoor and Indoor Air. Environ. Sci. Technol., 22 11 November 1988), 13811388 , 0001-3936X

147. F. H. Shair, K. L. Heitner, 1974 Theoretical Model for relating indoor pollution concentrations to those outside. Environ. Sci. Technol, 8 5 May 1974), 444451 , 0001-3936X

148. J. L. Shepherd, R. L. Corsi, J. Kemp, 1996 Chloroform in Indoor Air and Wastewater: The Role of Residential Washing Machines. J. Air Waste Manage. Assoc., 46 July 1996), 631642 , 1047-3289

149. M. H. Sherman, 1981 Air Infiltration in Buildings. Berkeley, Cal.: University of California, Ph.D. Dissertation,. (published as Lawrence Berkeley Laboratory Report LBL-1U712)

150. H. C. Shields, C. J. Weschler, (1992, (1992).Volatile Organic Compounds Measured at a Telephone Switching Center from 5/30/85-12/6/88: A Detailed Case Study. J. Air Waste Manage. Assoc., 42 42 6 June 1992), 792804 , 1047-3289

151. H. C. Shields, D. M. Fleischer, C. J. Weschler, 1996 Comparisons among VOCs Measured in Three Types of U.S. Commercial Buildings with Different Occupant Densities," Indoor Air, 6 217 , 0905-6947

152. N. Shinohara, Y. Kai, A. Mizukoshi, M. Fujii, K. Kumagai, Y. Okuizumi, M. Jona, Y. Yanagisawa, 2009 On-site passive flux sampler measurement of emission rates of carbonyls and VOCs from multiple indoor sources. Building Environ, 44 859863 , 0360-1323

153. S. Short, 2001 The issues and implications of setting and applying Indoor Air Quality Guidelines. IEH Seminar, University of Leicester,http://www.le.ac.uk/ieh/pdf/IAQmtgrep.

154. P. A. Siskos, K. E. Bouba, A. P. Stroubou, 2001 Determination of selected pollutsnts and measurements of physical parameters of the evaluation of indoor air quality in school buildings in Athens, Greece. Indoor Built. Environ., 10 185192

155. P. A. Siskos, 2003 An experimental Investigation of the Indoor Air Quality in Fifteen School Buildings in Athens, Greece. International Journal of Ventilation, 2 3 185201 , 1473-3315

156. A. P. Soldatos, E. B. Bakeas, P. A. Siskos, 2003 Occupational exposure to btex of workers in car parkings and gasoline service stations in Athens, Greece. Fresenius Environmental Bulletin, 12 9 10641070

157. S. Sollinger, K. Levsen, G. Wfinsch, 1993 Indoor Air Pollution by Organic Emissions from Textile Floor Coverings. Climate Chamber Studies under Dynamic Conditions. Atrnos. Environ., 27B 183192 , 1352-2310

158. S. Sollinger, K. Levsen, G. Wiinsch, 1994 Indoor Pollution by Organic Emissions from Textile Floor Coverings: Climate Test Chamber Studies under Static Conditions. Atrnos. Environ., 28 23692378 , 1352-2310

159. J. D. Spengler, D. W. Dockery, W. A. Turner, J. M. Wolfson, B. G. Ferris, 1981 Long-Term Measurements of Respirable Sulfates and Particles inside and outside Homes. Atrnos. Environ., 15 2330 , 1352-2310

160. J. D. Spengler, M. Brauer, J. M. Samet, W. E. Lambert, 1993 Nitrous Acid in Albuquerque, New Mexico, Homes. Environ. Sci. Technol., 27 5 May 1993), 841845 , 0001-3936X

161. J. Spengler, M. Schwab, P. B. Ryan, S. Colome, A. L. Wilson, I. Billick, E. Becker, 1994 Personal Exposure to Nitrogen Dioxide in the Los Angeles Basin. J. Air Waste Manage. Assoc., 44 January 1994), 3947 , 1047-3289

162. C. W. Spicer, D. V. Kenny, G. F. Ward, I. H. Billick, 1993 Transformations, Lifetimes, and Sources of 2 HONO, and HNO 3 in Indoor Environments. J. Air Waste Manage. Assoc., 43 November 1993), 14791485 , 1047-3289

163. T. H. Stock, 1987 Formaldehyde Concentrations inside Conventional Housing. JAPCA, 37 913918 , 0894-0630

164. G. A. Swedjemark, L. Mjönes, 1984 Radon and radon daughter concentration in Swedish Homes, Radiat. Prot. Dosim. 7 1-4), 341345

165. G. T. Tamura, 1975 Measurement of air leakage characteristics of house enclosures. ASHRAE Trans, 81 Pt. 1), 202211

166. C. V. Thompson, R. A. Jenkins, C. E. Higgins, 1989 A Thermal Desorption Method for the Determination of Nicotine in Indoor Environments. Environ. Sci. Technol., 23 4 April 1989), 429435 , 0001-3936X

167. B. A. Tichenor, L. A. Sparks, J. B. White, M. D. Jackson, 1990 Evaluating Sources of Indoor Air Pollution. J. Air Waste Manage. Assoc., 40 4 April 1990), 487492 , 1047-3289

168. G. W. Traynor, M. G. Apte, H. A. Sokil, J. C. Chuang, W. G. Tucker, J. L. Mumford, 1990 Selected Organic Pollutant Emissions from Unvented Paraffin Space Heaters. Environ. Sci. Technol., 24 9 September 1990), 12651270 , 0001-3936X

169. M. Tsakas, P. Siskos, 2010 Indoor Air Quality in the Control Tower of Athens International Airport, Greece. Indoor and Built Environment, 0 0106 , 0142-0326X

170. J. L. Tyson, P. W. Fairey, C. R. Withers, 1993 Elevated radon levels in ambient air, INDOOR AIR '93, Proceedings of the 6th International Conference on Indoor Air Quality and Climate,, 1 4, 443448 . Helsinki, Finland, July 4-8,1993

171. UNSCEAR (United Nations Scientific Committee on the Effects of Atomic Radiation) 1977 Sources and effects of ionizing radiation, United Nations ed., New York, E.77.IX.1.

172. UNSCEAR (United Nations Scientific Committee on the Effects of Atomic Radiation) 1993 Sources and effects of ionizing radiation, United Nations ed., New York, E.94.IX.2.

173. W. A. Wade, W. A. Cote, J. E. Yocom, 1975 A Study of Indoor Air Quality. JAPCA, 25 933939 , 0894-0630

174. J. Waldman, P. Jenkins, 2004 Indoor air exposure research and reducing indoor air exposures in California. In: Krzyzanowski, M., Jantunen, M., Bartonova, A., Oglesby, L., Kepahalopoulos, S., Kotzias D, (Eds.). Role of Human assessment in air quality management. Report EUR 21052, European Communities, Luxembourg, 6671

175. J. C. Wallace, L. P. Brzuzy, S. L. Simonich, S. M. Visscher, R. A. Hites, 1996 Case Study of Organochlorine Pesticides in the Indoor Air of a Home. Environ. Sci. Technol., 30 9 August 1996), 27152718 , 0001-3936X

176. L. Wallace, 1996 Indoor Particles: A Review. J. Air Waste Manage. Assoc., 46 February 1996), 98126 , 1047-3289

177. N. V. Waubke, W. Kusterle, 1990 Mould infestations in residential buildings. Paper presented at the 1st symposium on mould infestations, Innsbruck, January 1990

178. C. J. Weschler, H. C. Shields, D. V. Naik, 1989 Indoor Ozone Exposures. JAPCA, 39 15621568 , 0894-0630

179. C. J. Weschler, H. C. Shields, D. Rainer, 1990 Concentrations of Volatile Organic Compounds at a Building with Health and Comfort Complaints. Am. Ind. Hyg. Assoc. J., 51 261268 , 0002-8894

180. C. J. Weschler, H. C. Shields, D. V. Naik, 1994 Indoor Chemistry Involving O3, NO, and 2 as Evidence by 14 Months of Measurements at a Site in Southern California. Environ. Sci. Technol., 28 No. 12, 21202132 , 0001-3936X

181. C. J. Weschler, H. C. Shields, 1996 Chemical Transformations of Indoor Air Pollutants, in Indoor Air '96 (S. Yoshizawa, K. Kimura, K. Ikeda, S. Tanabe, and T. Iwata, Eds.). Organizing Committee of the 7th

International Conference of Indoor Air Quality and Climate, 1 919924 , Nagoya, Japan

182. D. L. West, 1977 Contaminant dispersion and dilution in a ventilated space. ASHRAE Trans, 83 Pt. 1), 125140

183. R. F. White, 1995 Landscape development and natural ventilation. Effect of moving air on buildings and adjacent areas. Landscape Archit, 45 7281 , 0023-8031

184. P. Wolkoff, 1999 Photocopiers and Indoor Air Pollution. Atmos. Environ., 33 21292130 , 1352-2310

185. World Health Organization. 1986 Regional Seminar on Human Exposure to Outdoor and Indoor Air Pollution and its Effects on Health. PEPAS, Kuala Lumpur

186. World Health Organization 1989 Indoor air quality: Organic Pollutants. Copenhagen, EURO 111

187. J. Zhang, Q. He, P. J. Lioy, 1994 Characteristics of Aldehydes: Concentrations, Sources, and Exposures for Indoor and Outdoor Residential Microenvironments. Environ. Sci. Technol., 28 1 January 1994), 146152 , 0001-3936X

188. L. Z. Zhang, J. L. Niu, 2004 Modeling VOCs emissions in a room with a single-zone multi-component multi-layer technique. Building Environ, 39 523531 , 0360-1323

CITATION

CHAPTER 1

L. Wang, J. Yang, P. Zhang, X. Zhao, Z. Wei, F. Zhang, J. Su and C. Meng, "A Review of Air Pollution and Control in Hebei Province, China," *Open Journal of Air Pollution*, Vol. 2 No. 3, 2013, pp. 47-55. doi:10.4236/ojap.2013.23007.

CHAPTER 2

Detlef Laussmann and Dieter Helm (2011). Air Change Measurements Using Tracer Gases: Methods and Results. Significance of air change for indoor air quality, Chemistry, Emission Control, Radioactive Pollution and Indoor Air Quality, Dr. Nicolas Mazzeo (Ed.), ISBN: 978-953-307-316-3, InTech, DOI: 10.5772/18600.

CHAPTER 3

M.D. Larrañaga, E. Karunasena, H.W. Holder, E.D. Althouse and D.C. Straus (2011). Statistical Considerations for Bioaerosol Health-Risk Exposure Analysis, Chemistry, Emission Control, Radioactive Pollution and Indoor Air Quality, Dr. Nicolas Mazzeo (Ed.), ISBN: 978-953-307-316-3, InTech, DOI: 10.5772/16323.

CHAPTER 4

David Straus, M.D. Larranaga, Enusha Karunasena, H.W. Holder and M.G. Beruvides (2011). Improving the Quality of the Indoor Environment Utilizing Desiccant-Assisted Heating, Ventilating, and Air Conditioning

Systems, Chemistry, Emission Control, Radioactive Pollution and Indoor Air Quality, Dr. Nicolas Mazzeo (Ed.), ISBN: 978-953-307-316-3, InTech, DOI: 10.5772/19776.

CHAPTER 5

Moinuddin Sarker (2011). Municipal Waste Plastic conversion into Different Category Liquid Hydrocarbon Fuel, Chemistry, Emission Control, Radioactive Pollution and Indoor Air Quality, Dr. Nicolas Mazzeo (Ed.), ISBN: 978-953-307-316-3, InTech, DOI: 10.5772/16276.

CHAPTER 6

F. Javier Álvarez-Hornos, Feliu Sempere, Marta Izquierdo and Carmen Gabaldón (2011). Lab-scale Evaluation of Two Biotechnologies to Treat VOC Air Emissions: Comparison with a Pilot Unit Installed in the Plastic Coating Sector, Chemistry, Emission Control, Radioactive Pollution and Indoor Air Quality, Dr. Nicolas Mazzeo (Ed.), ISBN: 978-953-307-316-3, InTech, DOI: 10.5772/17234.

CHAPTER 7

José Orosa (2011). One Way ANOVA Method to Relate Microbial Air Content and Environmental Conditions., Chemistry, Emission Control, Radioactive Pollution and Indoor Air Quality, Dr. Nicolas Mazzeo (Ed.), ISBN: 978-953-307-316-3, InTech, DOI: 10.5772/17927.

CHAPTER 8

Alessandro Bacaloni, Susanna Insogna and Lelio Zoccolillo (2011). Indoor Air Quality. Volatile Organic Compounds: Sources, Sampling and Analysis, Chemistry, Emission Control, Radioactive Pollution and Indoor Air Quality, Dr. Nicolas Mazzeo (Ed.), ISBN: 978-953-307-316-3, InTech, DOI: 10.5772/21645.

CHAPTER 9

Marios P. Tsakas, Apostolos P Siskos and Panayotis Siskos (2011). Indoor Air Pollutants and the Impact on Human Health, Chemistry, Emission Control, Radioactive Pollution and Indoor Air Quality, Dr. Nicolas Mazzeo (Ed.), ISBN: 978-953-307-316-3, InTech, DOI: 10.5772/18806.

INDEX

A

Air change 17, 18, 19, 20, 21, 22, 23, 24,
    25, 27, 29, 30, 31, 32, 34, 38, 39,
    40, 41, 42, 45, 46, 48, 50, 52, 54,
    56, 59, 60, 61, 62, 64, 65, 67, 319
air change rate (ACR) 18, 25, 31
air handling units (AHUs) 70
Air Pollution Index (API) 1
American Conference of Governmental
    Industrial Hygienists (ACGIH)
    69
American Industrial Hygiene Associa-
    tion (AIHA) 71

B

best available technologies (BATs) 200
biofilters (BF) 201, 202
bioscrubbers (BS) 201
biotrickling filters (BTF) 201, 202
Building-Related Environmental Com-
    plaints (BREC) 246
Building Related Health Complaints
    (BRC) 246
building-related illness (BRI) 118, 125,
    284
Building-Related Illness (BRI) 247

Building-Related Symptoms (BRS) 246

C

carbon dioxide (CO2) 226
Carbon dioxide (CO2) 29
Centers for Disease Control (CDC) 70

D

desiccant based air conditioning (DBAC)
    127

E

elimination capacity (EC) 202, 212
empty bed residence time (EBRT) 202
environmental tobacco smoke (ETS) 296
European Collaborative Action (ECA)
    268, 305

F

fresh air supply 17, 24, 52

G

Gas Chromatography & Mass Spectrom-
    eter (GC/MS) 162
Gross Domestic Product (GDP) 3, 121